International Union of Theoretical
and Applied Mechanics

H0080327

Buckling
of Structures

Symposium Cambridge/USA
June 17–21, 1974

Editor
Bernard Budiansky

Springer-Verlag
Berlin Heidelberg New York 1976

Professor Bernard Budiansky
Harvard University
Cambridge, Massachusetts, U.S.A.

With 214 Figures

ISBN 978-3-642-50994-0   ISBN 978-3-642-50992-6 (eBook)
DOI 10.1007/978-3-642-50992-6

Library of Congress Cataloging in Publication Data: Symposium on Buckling of Structures, Harvard University, 1974. Buckling of structures. At head of title: International Union of Theoretical and Applied Mechanics. Bibliography: p. Includes index. 1. Buckling (Mechanics) — Congresses. 2. Structural stability — Congresses. I. Budiansky, Bernard. II. International Union of Theoretical and Applied Mechanics. III. Title. TA656.2.S95. 1974. 624'.176. 75-31726.

Preface

This volume contains the written texts of the papers presented at a Symposium on Buckling of Structures held at Harvard University in June 1974. This symposium, one of several on various topics sponsored annually by the International Union of Theoretical and Applied Mechanics (IUTAM), was organized by a Scientific Committee consisting of B. Budiansky (Chairman), A. H. Chilver, W. T. Koiter, and A. S. Vol'-mir. Participation was by invitation of the Scientific Committee, and specific lecturers were invited to speak in the areas of experimental research, buckling and post-buckling calculations, post-buckling mode interaction, plasticity and creep effects, dynamic buckling, stochastic problems, and design. A total of 29 lectures were delivered, including a general opening lecture by Professor Koiter, and there were 93 registered participants from 16 different countries.

Financial support for the symposium was provided by IUTAM, in the form of partial travel support for a number of participants, and also by the National Science Foundation, the National Aeronautics and Space Administration, and the Air Force Office of Scientific Research, for additional travel support and administrative expenses. Meeting facilities and services were efficiently provided by the Science Center of Harvard University, and administrative support was generously provided by the Division of Engineering and Applied Physics of Harvard University. The scientific chairman enjoyed the invaluable assistance of his colleagues Professors J. W. Hutchinson and J. L. Sanders in making local arrangements for the symposium; and, finally, the dedicated secretarial and administrative services provided by Marion Remillard and Parian Temple, without whom nothing would have worked, are most gratefully acknowledged.

Cambridge, Mass., October 1975

Bernard Budiansky

Participants

(Authors denoted by asterisks)

Akkas, N.	Ankara, Turkey
*Almroth, B. O.	Palo Alto, California, U.S.A.
*Amazigo, J. C.	Troy, New York, U.S.A.
*Anderson, M. S.	Hampton, Virginia, U.S.A.
*Arbocz, J.	Pasadena, California, U.S.A.
*Augusti, G.	Florence, Italy
*Babcock, Jr., Ch. D.	Pasadena, California, U.S.A.
Basdekas, N.	Arlington, Virginia, U.S.A.
Batdorf, S. B.	Los Angeles, California, U.S.A.
Bauld, N. R.	Clemson, South Carolina, U.S.A.
Besseling, J. F.	Delft, Netherlands
Billington, D. P.	Princeton, New Jersey, U.S.A.
Blaauwendraad, J.	Gouda, Netherlands
Brown, E. H.	London, England
Brush, D. O.	Davis, California, U.S.A.
*Budiansky, B.	Cambridge, Massachusetts, U.S.A.
Bufler, H.	Stuttgart, Germany
Bushnell, D.	Palo Alto, California, U.S.A.
Calladine, C. R.	Cambridge, England
Ceradini, G.	Rome, Italy
*Chilver, A. H.	Cranfield, England
Como, M.	Cosenza, Italy
Danielson, D. A.	Charlottesville, Virginia, U.S.A.
Dickie, J. F.	Manchester, England
Dill, E. H.	Seattle, Washington, U.S.A.
*Duszek, M. K.	Warsaw, Poland
Dym, C. L.	Cambridge, Massachusetts, U.S.A.
Ebner, H.	Aachen, Germany
*Esslinger, M.	Braunschweig, Germany
*Gallagher, R. H.	Ithaca, New York, U.S.A.
Galletly, G. D.	Liverpool, England
*Hansen, H. R.	Oslo, Norway
*Hayman, B.	Leicester, England
Hedgepeth, J. M.	Santa Barbara, California, U.S.A.
Herrmann, G.	Stanford, California, U.S.A.
*Hoff, N. J.	Stanford, California, U.S.A.
Huang, N. C.	Notre Dame, Indiana, U.S.A.
Huseyin, K.	Waterloo, Ontario, Canada
*Hutchinson, J. W.	Cambridge, Massachusetts, U.S.A.
*Johns, K. C.	Sherbrooke, Quebec, Canada

Jones, N.	Cambridge, Massachusetts, U.S.A.
Kalnins, A.	Bethlehem, Pennsylvania, U.S.A.
Kempner, J.	Brooklyn, New York, U.S.A.
Khot, N. S.	Dayton, Ohio, U.S.A.
Knets, I.	Riga, U.S.S.R.
*Koiter, W. T.	Delft, Netherlands
*Kozarov, M. M.	Sofia, Bulgaria
*Leckie, F. A.	Leicester, England
*Leipholz, H. H. E.	Waterloo, Ontario, Canada
*Leonard, R. W.	Hampton, Virginia, U.S.A.
Libove, C.	Syracuse, New York, U.S.A.
Łukasiewicz, S.	Warsaw, Poland
McIvor, I. K.	Ann Arbor, Michigan, U.S.A.
*Massonnet, Ch.	Liège, Belgium
Masur, E. F.	Chicago, Illinois, U.S.A.
Nachbar, W.	La Jolla, California, U.S.A.
*Needleman, A.	Cambridge, Massachusetts, U.S.A.
*van der Neut, A.	Delft, Netherlands
Niordson, F. I.	Lyngby, Denmark
Ohtsubo, H.	Tokyo, Japan
*Pedersen, P. T.	Lyngby, Denmark
Pian, T. H. H.	Cambridge, Massachusetts, U.S.A.
*Pignataro, M.	Rome, Italy
Plaut, R. H.	Providence, Rhode Island, U.S.A.
Pomerantz, J.	Arlington, Virginia, U.S.A.
Rehfield, L. W.	Atlanta, Georgia, U.S.A.
Reissner, E.	La Jolla, California, U.S.A.
Rhodes, J.	Glasgow, Scotland
Riks, E.	Amsterdam, Netherlands
Roorda, J.	Waterloo, Ontario, Canada
*Røren, E. M. Q.	Oslo, Norway
Samuelson, L. A.	Bromma, Sweden
Sanders, J. L., Jr.	Cambridge, Massachusetts, U.S.A.
Seide, P.	Los Angeles, California, U.S.A.
*Sewell, M. J.	Reading, England
Simitses, G. J.	Atlanta, Georgia, U.S.A.
Simmonds, J. G.	Charlottesville, Virginia, U.S.A.
*Singer, J.	Haifa, Israel
Stein, M.	Hampton, Virginia, U.S.A.
Supple, W. J.	Guildford, England
*Tennyson, R. C.	Downsview, Ontario, Canada
*Thompson, J. M. T.	London, England
Thurston, G. A.	Denver, Colorado, U.S.A.
Tillman, S. C.	Manchester, England
*Tulk, J. D.	London, England
*Tvergaard, V.	Lyngby, Denmark
Uetani, K.	Kyoto, Japan
Valid, R.	Châtillon, France
de Veubeke, B. F.	Liège, Belgium
Wierzbicki, T.	Warsaw, Poland
*Williams, F. W.	Birmingham, England
*Yamaki, N.	Sendai, Japan
Zyczkowski, M.	Kraków, Poland

Contents

VIII. Design

Current Trends in the Theory of Buckling

W. T. Koiter

Technische Hogeschool, Delft, Netherlands

Abstract

Some current trends in buckling theory are reviewed. Particular attention is given to multi-mode bifurcation buckling, the associated imperfection-sensitivity and the correlation of experimental evidence with theory.

1. Introduction

The task of a speaker in an opening lecture at a meeting of selected experts in a particular field of science is hardly to be envied. In most aspects of his topic the audience comprises members with far more specialized expertise than the lecturer has at his command, and the easy way out, a retreat into generalities, is not likely to be welcomed. Having accepted the challenge, however, I shall try and review some current trends in the theory of buckling as I see them. I had the good fortune to read the abstracts of your papers which our chairman kindly made available to me, and this may reduce to some extent the inevitable personal bias in my talk.

The simplest possible model of a column, depicted in Fig. 1a, consists of a rigid bar supported on an elastic hinge. This simple model is employed in several elementary text-books, including the author's [28] which, incidentally, enjoys sales comparable to a volume of poetry. The perfect column, with zero eccentricity of the applied load ($e = \varkappa L = 0$), loses its stability and buckles by bifurcation at the critical load $F_1 = D/L$. The post-buckling behaviour depends on nonlinearities in the spring characteristic. Various possible load-deflection curves are shown in Fig. 1b, and the relationship between the load and the overall shortening is exhibited in Fig. 1c. In cases of an unstable descending post-buckling path ($\alpha \neq 0$ or $\alpha = 0, \beta < -1/6$) the column is sensitive to geometric imperfections or load eccentricity. The associated drop in critical load is shown in Fig. 1d, and the buckling load F_1 at the bifurcation point is replaced by the snap load at a limit point. In the absence of a descending post-buckling path of the perfect column, however, no loss of stability occurs in the column with imperfections or a load eccentricity. The presence of the buckling load F_1 of the perfect column appears here only in the form of a marked increase of the deflection when the load F approaches and passes the value F_1.

Post-buckling equilibrium paths and the influence of imperfections

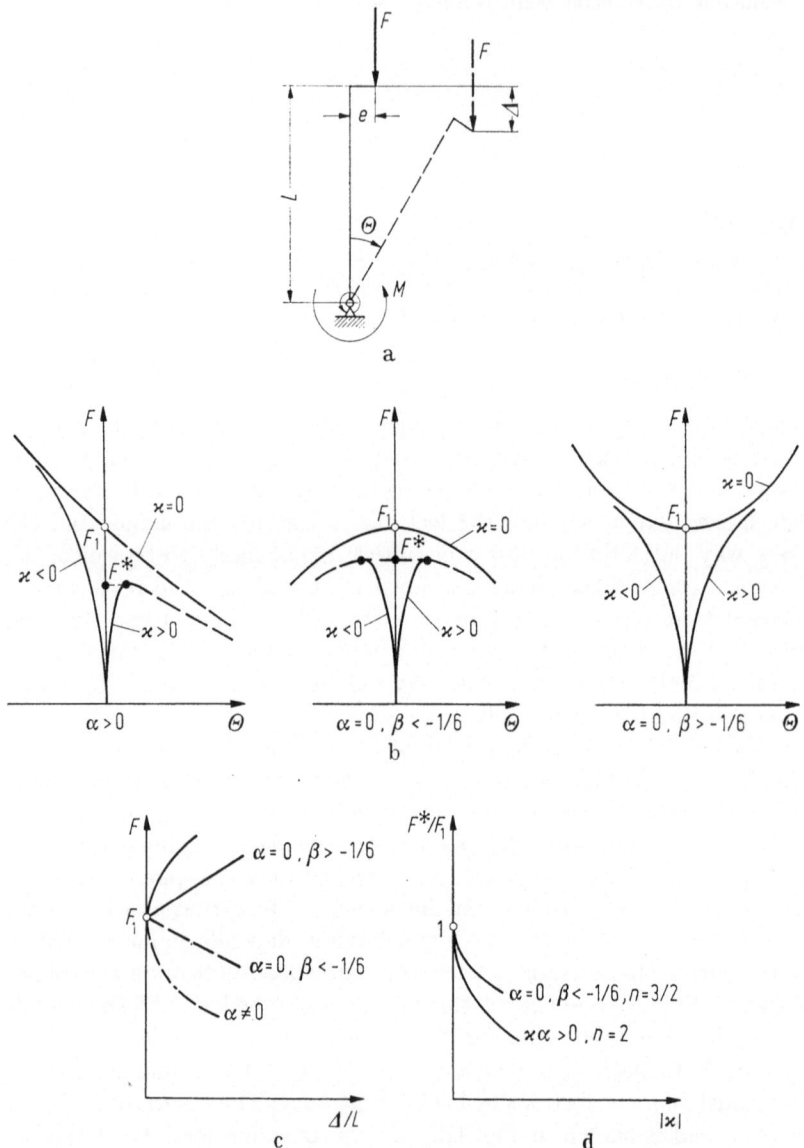

Fig. 1. a) Simplest column model eccentricity $e = \varkappa L$, spiral spring, $M = D[\theta - \alpha\theta^2 + \beta\theta^3]$; b) Load-deflection curves; c) Load-shortening curve for perfect column. d) Imperfection-sensitivity $(1 - F^*/F_1)^n = \text{const} \cdot |\varkappa| \, F/F_1$.

and load eccentricities may be analyzed directly by means of the equations of equilibrium or indirectly with the aid of the energy method. We prefer the latter approach because it delivers a verdict on stability at the same time. In this connection, however, we have to caution against a facile generalization of experience with the column model of Fig. 1a. Here it appears that a negative second variation of the potential energy of the external loads is the cause of a loss of stability, but this state of affairs is due to our simplifying assumption of an incompressible rod, and we may not generalize this experience to other cases, as still happens only too frequently. In fact, in the case of elastic structures under dead loads the potential energy of the external loads does not enter at all into the stability condition. Loss of stability of elastic structures is always due to an internal exchange of energy.

The literature abounds in further more or less simple examples of structural models which lose their stability when a certain critical value of the load factor is attained. In the case of elastic structures under conservative loads the critical load factor always corresponds to either a bifurcation point or a limit point. Moreover, the degeneration of a bifurcation point into a limit point due to the presence of imperfections, as in the example of Fig. 1, suggests that bifurcation is an exception rather than the rule. In spite of this the literature on elastic stability and buckling of structures deals preponderantly with bifurcation buckling problems.

Human frailty is, perhaps, the main reason for this state of affairs. Bifurcation problems are far more amenable to analysis because they often permit a linearization of the equations for the pre-buckling fundamental state. A similar simplification is impossible in a direct analysis of snapping problems. Modern computing techniques, and in particular the development of high-speed computers, have alleviated the difficulties of a nonlinear analysis ab initio, and significant results will also be reported by several speakers at this meeting.

Fortunately, there is also a more factual reason why the investigation of bifurcation phenomena has been popular and will remain highly valuable for many applications. Thin-walled structures are most efficient, if they carry their service loads in such a way that a membrane state of stress is approximated as closely as possible. In the ideal situation such structures are liable to bifurcation buckling rather than snap buckling. Unavoidable imperfections lead again to a snap buckling behaviour of the actual structure, but this phenomenon is usually more readily accessible to analysis once the bifurcation problem for the ideal structure is fully understood. Several papers at this meeting will testify to the effectiveness of this approach.

The theory of elastic stability presupposes an elastic material behaviour, and it breaks down when the stresses in the structure exceed the proportionality limit. A new phenomenon in the theory of elastic-plastic stability is that bifurcation may occur without a simultaneous loss of stability, and the associated post-buckling path is initially rising and stable. The rapid decrease of the tangent modulus with increasing deformation, however, leads to the occurrence of a limit point at a load only slightly above the bifurcation load. Initial imperfections have again often most serious detrimental consequences. The present situation with respect to plastic buckling and the associated imperfection-sensitivity is reviewed critically by Hutchinson in a forthcoming survey [17]. We also refer to Sewell's article with a detailed bibliography [35]. A more recent aspect of inelastic effects on the stability of structures is reflected in the topic creep buckling. A number of papers at our symposium are devoted to all these inelastic phenomena.

In the past the theory of structural stability has been restricted conventionally to conservative loading conditions, usually dead loads or pressure loads. Instability phenomena under these conditions are always of the so-called divergence type, a name borrowed from aircraft engineering. This name is employed in order to make a clear distinction with a second highly dangerous instability phenomenon called flutter. Dynamic interaction between a fluid flow and a solid structure is characteristic for the flutter problem, and the loads exerted on the structure depend both on its deformation and the time rate of deformation. Considerable attention has been given in recent years to the stability of structures under configuration-dependent non-conservative loads, in particular socalled follower forces, in an attempt to separate the effect of the action on the structure by its surroundings from the reaction on the surroundings by the structure. As in environmental engineering, it is open to some question whether such a separation is really feasible in many cases. The problem of structures under configuration-dependent non-conservative loads is further complicated by its sensitivity to small viscous (internal) damping.

Time-dependent loads are by nature non-conservative, and as in the case of configuration-dependent loading, a stability analysis can only be made on the basis of the equations of motion. Parametric excitation by loads of type $p = p_0 + p_1 \cos \omega t$ have been discussed by many authors, and a further example is to be expected at this meeting. The analysis of stability may normally be carried out on the basis of linear theory, and the stability diagrams are then of Mathieu- or Hill-type. Impulsive loading presents a more intractable problem because even a clear-cut definition of buckling is not at all obvious. We refer to [3, 4] for more detailed information.

In the sequel we shall confine our attention to the bifurcation theory of elastic stability and the associated influence of geometric and other imperfections, both for the human and factual reasons referred to before. It is indeed our firm belief that an adequate understanding of buckling phenomena cannot be achieved without proper knowledge of bifurcation theory. In our discussion we shall also dwell on some recent developments in the application of the theory for multi-mode buckling, and on its significance in the correlation of experimental evidence with the results of theory. We conclude the paper with a brief examination of the physical background of the energy criterion and the open mathematical questions in its application.

2. Résumé of General Bifurcation Theory[1]

The (conservative) loads on the structure are represented by the product of a unit load system and a non-dimensional load factor λ. Without loss in generality we may identify the unit load system with the loads on the structure at its critical bifurcation point. The critical load factor λ_1 thus equals unity. A particle of the body is identified by its radius vector x in the fundamental state of equilibrium at the critical factor $\lambda_1 = 1$. This fundamental state at an arbitrary load factor λ is described by the displacement vector field $U(x; \lambda)$ from the undeformed configuration to the fundamental pre-buckling configuration. It is stable for $0 \leqq \lambda < 1$ and unstable for $\lambda > 1$.

An arbitrary configuration II at the same load factor is described by the displacement field from the undeformed configuration $U(x; \lambda)$ $+ u(x)$, where $u(x)$ is the additional displacement vector field from the fundamental state I. Since we suppose the fundamental pre-buckling state I to be known, the increment in potential energy in the transition from the fundamental state to configuration II is a functional $P[u(x); \lambda]$ of the field $u(x)$ with the load factor as a parameter.

We expand the potential energy functional with respect to the (additional) displacement field $u(x)$. The linear term vanishes identically since the fundamental state is a configuration of equilibrium, and we have

$$P[u(x); \lambda] = P_2[u(x); \lambda] + P_3[u(x); \lambda] + P_4[u(x); \lambda] + \cdots. \tag{1}$$

The so-called second variation $P_2[u(x); \lambda]$ is positive definite for $0 \leqq \lambda < 1$, positive semi-definite for $\lambda = 1$, and normally indefinite for $\lambda > 1$.

Let $u_h(x)$, $h = 1, 2, \ldots, m$, denote the complete set of linearly independent buckling modes at $\lambda = 1$ for which the second variation

---

[1] [19—21].

is equal to zero. We orthonormalize these buckling modes in such a way that we may write the second variation, evaluated for a linear combination of buckling modes $a_h u_h(x)$ at a load factor λ near the critical value 1, in the form

$$P_2[a_h u_h(x); \lambda] = (1 - \lambda)\, a_h a_h [1 + O(1 - \lambda)]. \tag{2}$$

Here and in the sequel we employ the summation convention for a repeated subscript.

An arbitrary (additional) displacement field $u(x)$ may always be decomposed into a linear combination of buckling modes and a residual field $v(x)$ orthogonal to all buckling modes.

$$u(x) = a_h u_h(x) + v(x). \tag{3}$$

The second variation $P_2[v(x); \lambda]$ is positive definite (under the orthogonality conditions for $v(x)$) for $0 \leqq \lambda < \lambda_2$, where $\lambda_2 > 1$. We may therefore minimize the functional (1) at $\lambda = 1$ with respect to $v(x)$, at least for sufficiently small (fixed) values of all amplitudes a_h of the buckling modes. This minimizing field may then be written in the form

$$[v(x)]_{\min} = a_h a_k v_{hk}(x) + O[(a_h a_h)^{3/2}], \tag{4}$$

where the fields $v_{hk}(x) = v_{kh}(x)$ are unique.

The simplest case of so-called cubic systems (the general case in a mathematical sense) arises, if the cubic form

$$A_{ijk} a_i a_j a_k \overset{\text{def}}{=} P_3[a_h u_h(x); 1] \tag{5}$$

does not vanish identically. In that case we have near the bifurcation point

$$\underset{(a_h=\text{const})}{\text{Min}} P[a_h u_h(x) + v(x); \lambda] = (1 - \lambda)\, a_i a_i [1 + O(1 - \lambda)] +$$
$$+ A_{ijk} a_i a_j a_k + O[(1 - \lambda)(a_h a_h)^{3/2} + (a_h a_h)^2], \tag{6}$$

independent of the field (4).

The case of so-called quartic-systems arises, if the cubic form (5) vanishes identically. In this case we may define a quartic form by

$$A_{ijkl} a_i a_j a_k a_l \overset{\text{def}}{=} P_4[a_h u_h(x); 1] - P_2[a_h a_k v_{hk}(x); 1]. \tag{7}$$

Near the bifurcation point we have now

$$\underset{(a_h=\text{const})}{\text{Min}} P[a_h u_h(x) + v(x); \lambda] = (1 - \lambda)\, a_i a_i [1 + O(1 - \lambda)] +$$
$$+ A_{ijkl} a_i a_j a_k a_l + O[(1 - \lambda)(a_h a_h)^2 + (a_h a_h)^{5/2}]. \tag{8}$$

The post-buckling behaviour near the bifurcation point is now described to a first approximation by the function

$$F(a_h; \lambda) = (1 - \lambda)\, a_i a_i + \frac{A_{ijk} a_i a_j a_k}{A_{ijkl} a_i a_j a_k a_l}, \tag{9}$$

where the upper and lower expressions hold for cubic and quartic systems respectively. The equilibrium configurations are characterized by stationary values of (9) with respect to the amplitudes a_h, and they are stable if and only if these stationary values are proper (relative) minima.

A significant quantity, not considered explicitly in [19], is the generalized additional deflection associated with the load factor in a post-buckling path, the generalization of $F_1 \Delta$ in the example of Fig. 1 c. This additional generalized deflection $\Delta\varepsilon(\lambda)$ is defined by the equation

$$\int_{\lambda}^{1} \Delta\varepsilon(\lambda)\, \mathrm{d}\lambda = F(a_h; \lambda), \qquad (10)$$

where a_h represents the equilibrium values of the amplitudes at the load factor λ. We have therefore Budiansky's general result $\Delta\varepsilon(\lambda) = a_i a_i$ [5].

The influence of initial imperfections in the structure, of any type, manifests itself primarily by the fact that the fundamental state of the perfect structure $U(x; \lambda)$ is no longer an equilibrium configuration for the imperfect structure. This implies that the potential energy functional (1) has to be modified, first of all by the addition of a term which is linear in the (additional) displacement field $u(x)$. For small imperfections this additional linear term will also be linear in the imperfections, and for small deflections this bilinear term is certainly the dominant influence of the imperfections. The result of this simple argument is that the function (9) is replaced by

$$F^*(a_h; \lambda) = F(a_h; \lambda) - 2\varkappa B_i a_i, \qquad (11)$$

where B_i depends on the type of imperfections, and $\varkappa$ is a nondimensional parameter which measures their magnitude. The numerical factor (-2) has only been included for convenience.

Considerable simplifications are possible, if the pre-buckling fundamental state of the perfect structure is governed, with adequate accuracy, by the linearized equations of classical elasticity theory. Whereas the function (9) is an adequate approximation in the general theory, if and only if both $(1 - \lambda)$ and all amplitudes a_h are small in magnitude, the first restriction does not apply in the case of a linear pre-buckling state. Moreover, in the most important case in practice of purely geometric imperfections $u^0(x) = \varkappa[a_h^0 u_h(x) + v^0(x)]$ the function (11) is simplified to

$$F^*(a_h; \lambda) = F(a_h; \lambda) - 2\lambda\varkappa a_i^0 a_i, \qquad (12)$$

whose validity is again subject to the single restriction that all amplitudes a_h must be small in magnitude, as well of course as the imperfection amplitudes $\varkappa a_h^0$.

The general theory of bifurcation buckling and of the influence of imperfections outlined above, and elaborated in a little detail for simultaneous buckling modes in the next section, has been rather slow in gaining wider acceptance and application. In the past ten to fifteen years, however, the relatively few examples in [19—21], only four in number, have been augmented significantly, in particular by the Harvard group of solid mechanics and by the similar group at University College in London. We may refer to a survey article [16] and to the numerous references in the excellent monograph by Thompson and Hunt [36]. Equivalent alternative formulations of the general theory have been given, e.g. by Sewell [33, 34] in the form of a perturbation analysis, and in particular by Budiansky [5] with a most effective and lucid employment of the concepts of functional analysis. An account of the theory is also contained in Vol'mir's monumental treatise on the stability of deformable systems [38].

The general theory is exact in an asymptotic sense. Its range of validity may be quite small, in particular if the fundamental state has higher bifurcation points λ_2, λ_3 etc. close to $\lambda_1 = 1$, whose associated buckling modes couple with the critical modes at $\lambda_1 = 1$ in the cubic part $P_3[u(x); \lambda]$ of the energy functional. A first step to overcome this difficulty was suggested in [19] in the replacement of $P_2[v(x); 1]$ by $P_2[v(x); \lambda]$ when the energy functional is minimized with respect to the field $v(x)$ orthogonal to the critical buckling modes. The result of this modification is a replacement of the function $F(a_h; \lambda)$ by another function of degree 4 in the amplitudes, with a more complicated dependence on the load factor λ. Unfortunately, no general criterion seems to be available to assess the accuracy of this refinement. The method has been applied, however, to the buckling problem of a complete spherical shell under external pressure where the difficulty occurs in an extreme form. In this case we have a cluster of bifurcation points just above the critical load factor $\lambda_1 = 1$, and the associated buckling modes exhibit a strong coupling with the critical modes in the cubic energy term. The results of the modified analysis are indeed spectacular in this case [39], even if we have no criterion to assess its range of validity.

3. Some General Theorems for Multi-Mode Buckling

We may always divide the functions $F(a_h; \lambda)$ and $F^*(a_h; \lambda)$ by their dimensional factor. Without any loss in generality we shall therefore now assume that these functions, as well as all quantities appearing in them, are non-dimensional. The amplitudes a_i are regarded as components of a vector a in Euclidean m-space, and we write $a = ae$, where e is a unit vector in the direction of the post-buckling path and a is the magnitude of the deflection. The generalized deflection associated with the load is a^2. Again without loss in generality we may take B_i in (11) and a_i^0 in (12) as the components of unit imperfection vectors B and a^0 in m-space.

The equations of equilibrium and the stability condition for the imperfect structure, associated with the energy function (11) are

$$2(1 - \lambda)\, ae_i + \frac{3a^2 A_{ijk}e_je_k}{4a^3\, A_{ijkl}e_je_ke_l} - 2\varkappa B_i = 0, \quad i = 1, 2, \ldots, m, \tag{13}$$

$$2(1 - \lambda) + \frac{6a\, A_{ijk}b_ib_je_k}{12a^2\, A_{ijkl}b_ib_je_ke_l} \geqq 0, \tag{14}$$

where the latter condition must hold for all unit vectors b.

For the perfect structure ($\varkappa = 0$) the directions e of the post-buckling paths evidently coincide with the unit vectors t for which the form

$$A_{ijk}t_it_jt_k \quad \text{or} \quad A_{ijkl}t_it_jt_kt_l \tag{15}$$

takes a stationary value on the unit sphere $|t| = 1$. Let $A_3(t)$ or $A_4(t)$ denote such a stationary value. The magnitude of the deflection a in the post-buckling path in the stationary direction t is given by

$$a = \frac{2(\lambda - 1)}{3A_3(t)} \quad \text{or} \quad a = \left[\frac{\lambda - 1}{2A_4(t)}\right]^{1/2}, \tag{16}$$

where the sign of $(\lambda - 1)$ coincides with the sign of $A_3(t)$ or $A_4(t)$. It is easily verified from a violation of the stability condition (14) for $b = t$ in descending post-buckling paths ($\lambda < 1$ for $A_3(t) < 0$ or $A_4(t) < 0$) that these descending paths are necessarily unstable. The generalized additional deflection associated with the load (a nondimensional measure for the end shortening in the example of Fig. 1) is given by

$$\Delta\varepsilon(\lambda) = \frac{4(\lambda - 1)^2}{9A_3^2(t)}, \quad \Delta\varepsilon(\lambda) = \frac{|\lambda - 1|}{2A_4(t)}. \tag{17}$$

Of some special interest is the post-buckling path for which the cubic or quartic form (15) takes its absolute minimum on the unit sphere. Let t^* denote the minimizing direction. The associated minimum for a cubic system $A_3(t^*) = A_3^*$ is always negative, and it corresponds by (16) and (17) to the post-buckling path of steepest descent. For a quartic system the minimum $A_4(t^*) = A_4^*$ may have either sign. The associated post-buckling path is stable for $A_4^* > 0$, the path of slowest ascent. On the other hand, for $A_4^* < 0$ the post-buckling path in the direction t^* is again the (unstable) path of steepest descent.

The analysis of imperfect structures is nearly equally simple, if the imperfection vector B coincides with the direction t of a post-buckling path. The direction e of the deflection $a = ae$ also coincides with t, and the magnitude a is the root of the equation

$$2(1 - \lambda)\, a + \frac{3A_3(t)\, a^2}{4A_4(t)\, a^3} - 2\varkappa = 0 \tag{18}$$

which tends to zero for $\varkappa \to 0$. For negative values of $A_3(t)$ or $A_4(t)$ this so-called natural branch has a limit point at a value $\lambda < 1$ specified by the equation

$$(1 - \lambda)^2 = -6\varkappa A_3(t) \quad \text{or} \quad (1 - \lambda)^{3/2} = (3/2)\,\varkappa[-6A_4(t)]^{1/2}. \tag{19}$$

This limit load factor is an upper bound for the stability of equilibrium on the natural branch. Of particular interest here are imperfections in the direction of the post-buckling path of steepest descent $B = t^*$ and the associated limit load factor λ^* is the root $\lambda^* < 1$ of the equation

$$(1 - \lambda^*)^2 = -6\varkappa A_3^* \quad \text{or} \quad (1 - \lambda^*)^{3/2} = (3/2)\,\varkappa(-6A_4^*)^{1/2}. \tag{20}$$

In cases where a linearized pre-buckling analysis is sufficiently accurate, the asymptotic formulae (18), (19) and (20) for small imperfection parameters $\varkappa$ may often be rendered more accurate by the replacement of $\varkappa$ by $\lambda\varkappa$ (cf. (11) and (12)).

In connection with (19) we have noted that this limit load factor constitutes an upper bound for the critical load factor of an imperfect structure with an imperfection vector $\varkappa t$, where t is a descending post-buckling path of the perfect structure. The natural branch of the imperfect structure may indeed reach a critical load factor by bifurcation before the limit load factor is attained. For cubic systems, however, it has recently been proved by D. Ho [14] that the limit load factor λ^*, defined by (20), associated with an imperfection vector in the direction of the post-buckling path of steepest descent, is a lower bound for the stability limit of the imperfect structure with an arbitrary direction of the imperfection vector. In other words, the most detrimental imperfections are those in the direction of the post-buckling path of steepest descent. Ho has also derived a lower bound for the stability limit of imperfect structures in the case of quartic systems, but this result is far less significant because it also involves the maximum value of the quartic form (15).

Ho's theorem for cubic systems is based on some simple properties of homogeneous cubic forms on the unit sphere. We first define the quantities

$$A_3(e) = A_{ijk}e_i e_j e_k, \quad C_3(e) = \operatorname*{Min}_{e = \text{const}} A_{ijk}b_i b_j e_k, \tag{21}$$

where both e and b are unit vectors. The properties required for our purpose are now

$$A_3(e) \geqq C_3(e) \geqq A_3(t^*) = A_3^*. \tag{22}$$

The first part of this continued inequality is an immediate consequence of the definitions (21). The second part is a consequence of the theorem that the minimum value of a trilinear form $A_{ijk}a_i b_j c_k$ in three unit vectors a, b, c occurs when these vectors have either the same or opposite directions.

In our simplified proof of Ho's theorem we note first that on the natural branch the stability condition (14) is always satisfied for sufficiently small values of the load factor. The stability limit is reached, therefore, when the equality sign holds for the critical unit vector b. At the critical load factor λ we have therefore

$$a = -\frac{1-\lambda}{3C_3(e)}. \tag{23}$$

Since a is necessarily positive, and because we are interested in critical load factors less than unity, we need only consider cases where $C_3(e)$ is negative at the critical load factor.

Multiplying both members of Eq. (13) by e_i and summing, we obtain

$$2(1-\lambda)a + 3a^2 A_3(e) - 2\varkappa B = 0, \tag{24}$$

where $B = B_i e_i$ is the scalar product of the unit vectors in the direction of the imperfection and the deflection. On the natural branch we have $B > 0$ for sufficiently small loads. Inequality (22) then ensures that B remains positive at least up to the critical value of a specified by (23). Substituting now from (23) into (24) we arrive at the equation for the critical load factor

$$(1-\lambda)^2 \left[2 - \frac{A_3(e)}{C_3(e)} \right] = -6\varkappa B C_3(e). \tag{25}$$

Since $0 < B \leqq 1$ Ho's theorem $\lambda^* \leqq \lambda < 1$ is now easily verified by a comparison between (20) and (25), and with the aid of inequalities (22).

4. Experimental Evidence and Its Correlation to Theory

To put it mildly, buckling theory and experiments have not always co-existed in harmony. The cold war before the end of last century arose out of the circumstance that columns were hardly ever slender enough to allow the application of Euler's formula. Agreement between theory and experiment was hardly better for the buckling of plates and shells in the twenties and thirties of the present century. As usual, both sides were again to blame for their disagreement. Ill-conceived applications of linear theory to essentially nonlinear problems can hardly be expected to agree with the results of badly controlled experiments, no matter how much money is spent on both.

The application of nonlinear analysis occurred first to buckling problems of thin flat plates which have a rewarding stable post-buckling behaviour with a significant load-carrying capacity. The pioneering investigations of nonlinear imperfection-sensitive column models by Cox [8] and von Kármán, Dunn and Tsien [18] spawned many further researches, including the general theory in [19]. Considerable experimental difficulties had still to be overcome in the buckling of shells, and the first reliable verification of theory was achieved by Roorda [31, 32, 24], somewhat ironically on a simple model which might also have been investigated a long time ago. Time does not permit a detailed discussion of the many careful experiments on shell buckling, carried out in recent years in many institutes, but a number of speakers at our meeting will report on this work. It would seem that a reasonable measure of agreement between theory and experiment has now been achieved, at least to the extent that theorists and experimentalists can safely attend the same meeting. I should like to add only a few comments on aspects which I have noted recently.

In an extensive series of experiments on the buckling of cylindrical shells under axial compression at the California Institute of Technology, in which the geometric imperfections of the shells were measured quite carefully before the test, it was found that asymmetric imperfections with a long axial wave length dominate. The calculated critical stress on the basis of these long-wave imperfections alone is unfortunately still considerably larger than the experimental buckling stress. On the other hand, if one assumes the simultaneous presence of very small imperfections in the short-wave axisymmetric and asymmetric modes which couple with the long-wave mode in the post buckling behaviour, the influence of the imperfections is multiplied by a factor 4! Small irregularities in the shell thickness, of a magnitude likely to occur (on the basis of a few measurements of the thickness distribution), may already provide the required short-wave imperfections.

Most calculations of imperfection-sensitivity have been carried out for simple shapes of the imperfection distribution, selected for the convenience of analysis. It is generally realized, of course, that actual imperfections are unlikely to follow this regular pattern. Some information is available about more or less localized imperfections [1, 2], and several computer programs are available for a detailed analysis of arbitrary imperfections, be it at the expense of considerable computer time. A somewhat different approach is suggested in a note [29] submitted for discussion next Friday, which suggests that the effect of more or less localized imperfections is not much less severe than the regular distribution assumed in earlier calculations. It should be acknowledged that the convenience of the analyst again guided the selection of the shape of localized imperfections: the evaluation does not require any calculation at all.

5. Physical Background and Mathematical Difficulties of the Energy Criterion

The subservience of continuum mechanics to thermodynamics is rightly ignored in problems of structural analysis dealing with equilibrium situations. It is not always realized, however, that stability is essentially a dynamic concept, even if its analysis in the theory of structures is usually performed by means of equilibrium considerations or by applying the quasi-static energy criterion. A proper justification of these methods cannot be obtained, however, without recourse to the theory of continuum thermodynamics.

The foundations of a thermodynamic theory of elastic stability were laid by Gibbs [12], and the theory was elaborated considerably by Duhem in his monumental treatise [9]. The energy criterion of stability in its purely mechanical form, however, did not emerge before Ericksen succeeded in separating the thermal effects from the mechanical aspects [10, 11].

In our discussion of the thermodynamic background of the theory of elastic stability [26, 27] we have widened the concept of an elastic body in the following sense. A body is called elastic, if there exists a reversible, infinitely slow path from one configuration to another configuration, but actual deformations with nonvanishing strain rates need not be reversible.[1] The elastic energy density, to be employed in the quasi-static energy criterion, is defined by the free energy density at the uniform equilibrium temperature of the pre-buckling fundamental state. Following in Ericksen's footsteps it is then shown that a positive definite total potential energy is a sufficient condition for dynamic stability. Likewise it has been shown that the presence of a positive definite energy dissipation for non-vanishing strain rates implies instability in the case of an indefinite potential energy.

Whereas the physical background of the quasi-static energy criterion of elastic stability seems now to be well-established, we have to recognize a lack of mathematical foundations for the conventional application of this criterion. This application is always based, tacitly or explicitly, on the replacement of the requirement of a positive definite potential energy functional by the less exacting requirement that this functional must have a weak proper relative minimum in the equili-

---

[1] Our widened concept of an elastic material appears to be essentially equivalent to Coleman's material with fading memory, at least in its application to elastic stability theory [6, 7].

brium configuration. There is no question (in the presence of dissipation) about the necessity of the stability conditions resulting from the conventional energy criterion, but no rigorous general proof exists of their sufficiency.

The mathematical difficulty in question was noted already early in this century [13], and the reason for its absence in the simple case of Euler's elastica problem, viz. the occurrence of a positive definite term in the highest derivative of the normal deflection, was already recognized at that time. It has been shown recently that the difficulty may also be overcome in the theory of stability of flat plates and shallow shells, again as a result of the presence of a positive definite term in the second derivatives of the normal deflection [25]. In the three-dimensional theory we are not aware of any significant result, except when the somewhat artificial assumption is made that the elastic energy density includes a positive definite term in the strain gradients, in addition to the classical expression in terms of the strains themselves [22]. The recent advance claimed in [30] is spurious because assumption (c) in the basic theorem 4.1 of [30] is equivalent to the assumption that stability is governed by a weak minimum of the potential energy.

6. Concluding Remarks

In spite of the various open questions of a mathematical character, referred to in the previous section, I do not share the pessimistic and derogatory view, expressed so eloquently by some critics [37], that the vast effort devoted to our field in the past and now would be to little purpose. I have refuted these malignant attacks already on a previous occasion [23]. The proof of the pudding in any theory of engineering science lies in the eating, the practical application to pertinent problems, and I believe the theory of buckling of structures has passed this crucial test with honours. I have no doubt that the many papers at the present IUTAM-symposium will confirm my confidence in the theory of buckling as a healthy, virile and rewarding field of engineering science.

References

1. Amazigo, J. C.: Buckling under External Pressure of Cylindrical Shells with Dimple Shaped Initial Imperfections. Int. J. Solids and Structures **7**, 883—900 (1971).
2. Amazigo, J. C.; Budiansky, B.: Asymptotic Formulas for the Buckling Stresses of Axially Compressed Cylinders with Localized or Random Axisymmetric Imperfections. J. Appl. Mech. **39**, 179—184 (1972).
3. Budiansky, B.; Hutchinson, J. W.: Dynamic Buckling of Imperfection-Sensitive Structures. Proc. 11th Int. Congr. Appl. Mech., München 1964. Berlin, Heidelberg, New York: Springer 1966, pp. 636—651.

4. Budiansky, B.: Dynamic Buckling of Elastic Structures: Criteria and Estimates. In: Dynamic Stability of Structures. Pergamon Press 1965, pp. 83—106.
5. Budiansky, B.: Theory of Buckling and Post-Buckling Behavior of Elastic Structures. Advances in Applied Mechanics 14, 1—65 (1974).
6. Coleman, B. D.: On the Energy Criterion for Stability. In: Nonlinear Elasticity. New York: Academic Press 1973, pp. 31—55.
7. Coleman, B. D.; Dill, E. H.: On Thermodynamics and the Stability of Motions of Materials with Memory. Arch. Rat. Mech. Anal. 51, 1—53 (1973).
8. Cox, H. L.: Stress Analysis of Thin Metal Construction. J. Roy. Aer. Soc. 44, 231 (1940).
9. Duhem, P.: Traité d'énergétique. 2 vols. Paris 1911.
10. Ericksen, J. L.: A Thermo-Kinetic View of Elastic Stability Theory. Int. J. Solids and Structures 2, 573—580 (1966).
11. Ericksen, J. L.: Thermoelastic Stability. Proc. 5th U.S. National Congr. Appl. Mech., 187—193 (1966).
12. Gibbs, J. W.: On the Equilibrium of Heterogeneous Substances. Trans. Connecticut Acad. Arts and Sci. 3, 108—248, 343—524 (1875—1878). Also in Collected Works, Vol. 1, 1931.
13. Hellinger, E.: Allgemeine Ansätze der Mechanik der Kontinua. Enz. math. Wiss. VI-4, 601—694, in particular 653—654 (1914).
14. Ho, D.: Buckling Load of Nonlinear Systems with Multiple Eigenvalues. Int. J. Solids and Structures 10, 1315—1330 (1974).
15. Hutchinson, J. W.; Amazigo, J. C.: Imperfection-Sensitivity of Eccentrically Stiffened Cylindrical Shells. AIAA Journal 5, 392—401 (1967).
16. Hutchinson, J. W.; Koiter, W. T.: Postbuckling Theory. Appl. Mech. Revs. 23, 1353—1366 (1970).
17. Hutchinson, J. W.: Plastic Buckling. Advances in Appl. Mech. 14, 67—144 (1974).
18. von Kármán, T.; Dunn, L. G.; Tsien, H. S.: The Influence of Curvature on the Buckling Characteristics of Structures. J. Aer. Sci. 7, 276 (1940).
19. Koiter, W. T.: Over de stabiliteit van het elastisch evenwicht. Thesis Delft. Amsterdam: H. J. Paris 1945. English translations issued as NASA TT F10, 833 (1967) and AFFDL TR 70-25 (1970).
20. Koiter, W. T.: Buckling and Post-Buckling Behaviour of a Cylindrical Panel under Axial Compression. NLR Report S476, Reports and Trans. Nat. Aero. Res. Inst. 20, Amsterdam (1956).
21. Koiter, W. T.: Elastic Stability and Post-Buckling Behaviour. Proc.. Symp. Nonlinear Problems. Madison: Univ. of Wisconsin Press 1963, pp. 257—275.
22. Koiter, W. T.: The Energy Criterion of Stability for Continuous Elastic Bodies. Proc. Kon. Ned. Ak. Wet. B 68, 178—202 (1965).
23. Koiter, W. T.: Purpose and Achievements of Research in Elastic Stability. Proc. 4th Techn. Conf., Soc. Eng. Sci., North Carolina State Univ., Raleigh, N.C. (1966).
24. Koiter, W. T.: Postbuckling Analysis of a Simple Two-Bar Frame. Recent Progress in Applied Mechanics. Stockholm: Almquist and Wicksel 1966, pp. 337—354.
25. Koiter, W. T.: A Sufficient Condition for the Stability of Shallow Shells. Proc. Kon. Ned. Ak. Wet. B 70, 367—375 (1967).
26. Koiter, W. T.: On the Thermodynamic Background of Elastic Stability Theory. In: Problems of Hydrodynamics and Continuum Mechanics (L. I. Sedov Anniversary Volume) SIAM, Philadelphia 1969, pp. 423—433.

27. Koiter, W. T.: Thermodynamics of Elastic Stability. Proc. 3rd Can. Congr. Appl. Mech. Calgary 1971, pp. 29—37.
28. Koiter, W. T.: Stijfheid en Sterkte I, Grondslagen. Deel 4 van de reeks Mechanica. Haarlem: Scheltema en Holkema 1972.
29. Koiter, W. T.: The Influence of More or Less Localized Short-wave Imperfections on the Buckling of Circular Cylindrical Shells under Axial Compression (1974-05-15). Note submitted for general discussion session.
30. Naghdi, P. M.; Trapp, J. A.: On the General Theory of Stability for Elastic Bodies. Arch. Rat. Mech. Anal. 51, 165—191 (1973).
31. Roorda, J.: The Buckling Behavior of Imperfect Structural Systems. J. Mech. Phys. Solids 13, 267—280 (1965).
32. Roorda, J.: Stability of Structures with Small Imperfections. J. Eng. Mech. Div. ASCE 91, 87—106 (1965).
33. Sewell, M. J.: The Static Perturbation Technique in Buckling Problems. J. Mech. Phys. Solids 13, 247—264 (1965).
34. Sewell, M. J.: A General Theory of Equilibrium Paths through Critical Points, I, II. Proc. Roy. Soc. London A 306, 201—223, 225—238 (1968).
35. Sewell, M. J.: A Survey of Plastic Buckling. In: Stability, edited by H. Leipholz 1972, ch. 5, pp. 85—197.
36. Thompson, J. M. T.; Hunt, G. W.: A General Theory of Elastic Stability. London: John Wiley & Sons 1973.
37. Truesdell, C.; Noll, W.: The Nonlinear Field Theories of Mechanics. Handbuch der Physik, Bd. III/3. Berlin, Heidelberg, New York: Springer 1965. In particular section 68 b.
38. Vol'mir, A. S.: Stability of Deformable Systems (in Russian), Moscow (1967). An earlier edition under the title "Stability of Elastic Systems" (1963) was issued in an English translation as FTD-MT64-335 (1965).
39. Koiter, W. T.: The Nonlinear Buckling Problem of a Complete Spherical Shell under Uniform External Pressure. Proc. Kon. Ned. Ak. Wet. B 72, 40—123 (1969), in particular sections 8—10, pp. 80—110.

Critical Buckling Loads
of some Prismatic Plate Assemblies

F. W. Williams, W. H. Wittrick, R. J. Plank

University of Birmingham, Birmingham, England

Abstract

Prismatic plate assemblies include closed and open section struts, stiffened panels with open or closed section longitudinal stiffeners and longitudinally stiffened tubes. Recent theoretical advances have made possible the computation of critical buckling loads (or natural frequencies) for such assemblies. The answers obtained are "exact", in the sense that only the usual thin plate assumptions are made, so long as the ends of each component plate are simply supported or the longitudinal half-wavelength of the buckling mode is short compared with the length of the assembly. Initially, the theory only applied to assemblies of uniformly compressed, flat, isotropic plates. However, results can now be computed for assemblies which include curved anisotropic plates which carry in-plane bending and uniform in-plane shear and transverse stresses.

Existing computer programs enable one to investigate the critical buckling behaviour of a vast range of interesting plate assemblies. So far, results obtained cover only a relatively small part of this range. In this paper, a summary is given of the more interesting conclusions drawn from these results. The results include (a) comparisons with results from simpler, approximate analyses, (b) investigations of novel, advantageous structural forms, (c) examples of the effects of idealisations such as ignoring eccentricities between connected plates and finite radii of curvature at bends, (d) interaction between shear, bending and longitudinal compression for stiffened panels and (e) critical loads other than the lowest.

1. Introduction

Seven years ago the second author began working on a theory giving "exact" critical buckling stresses (and natural frequencies of undamped vibration) of prismatic assemblies of thin, flat, rectangular plates [1]. This theory now includes the most general form of plate anisotropy for which it is practical to obtain analytical expressions for the stiffnesses of component plates [2], although numerical procedures can be used to handle all forms of anisotropy [3]. The authors' computer programs, unlike other programs for similar structures [3, 4], include an algorithm [5] which can be applied to any linearly elastic structure to ensure that no critical buckling loads are missed.

The aim herein is to bring together conclusions drawn from the authors' many published results. Thus these results are not repeated in detail and only essential theory is described. The theory assumes that component plates are rigidly connected together along common longitudinal edges and that buckling displacements vary sinusoidally along every longitudinal line with a common half-wavelength λ. Thus, plates must be simply supported at their ends or λ must be small compared with l, the length of the assembly. The plates can be isotropic or anisotropic (subject to some restrictions, e.g. see Section 8) and can carry in-plane longitudinal, transverse and shear forces per unit length, of magnitudes N_L, N_T and N_S respectively, which are longitudinally and transversely invariant, see Fig. 1. The theory pre-

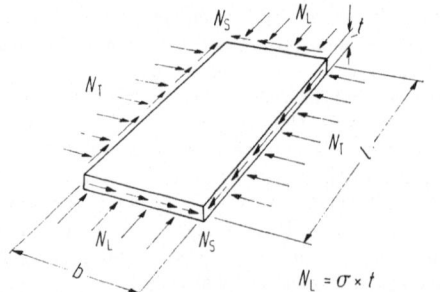

Fig. 1.
A component plate.

dicts all possible modes of buckling, does not involve the approximations of finite element analyses and permits considerable computational savings through sub-structuring [6].

The typical plate assembly cross-sections of Figs. 2a—g are for, respectively, panels with N unflanged and flanged integral stiffeners, panels with N trough and vee stiffeners, a corrugated core sandwich panel with N corrugation repetitions and unstiffened and stiffened N-sided regular polygonal tubes (with $N = 8$). Each structure is an assembly of sets of identical plates, e.g. the panel of Fig. 2b is assembled from $3N + 1$ plates of four different types, representing flanges, webs,

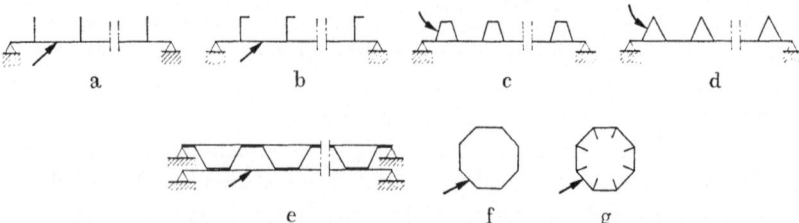

Fig. 2a—g. Some typical cross-sections of prismatic plate assemblies. Δ denotes a continuous longitudinal support which only prevents vertical motion.

inter-stiffener skin portions and the two outer skin portions. For all problems considered later, $b = b_1$ and $t = t_1$ for the arrowed plates on this and subsequent figures, where b_1/t_1 exceeds or equals b/t for all other plates in the assembly.

2. Presentation of Results

The assembly loading is defined by base values of N_L, N_T and N_S in each plate and a load factor F, by which they are all multiplied. Therefore initial buckling corresponds to F_C, the least value of F which causes critical buckling. However, the theory involves assuming successive values for λ and finding the corresponding values of F at buckling. This enables F at buckling to be plotted against λ, giving a curve applying to any panel with the specified cross-section, irrespective of its length, l. Figure 3 [7] gives such a curve for the panel of Fig. 2b, with $N = 12$, $t = t_1$ for all plates, $b_1/t_1 = 50$ and $b = b_1/2$ and $0.15b_1$, for the webs and flanges, respectively. The plates are isotropic, their base loads correspond to a uniform longitudinal stress of unity and E is Young's modulus. Thus here $F_C = \sigma_C$, the initial buckling stress of the panel in longitudinal compression. Strictly, Fig. 3 gives the buckling load factor for an infinitely long panel which is constrained to buckle with a half-wavelength of λ. However, for this and all other problems which do not involve sheared or anisotropic plates, the buckling mode at any value of λ is such that the translations perpendicular to all longitudinal lines are in-phase with each other [2]. Thus there are undistorted cross-sections at longitudinal intervals of λ, although points on these cross-sections translate longitudinally by

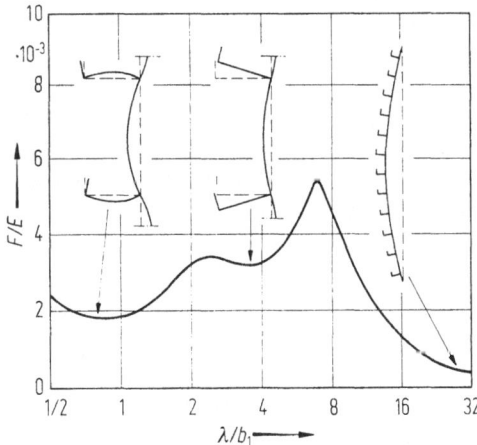

Fig. 3. Illustrative plot of F versus λ. The modes shown are roughly to scale.

varying amounts. Therefore, such modes also apply to panels of length l with simply supported ends, if $\lambda = l/i$ and i is an integer. Hence F_C is the lowest value of F given by Fig. 3 for $\lambda = l/i$, with $i = 1, 2, 3, \ldots$, and λ_C is defined as the corresponding value of λ.

This shows how the buckling load of any assembly with simply supported ends can be found from a figure like Fig. 3, so long as no plates are anisotropic or sheared. If the end constraints include all those implied by simple supports plus others (e.g. clamped ends) the same procedure will give a lower bound on F_C [8], since adding constraints to a structure can only move its critical loads upwards [5]. Likewise, it can be shown [8] that the procedure also gives a lower bound for assemblies with simply supported ends when sheared or anisotropic plates are present. Finally, all lower bounds are presumed to be close for $\lambda \ll l$.

Figure 3 includes the modes corresponding to the three concave portions into which the curve is divided by the convex portions around $\lambda/b_1 = 2.4$ and $\lambda/b_1 = 7$. From left to right the modes are local, with negligible plate junction translations, so-called torsional, primarily involving plate junction rotations and in-plane stiffener flange translations, and overall.

For convenience, σ and σ_C are used in place of F and F_C in Sections 3 to 6, since only uniform longitudinal compression is involved.

3. Comparisons with Approximate Results

Approximate methods of analysis frequently involve considering possible buckling modes separately, but their relative simplicity can still make them attractive.

Local buckling calculations often neglect plate junction translations and our results justify this approximation, e.g. the local buckling stress was less than 2% too high for a panel almost identical to that of Fig. 3 [7]. They also frequently assume mode repetition, at transverse intervals of $2b_1$, for the panels of Fig. 2a, b, implying $N = \infty$. Our results [7] show that if N is varied for the panel of Fig. 3, the buckling stress for $N = \infty$ under-estimates the values for $N = 3$, 5 and 8 by, respectively, about 8.6%, 3.5% and 1.5%. Such under-estimates are usually acceptable, except for N low.

Torsional buckling stresses can be over-estimated by 100%, in unfavourable situations, by assuming that the stiffener cross-sections of Fig. 2b rotate without distorting [7].

The overall buckling stress for one of the panels of Figs. 2a—e is often assumed to equal the Euler stress of a pin-ended strut of length l with cross-section identical to the repeated part of the panel cross-section. This approximation gave results which were about -9%, 10% and 41% too low at $\lambda/b_1 = 8$, 32 and 64, respectively, for the panel of Fig. 3, modified to give $b = 0.05b_1$ for the flanges [7]. Such accuracy is often acceptable, particularly for the most important range for this problem, $8 \leqq \lambda/b_1 \leqq 32$.

4. Further Comments on Modes

While Fig. 3 is typical for panels with flanged integral stiffeners many other panel cross-sections, e.g. see Fig. 2c, give curves with only two concave portions, corresponding to local and overall modes. In contrast, the curve for the panel of Fig. 4a has four portions. The outer portions

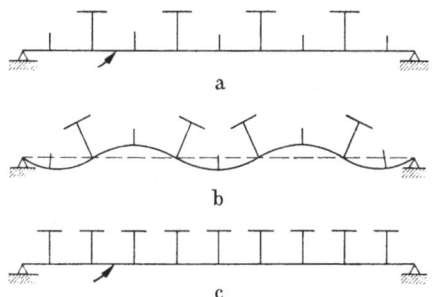

Fig. 4. a) and c) Two panel cross-sections with plate breadths approximately to scale and b_1/t_1 values of 29 and 30, respectively; b) Partial overall mode for the panel of a).

correspond to local and overall modes, the right-hand portion remaining corresponds to a torsional mode dominated by plate junction rotations and horizontal translations of stiffener flanges and the other portion corresponds to the partial overall mode sketched in Fig. 4b, which primarily involves plate junction rotations, horizontal flange translations and vertical translations of the small stiffeners. This panel was about 7% lighter than the comparable panel of Fig. 4c, which had the same length, breadth and buckling *load*, but only involved one stiffener size [9]. The weight saving emphasises the value of programs which enable novel panels to be analysed without prior knowledge of their possible modes.

5. Higher Critical Loads

Figure 5 demonstrates how the 10 lowest critical stresses vary with λ, for the symmetrical panel cross-section shown [8]. The panel has fully clamped longitudinal boundaries and b and t for the webs and flanges (which are $14°$ from the horizontal) are $0.372b_1$, $0.064b_1$, $0.937t_1$ and $2.25t_1$, respectively, where $b_1/t_1 = 54.6$.

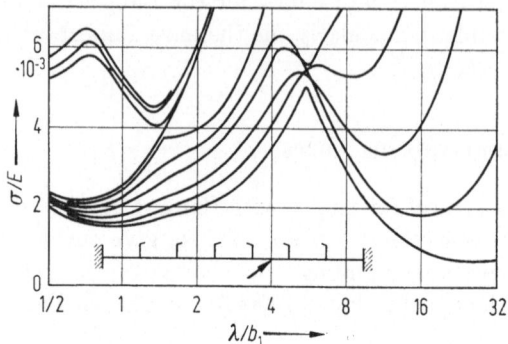

Fig. 5. The variation, with λ, of the 10 lowest critical stresses of the panel shown.

For $l = 16b_1$ the figure gives $\sigma_C/E \approx 0.95 \cdot 10^{-3}$ at $\lambda_C = l$ and the next critical stress with $\lambda = l$ is $\sigma/E \approx 1.8 \cdot 10^{-3}$. These two stresses are clearly separated by many (about 42) critical stresses [8] given by the three lowest curves with $\lambda = l/i$ and $8 < i < 32$.

The seven lowest curves bunch at $\lambda \approx b_1$ because b/t is highest for the seven identical plates forming the skin. Thus clamping the stiffener — skin junctions would make the first seven local buckling stresses coincide at σ_{CC}, the initial buckling stress of one skin-forming plate with its longitudinal edges clamped. Therefore, since removing clamps cannot raise critical stresses [5], the actual panel must have seven critical stresses at or below σ_{CC}.

Similarly, the lowest curves must be close at $\lambda \approx b_1$ for any of the cross-sections of Fig. 2. Thus the lowest short wavelength critical stresses will always be close together, and this may be significant when considering post-buckling behaviour.

6. Cross-Section Modelling Approximations

Herein, cross-sections were represented as assemblies of plates inter-connected along their longitudinal edge centre-lines. Thus Fig. 6b would represent the panel with bonded Z-section stiffeners of Fig 6a, involving the four following approximations:

1. Some material is omitted at web-flange junctions,
2. shaded material is counted twice,
3. the $2t_1$ thick plate is displaced by $t_1/2$ and
4. the small curved portions at web-flange junctions are ignored.

The first three approximations can be avoided by connecting plates eccentrically to appropriate longitudinal line junctions, see Fig. 6c.

Buckling stresses obtained, using Vipasa [2] for the alternative representations of Fig. 6b, c, agreed to within 2% for $\lambda/b_1 = 1, 2, 4, 8, 16$ and 32 [8]. Further results in reference [8] suggest that approximation

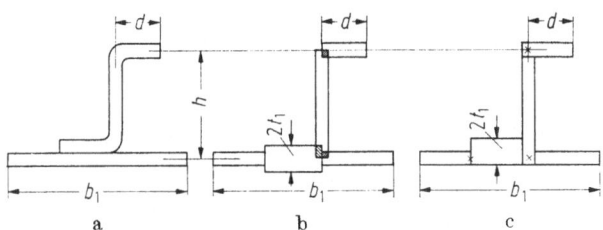

Fig. 6a—c. Two possible representations of a panel with Z-section stiffeners; b_1 is the stiffener pitch, $N = \infty$, $h = 0.5b_1$, $d = 0.3h$, $b_1/t_1 = 50$ and all plate thicknesses are t_1 except for the two plates of thickness $2t_1$ shown; crosses denote longitudinal line junctions.

(4) will also usually give acceptable accuracy. Thus the four approximations have probably affected the results presented herein very little, although 20% errors are possible for exceptionally unfavourable problems [8].

7. Interaction of Shear and Non-Uniform Compression

The results given are for the isotropic panels of Figs. 2a—e, with $N = 4$. $b_1/t_1 = 50$ for Figs. 2a, b, e, $= 41$ for Fig. 2c and $= 56$ for Fig. 2d. Component plate thicknesses are $2t_1$ for the skin in Figs. 2c, d, $t_1/2$ for the corrugation of Fig. 2e and t_1 elsewhere, while breadths are given approximately by scaling from Fig. 2. Each panel cross-section was loaded in shear and by a longitudinal compressive stress which varied linearly from σ_1 at the bottom to σ_2 at the top, with the greater of σ_1 and σ_2 denoted by σ^1. The shear stress was τ ($= N_S/t$ on Fig. 1) for each arrowed plate on Fig. 2 and the other plate shear stresses follow from the conditions of equilibrium at plate junctions and zero twist around closed cells. Thus the shear stresses in the skin and stiffeners are τ and zero, respectively, for the two panels of Figs. 2a, b.

Each point used when plotting Fig. 7 was found by obtaining an F versus λ curve (see Fig. 3), for the loading pattern defined by the σ^1/τ and σ_1/σ_2 values corresponding to the point, and picking the lowest point on the curve for which the mode was not overall. Thus, Fig. 7 only applies to panels for which overall buckling does not occur, and this implies limiting values of l (given in reference [8]) above which Fig. 7 cannot be used. Each curve on Fig. 7 has been made dimensionless using σ_0^1 and τ_0, the values at buckling of, respectively, σ^1 when

$\tau = 0$ and τ when $\sigma^1 = 0$. The values of σ_0^1 and τ_0 for each curve are given in reference [8].

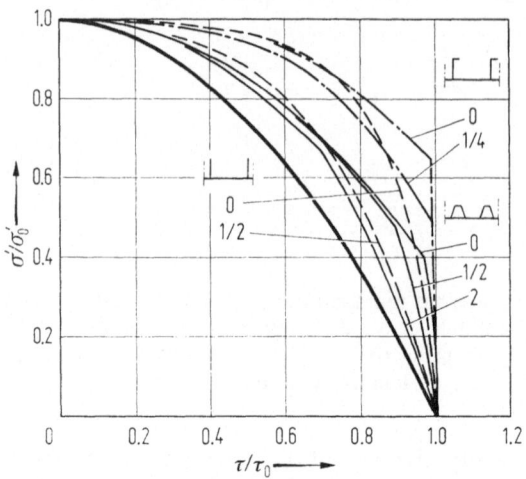

Fig. 7. Interaction curves, for the values of σ_1/σ_2 shown; dashed, chain-dotted and solid curves relate to Figs. 2a, b and c, respectively; the chain-dotted curves are approximate for $\tau/\tau_0 > 0.8$.

Reference [8] gives interaction curves with $\sigma_1/\sigma_2 = 0$, 1/4, 1/2, 1, 2, 4 and ∞ for the five panels and all 35 curves lie outside the parabola $\sigma^1/\sigma_0^1 + (\tau/\tau_0)^2 = 1$, which is the bold line of Fig. 7. Furthermore, the seven curves for each panel are nested in the order of their σ_1/σ_2 values, with the curve for $\sigma_1/\sigma_2 = \infty$ innermost. Most curves are within 3% of the parabola, in the sense that the distance from the origin to the curve along any straight line exceeds the distance to the parabola by less than 3%. The exceptions are the outer 4, 2 and 6 curves of Fig. 2a, b, and c, respectively, and Fig. 7 gives a representative sample of these curves, many of which are far from the parabola. The occasional discontinuities of slope correspond to mode changes.

8. Some Useful Approximations

The results of Section 7 were obtained using a special program which treated bending exactly [10, 8]. This program no longer exists. Available programs [2, 3], while being more powerful in most respects, do not include this facility. However, Vipasa [2] was used to check the values of σ_0^1 for the panel of Fig. 2a with $\sigma_2/\sigma_1 = 0$ and $\sigma_2/\sigma_1 = \infty$ [11], by treating each stiffener as j rigidly inter-connected plates of width $b_1/2j$, with the ith plate from the bottom $(i = 1, 2, ..., j)$ carrying a uniform stress of $\sigma_1 + (1/2j)\,(2i - 1)\,(\sigma_2 - \sigma_1)$. Thus, the linear stress variation was replaced by an approximately equivalent stepped variation. Calculations performed for $j = 2$, 4, 6 and 8 gave errors of about 10%, 2%, 1% and 0.5%, respectively, for $\sigma_2/\sigma_1 = \infty$, and 1%, 0.2%, 0.1% and 0.05% for $\sigma_2/\sigma_1 = 0$. Reference [11] gives further simulated bending results.

Encouraging comparisons with Flügge's results (reference [12], p. 427) for a uniformly longitudinally compressed isotropic cylinder of mean radius R and thickness T, with $R/T \approx 91.3$, were obtained when the cylinder was modelled as in Fig. 2f, with $t_1 = T$, $b_1 = 2\pi R/N$ and $N = 60$. Calculations performed for 15 typical values of λ in the range $0.05 \leq \lambda/R \leq 50$ agreed to within 2% with those given by Flügge's formula, the greatest difference being at $\lambda/R = 30$. This discrepancy was little changed when N was increased to 120 [11]. Similar agreement, this time with the exact results of reference [3], was obtained when curved anisotropic plates, loaded in either shear ($N_S \neq 0$) or longitudinal compression ($N_L \neq 0$) or hoop compression ($N_T \neq 0$), were represented by several flat plates [11].

Vipasa [2] can only handle anisotropic plates with uncoupled in- and out-of-plane behaviour, i.e. with $\boldsymbol{B} = \boldsymbol{0}$. However, the common composite plate with $\boldsymbol{B} \neq \boldsymbol{0}$ which occurs when two differing plates with $\boldsymbol{B} = \boldsymbol{0}$ are bonded together (e.g. the lower flange of a Z stiffener and the skin of a panel) can be handled approximately, by splitting each component plate into i portions of breadth b/i and rigidly interconnecting these portions with appropriate eccentricities, as shown in Fig. 8, where $i = 2$. About 4.5% and 0.3% agreement [11] with the

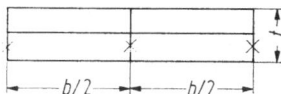

Fig. 8. Four anisotropic plates, connected together at the line junctions indicated by crosses, forming a single plate of width b and thickness t.

exact results [13] was obtained, with $i = 8$ and 32 respectively, for the buckling load, in pure longitudinal compression, of a plate which was simply supported along all four edges. This was a very demanding problem, with $b/t = 100$ and component plate thicknesses of $0.85t$ and $0.15t$. Smaller values of i (even $i = 1$) give adequate results for many problems [11]. The component plates in the above problem were assumed to be equally strained, both in the longitudinal and (in-plane) transverse directions. Thus, the two different plate portions carried differing longitudinal stresses plus transverse forces per unit length which were equal and opposite for the two portions.

9. Stability of Calculations

All the results for practical problems ever obtained by the authors have only needed single length arithmetic (11—13 significant figures, for the computers used). However, there are indications that polygonal representation of cylinders may sometimes become ill-conditioned at high, but practical, values of λ/R, e.g. those associated with Euler buckling.

References

1. Wittrick, W. H.: A Unified Approach to the Initial Buckling of Stiffened Panels in Compression. Aeronautical Quarterly. **XIX**, 265—283 (1968).
2. Wittrick, W. H.; Williams, F. W.: Buckling and Vibration of Anisotropic or Isotropic Plate Assemblies under Combined Loadings. Int. J. of Mech. Sci. **16**, 209—239 (1974).
3. Viswanathan, A. V.; Tamekuni, M.; Baker, L. L.; Elastic Stability of Laminated, Flat and Curved, Long Rectangular Plates Subjected to Combined In-Plane Loads. NASA-CR-2330, 1973.
4. Smith, C. S.: Bending, Buckling and Vibration of Orthotropic Plate-beam Structures. J. of Ship Research **12**, 249—268 (1968).
5. Wittrick, W. H.; Williams, F. W.: An Algorithm for Computing Critical Buckling Loads of Elastic Structures. J. of Structural Mechanics **1**, 497—518 (1973).
6. Williams, F. W.: Computation of Natural Frequencies and Initial Buckling Stresses of Prismatic Plate Assemblies. J. of Sound and Vibration **21**, 87—106 (1972).
7. Williams, F. W.; Wittrick, W. H.: Numerical Results for the Initial Buckling of Some Stiffened Panels in Compression. Aeronautical Quarterly **XXIII**, 24—40 (1972).
8. Plank, R. J.; Williams, F. W.: Critical Buckling of Some Stiffened Panels in Compression, Shear and Bending. Aeronautical Quarterly **XXV**, 165—179 (1974).
9. Williams, F. W.: Stiffened Panels with Varying Stiffener Sizes. J. of the Royal Aeronautical Society **77**, 350—354 (1973).
10. Wittrick, W. H.: General Sinusoidal Stiffness Matrices for Buckling and Vibration Analyses of Thin Flat-Walled Structures. Int. J. of Mech. Sci. **10**, 949—966 (1968)
11. Williams, F. W.: Approximations in Complicated Critical Buckling and Free Vibration Analyses of Prismatic Plate Structures. Aeronautical Quarterly **XXV**, 180—185 (1974).
12. Flügge, W.: Stresses in Shells. Berlin, Göttingen, Heidelberg: Springer 1960.
13. Jones, R. M.: Buckling and Vibration of Unsymmetrically Laminated Cross-Ply Rectangular Plates. American Institute of Aeronautics and Astronautics J. **11**, 1626—1632 (1973).

On the Collapse Load of Cylindrical Shells

P. T. Pedersen

The Technical University of Denmark, Lyngby, Denmark

Abstract

Bifurcation stresses and post-buckling behaviour of circular cylindrical shells under axial compression are analyzed. For shells with axisymmetrical, sinusoidal imperfections it is found that the advanced post-buckling behaviour can be such that higher loads than the bifurcation load can be supported before total collapse occurs. This holds good for relatively large imperfections (small bifurcation stresses), even in cases where the initial post-buckling behaviour indicates an unstable bifurcation.

An extension of the asymptotically exact initial post-buckling analysis into the advanced post-buckling region is presented. The advanced post-buckling behaviour is analyzed by means of the Galerkin method using as coordinate functions a set of initial post-buckling functions corresponding to a given axial wavelength.

Notation

a_j	Amplitudes of coordinate functions
A_{lj}, B_{lij}, C_{lijk}, D_{lj}	Coefficients in Galerkin equations (5.5)
$B = \beta^4 + 1 - 2\lambda\beta^2$	Classical buckling parameter
b_j	Amplitudes of nonaxisymmetric imperfection modes
$c = [3(1 - \nu^2)]^{1/2}$	
$D = Et^3/12(1 - \nu^2)$	Plate bending stiffness
E	Young's modulus
F	Stress function
$f = 2cF/Et^3$	
k	Longitudinal wavenumber
$q_0 = [12(1 - \nu^2)]^{1/4} [R/t]^{1/2}$	
R	Cylinder radius
s	Circumferential wavenumber
t	Shell wall thickness
U, V, W	Axial, circumferential, and radial displacements
W_0	Initial radial deflection (imperfection)
$w = W/t$	
w_u	Radial displacement in prebuckling state
w_j	Radial deflection function defined by Eq. (4.1)
X, Y	Longidutinal and circumferential coordinates
$x = Xq_0/R$	

28   P. T. Pedersen

$y = Yq_0/R$
β Wavenumber for imperfection
$\gamma = 2c\,\delta/t$
δ Axisymmetric imperfection amplitude
$\lambda = (Et/cR)$ Average axial compression
λ_c Value of λ at bifurcation point
ν Poisson's ratio
ξ Amplitude of initial buckling displacement

Operators

$$(\dot{\ }) = \frac{\partial(\ )}{\partial y}, \quad (\ )' = \frac{\partial(\ )}{\partial x}$$

ψ see Eq. (2.4)

1. Introduction

The present paper is devoted to a study of equilibrium paths in the advanced post-buckled region of imperfect circular cylindrical shells under axial compression.

In [1] the author studied the bifurcation loads and the initial post-buckling behaviour of axially compressed, unstiffened and ring-stiffened, circular cylindrical shells with axisymmetric, sinusoidal imperfection of arbitrary wavelength and amplitude. The method of analysis employed was that proposed by Budiansky and Hutchinson [2], where the bifurcation problem is formulated exactly and a complete family of bifurcation modes is identified. The initial post-buckling analysis is made within the context of Koiter's general theory [3].

The results of [1] show that when the imperfection amplitude is small, bifurcations from axisymmetric state into a non-axisymmetric state are initially unstable, but for larger imperfection amplitudes the bifurcations become stable so that loads above the bifurcation loads can be sustained. On the other hand, for long wavelengths of the imperfection this transition from unstable to stable bifurcations takes place at loads which are low compared to the reported experimental buckling loads for cylindrical shells. In [1] only the initial post-buckling behaviour was studied, and although the initial post-buckling behaviour indicates an unstable bifurcation, the possibility does exist that the general post-buckling behaviour is such that a higher load than the bifurcation load can be supported before total collapse occurs. This possibility is indicated in Fig. 1. Further, even if the initial post-buckling behaviour is stable it is possible that only a slightly higher load than the bifurcation load can be supported before collapse occurs. In this paper we will examine these possibilities by presenting results for the advanced post-buckling behaviour; that is, develop an extension of the asymptotically exact initial post-buckling behaviour determined in [1].

The large-displacement analysis giving the advanced post-buckling behaviour will here be based on a Galerkin method, where the coordinate functions are functions associated with the initial post-buckling

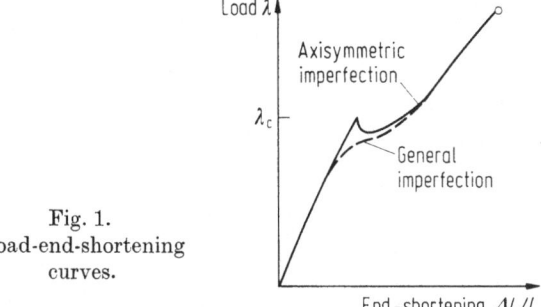

Fig. 1.
Load-end-shortening
curves.

analysis. Using such a series representation of large-deflection behaviour accurate results in the initial post-buckling region are obtained. In this respect the present analysis differs from the large-deflection analysis given by Hoff, Madsen and Mayers [4]. Their series representation is well suited for making real large-deflection calculations but lacks accuracy when applied to the initial post-buckling region. Also Hoff, Madsen and Mayers considered a perfect shell. We will here consider shells with initial imperfections.

2. Governing Equations

Figure 2 shows an infinitely long shell with thickness t. The middle surface of the shell is assumed to have an initial displacement W_0 in the radial direction from a perfect cylindrical surface of radius R.

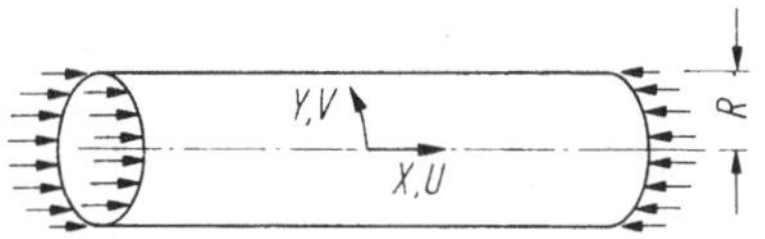

Fig. 2.
Part of cylindrical shell.

Points on the cylindrical surface are specified by the axial coordinate X and the circumferential coordinate Y. The initial imperfection of the shell is assumed to be composed of an axisymmetric part and a relatively smaller asymmetric part $\hat{W}$:

$$W_0 = -\delta \cos (p_x X/R) + \hat{W}(X, Y). \tag{2.1}$$

The behaviour of the elastic shell will be characterized by the radial displacement W and an Airy stress function F which gives the resulting membrane stresses as $N_{xx} = \partial^2 F/\partial Y^2$, $N_{yy} = \partial^2 F/\partial X^2$, and $N_{xy} =$

$-\partial^2 F/\partial X\,\partial Y$. Using the Karman-Donnell shell theory the non-dimensional equilibrium and compatibility equations take the form

$$\nabla^4 w + f'' = 2c\psi(f, w + w_0), \tag{2.2}$$

$$\nabla^4 f - w'' = -c\psi(w, w + 2w_0), \tag{2.3}$$

where ∇^4 is the two-dimensional biharmonic operator. With $q_0 = [12(1 - \nu^2)]^{1/4} [R/t]^{1/2}$, the independent variables are $x = Xq_0/R$ and $y = Yq_0/R$, and with $c = [3(1 - \nu^2)]^{1/2}$, the dependent variables are $f = [2c/Et^3]\,F$ and $w = W/t$. Also, $(\ )' \equiv \partial/\partial x\,(\ )$ and $(\dot{\ }) \equiv \partial/\partial y\,(\ )$. The operator ψ is given by

$$\psi(g_1, g_2) = g_1'' \ddot{g}_2 + \ddot{g}_1 g_2'' - 2\dot{g}_1' \dot{g}_2'. \tag{2.4}$$

The non-dimensional imperfection is assumed to be given by $w_0 = -\delta/t \cos \beta x + \hat{w}(x, y)$ where $\beta = p_x/q_0$ and $\hat{w} = \hat{W}/t$.

Now, it is well known that solutions to equations (2.2) and (2.3) do not necessarily fulfil the requirement that the tangential displacements are single-valued over any complete circuit of the shell. For the circumferential displacement V this condition is enforced if

$$\int_0^{2\pi q_0} \{f'' - \nu \ddot{f} - w - c(\dot{w})^2\}\,\mathrm{d}y = 0. \tag{2.5}$$

3. Axisymmetric Prebuckling State

Let us now assume that the initial imperfection is axisymmetric, i.e. $\hat{w}(x, y) \equiv 0$. Then the non-linear equations (2.2) and (2.3) admit the following axisymmetric solution for the unbuckled cylinder, which satisfies the condition (2.5) for single valued circumferential displacement:

$$w_{\mathrm{u}} = \frac{\nu\lambda}{c} - 2\,\frac{\delta}{t}\,\frac{\lambda\beta^2}{B}\cos\beta x, \quad f_{\mathrm{u}} = -\frac{\lambda y^2}{2c} + 2\,\frac{\delta}{t}\,\frac{\lambda}{B}\cos\beta x. \tag{3.1}$$

Here $B = \beta^4 + 1 - 2\lambda\beta^2$ and λ is the average axial compressive stress normalized by the classical buckling stress (Et/cR). The independent variable x is normalized so that the wavelength of the corresponding classical buckling mode is 2π.

For increasing compression λ, the axisymmetric solution (3.1) will be a stable equilibrium solution until a bifurcation point is reached where the shell bifurcates into a non-axisymmetric state.

4. Initial Post-Buckling Behaviour

As the bifurcation problem serves to generate coordinate functions for use in the analysis of the advanced post-buckling behaviour, we will, in this section, briefly present a solution to this problem. We will determine the compressive stress λ_c at which the bifurcation takes place and determine the functions w, f, which govern the initial post-buckling behaviour.

In [1] it was shown that the post-buckling behaviour in the vicinity of the bifurcation point λ_c can be expanded in the form

$$\begin{Bmatrix} w \\ f \end{Bmatrix} = \begin{Bmatrix} w_u(\lambda) \\ f_u(\lambda) \end{Bmatrix} + \xi \cos sy/2 \begin{Bmatrix} w_a \\ f_a \end{Bmatrix} + \xi^2 \left(\cos sy \begin{Bmatrix} w_b \\ f_b \end{Bmatrix} + \begin{Bmatrix} v_b \\ g_b \end{Bmatrix} \right) + \cdots,$$
(4.1a)

$$\lambda/\lambda_c = 1 + b\xi^2 + \cdots,$$
(4.1b)

where $\{w_a(x), f_a(x)\}$ is the buckling mode normalized so that $|w_a|_{max} = 1$, ξ is a scalar parameter, s is a circumferential wave parameter, which, in order to satisfy periodicity, must fulfil the condition that $q_0 s/2$ is an integer, and b, a post-buckling coefficient.

The buckling mode in the expansion (4.1a) can be written in the form

$$w_a = \mathrm{Re}\,[w_{11}(x)\,e^{ikx/2}]; \quad f_a = \mathrm{Re}\,[f_{11}(x)\,e^{ikx/2}],$$
(4.2)

and the post-buckling mode, in the form

$$w_b = 2cs^2\,\mathrm{Re}\,\{w_{02} + w_{22}\,e^{ikx}\}, \quad v_b = 2cs^2\,\mathrm{Re}\,\{w_{00} + w_{20}\,e^{ikx}\},$$
$$f_b = 2cs^2\,\mathrm{Re}\,\{f_{02} + f_{22}\,e^{ikx}\}, \quad g_b = 2cs^2\,\mathrm{Re}\,\{f_{00} + f_{20}\,e^{ikx}\},$$
(4.3)

where the functions (w_{pq}, f_{pq}) are complex periodic functions in x with period $2\pi/\beta$. The parameter k is a free axial wave parameter, which can be restricted to the interval $0 \leq k \leq \beta$ and still give a complete family of bifurcation modes. Note that when k assumes a value given by $k = \beta l/m$, where l and m are integers, the functions (w, f) will be periodic with period $4\pi m/\beta$.

The bifurcation load λ_c is determined together with the functions (w_{11}, f_{11}) by a linear eigenvalue problem. For given values of the axial waveparameter k and of the axisymmetric imperfection parameters $\gamma = 2c\,\delta/t$ and β, the bifurcation load λ_c is determined numerically by a trial and error method as the smallest value that makes the determinant of the homogeneous difference equations corresponding to the linear eigenvalue problem equal to zero. The circumferential wavenumber s is chosen such that the eigenvalue λ_c attains a minimum with the restriction that $sq_0/2$ is an integer.

When the bifurcation stress λ_c and the corresponding normalized eigenfunctions (w_{11}, f_{11}) have been determined it is a straight-forward task to determine the second order functions (w_{pq}, f_{pq}); $(p, q) = (0, 2)$, $(2, 2)$, $(0, 0)$, and $(2, 0)$ by a finite difference method from a set of four inhomogeneous linear ordinary differential equations.

In the advanced post-buckling analysis presented in the next section we will only use sets of first and second order expansion functions as

coordinate functions. Therefore, solutions of the above-mentioned eigenvalue problem and the four boundary value problems are sufficient to
yield these coordinate functions.

The initial relationship between the load and the bifurcation displacement amplitudes hinges on the sign and the magnitude of the
post-buckling coefficient b in Eq. (4.1b). If b is negative, the load-
carrying capacity will diminish following bifurcation and the bifurcation point will be unstable, while if b is positive, the shell with retain
some ability to support increased loads after bifurcation. More details
about the calculation of the coefficient b and a discussion of the initial
post-buckling behaviour of these shells with axisymmetric imperfections
can be found in [1].

5. Advanced Post-Buckling Behaviour

For each value of the axial wave parameter k a bifurcation load and
an associated set of bifurcation functions exist, which determine the
initial slope of the load-end deflection equilibrium path in the post-
buckled region. By varying k in the closed interval $0 \leq k \leq \beta$, we can
obtain a complete family of bifurcation loads and associated bifurcation functions. Normally it is found that the bifurcation stresses λ_c
are bound by those obtained from $k = 0$ and $k = \beta$, as indicated in
Fig. 3. Whether it is the case $k = 0$ or the case $k = \beta$ that corresponds
to the smallest bifurcation load λ_c depends on the imperfection parameters δ and β.

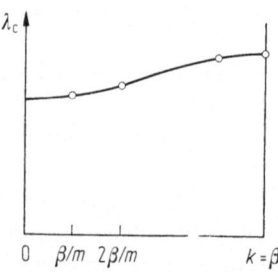

Fig. 3.
Bifurcation load as func-
tion of axial wave-para-
meter.

For the limiting case $k = 0$ it is possible to obtain an analytical
solution to the bifurcation problem, and the results provide a good
approximation for the modes associated with small values of k and
the circumferential wave parameter s. However, it must be noted that
the case $k = 0$ corresponds to infinitely long axial wavelengths, so,
strictly speaking, this case has no physical significance.

In order to determine the continuation of the equilibrium path into
the post-buckled region we will assume that the radial deflection w can

be approximated by linear combinations of the prebuckling solution $w_u(x)$ and m sets of first and second order initial post-buckling functions. The initial post-buckling functions will be determined for m different values of the axial wave parameter $k = \beta l/m$, $(l = 1, 2, \ldots, m)$, indicated by the dots in Fig. 3. The approximation takes the form

$$w = w_u(x, \lambda) + \sum_{j=1}^{2m} a_j(\lambda) \{w_j(x) \cos (s_j y) + v_j(x)\}, \qquad (5.1)$$

where a_j $(j = 1, 2, \ldots, 2m)$ are the amplitudes of the radial deflection modes to be determined, and

$$w_{2l-1}(x) = w_a(\varkappa x)|_{k=\beta l/m}, \quad v_{2l-1}(x) \equiv 0,$$

$$s_{2l-1} = s/2 \,|_{k=\beta l/m} - \mu/q_0,$$

$$w_{2l}(x) = w_b(\varkappa x)|_{k=\beta l/m}, \quad v_{2l}(x) = v_b(\varkappa x)|_{k=\beta l/m},$$

$$s_{2l} = s \,|_{k=\beta l/m} - 2\mu/q_0.$$

Here $\varkappa$ is an axial wave parameter, which, in order to satisfy periodicity in the axial direction, must fulfil the condition that $\varkappa = 2m/n$, where n is an integer, and μ is a circumferential wave parameter, which must assume integer values in order to satisfy periodicity in the circumferential direction.

It is hoped that the representation (5.1) has enough freedom to characterize a sufficiently general deflection pattern to describe the post-buckling behaviour in a region relatively close to the bifurcation point. Obviously, an exact representation of the initial post-buckling behaviour is possible with these coordinate functions. For example, suppose that among the m bifurcation loads corresponding to these initial post-buckling functions, the one corresponding to $k = \beta/m$ is the lowest, then, in the vicinity of the bifurcation point, the $a_2(\lambda)$ will be nearly proportional to $a_1^2(\lambda)$, $a_3, \ldots, a_{2m}$ will be close to zero, and the wave parameters $\varkappa$ and μ will be 1 and 0, respectively.

In order to investigate the influence of relatively small additional imperfections we will assume that these can be expressed as a linear combination of the chosen coordinate functions so that the total initial imperfection can be expressed as

$$w_0 = -\delta/t \cos \beta x + \sum_{j=1}^{2m} b_j\{w_j(x) \cos s_j y + v_j(x)\}, \qquad (5.2)$$

where b_j are known constants.

With these assumptions approximate solutions to the full non-linear equations (2.2) and (2.3), with the additional condition (2.5) in the advanced post-buckled region, can be obtained in the following manner. First the compatibility equation is solved exactly for the stress function f in terms of the assumed radial deflection w. Substituting (5.1) and (5.2) into Eq. (2.3) we obtain the stress function f in the form

$$f(x, y) = f_u(x, y) + \sum_{j=1}^{2m} [b_j \Phi_{01j} \cos (s_j y) + a_j \{\Phi_{02j} + \Phi_{03j} \cos (s_j y)\}] +$$

$$+ \sum_{i,j=1}^{2m} \{a_j a_i + 2b_i a_j\} \{\Phi_{1ij} \cos (s_i + s_j) y + \qquad (5.3)$$

$$+ \Phi_{2ij} \cos (s_i - s_j) y + \Phi_{3ij} \cos (s_i y) + \Phi_{4ij} \cos (s_j y)\}.$$

Due to the condition (2.5) it is found that $\Phi''_{02j}(x) = v_j(x)$, and the remaining functions $\Phi_{lij}(x)$ are determined numerically from ordinary differential equations of the type

$$\Phi^{IV}_{lij}(x) + E_{lij} \Phi''_{lij}(x) + F_{lij} \Phi_{lij}(x) = G_{lij}(x),$$

where the boundary conditions are obtained from

$$\Phi_{lij}(-x) = \Phi_{lij}(x) \quad \text{and} \quad \Phi_{lij}[(\pi n/\beta) - x] = \Phi_{lij}[(\pi n/\beta) + x].$$

An approximate solution to the equilibrium equation is obtained by the Galerkin procedure, which, in this case, can be expressed as

$$\int_0^{2\pi q_0} \int_{-\pi n/\beta}^{\pi n/\beta} \{w_l \cos (s_l y) + v_l\} \{\nabla^4 w + f'' - 2c\psi(f, w + w_0)\} \, dx \, dy = 0,$$

$$(l = 1, 2, ..., 2m). \qquad (5.4)$$

Substituting the expressions for the radial deflection w (5.1), the initial imperfection w_0 (5.2) and the corresponding stress function f (5.3) into the Eqs. (5.4), fairly lengthy calculations lead to a set of $2m$ non-linear equations for the $2m$ unknown buckling amplitudes $a_j(\lambda)$ in the form

$$\sum_{j=1}^{2m} A_{lj}(\lambda) a_j + \sum_{i,j=1}^{2m} B_{lij}(\lambda) a_i a_j + \sum_{i,j,k=1}^{2m} C_{lijk}(\lambda) a_i a_j a_k +$$

$$+ \sum_{j=1}^{2m} D_{lj}(\lambda) b_j = 0, \quad l = 1, 2, ..., 2m. \qquad (5.5)$$

The Eqs. (5.5) can be solved numerically by a Newton-Raphson iteration method as follows. Let us assume that among the m bifurcation loads associated with $2m$ coordinate functions used in the approximation (5.1), the smallest bifurcation load is the one corresponding to the axial wave-parameter $k = \beta l/m$. Then, a small value is specified for the amplitude a_{2l-1}, and using the Newton-Raphson iteration method, the corresponding load λ and the remaining amplitudes a_j are determined. Now, the value of a_{2l-1} is increased a little and the previous solution is used as a starting vector for a new iteration and so on. In this way it is possible to follow an equilibrium path in the post-buckled region.

The disappearance of the Jacoby determinant of the right-hand sides of the Eqs. (5.5) serves to locate secondary branching points.

In the numerical results presented no attempt is made to continue the post-buckling path beyond these secondary bifurcation points. However, such an analysis can be carried out using, for example, the general theory for branching analysis of discrete structural systems given by Thompson and Walker in [5].

The average axial shortening per unit length of the cylinder due to the axial compression can be defined as

$$\Delta L/L = -\lim_{L \to \infty} \frac{1}{2\pi RL} \int_0^L \int_0^{2\pi R} \left(\frac{\partial U}{\partial X}\right) dX\, dY. \qquad (5.6)$$

Using the Donnell strain relations (5.6) can be expressed as

$$\Delta L/L = \lim_{L \to \infty} \frac{t}{2\pi q_0 RL} \int_0^L \int_0^{2\pi q_0} [2c\{w'w_0' + w'^2/2\} - \ddot{f}]\, dx\, dy. \qquad (5.7)$$

Thus the average shortening can be calculated when the non-linear equations (5.5) have been solved.

6. Results and Discussions

In Fig. 4 the results of the advanced post-buckling analysis for a shell with a small axisymmetric imperfection are presented. The figure shows the load as a function of the end-shortening for a shell with a radius to thickness ratio equal to 757 ($q_0 = 50$), Poisson's ratio 0.3 and an imperfection in the shape of the classical axisymmetric buckling mode ($\beta = 1.0$), the amplitude of which is 0.0095 times the shell wall thick-

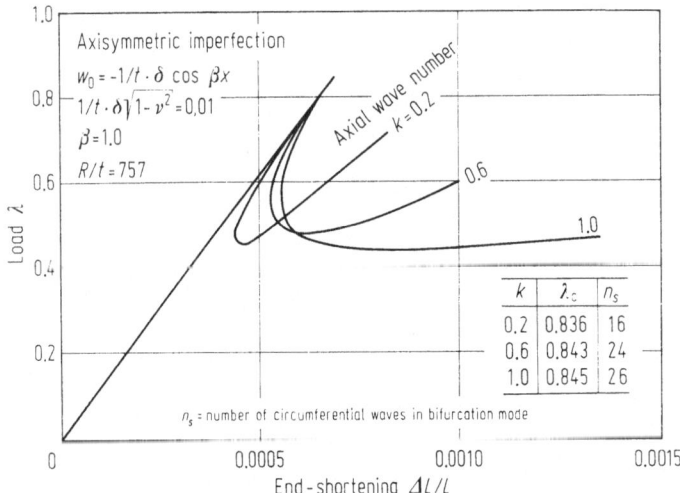

Fig. 4. Load-end-shortening equilibrium paths for different initial post-buckling modes (small imperfection).

ness. The post-buckling analysis is performed with ten coordinate functions in the approximation for the radial deflection ($m = 5$). The initial post-buckling analysis shows that, for this case, it is the bifurcation mode associated with the axial wave number $k = 0.2$ that corresponds to the lowest bifurcation load. Therefore, the iteration procedure is naturally started by specifying the amplitude of the bifurcation mode w_a corresponding to $k = 0.2$. Such a calculation results in the curve marked $k = 0.2$ in Fig. 4 when $\varkappa$ and μ are chosen so that, for a given value of the prescribed amplitude, the load λ assumes a minimum value. It is found that, at the minimum point of the load-end-shortening curve, the axial and circumferential wavelengths of the buckling pattern are considerably longer than those associated with the bifurcation point. This variation in wavelengths is in agreement with experimental observations. In order to determine other possible equilibrium paths in the post-buckled region, similar equilibrium paths starting from the bifurcation points corresponding to $k = 0.6$ and 1.0 are also determined. The resulting curves are shown in Fig. 4 as the curves marked $k = 0.6$ and $k = 1.0$. No attempt has been made to determine equilibrium paths that are not directly connected with these primary bifurcation points.

Figure 5 shows load end-shortening equilibrium paths starting from the lowest bifurcation load for shells with large axisymmetric imperfections. These curves are calculated using coordinate functions, where $\varkappa = 1$ and $\mu = 0$. Since the initial post-buckling analysis in [1] shows that it is the relatively long wavelength of the imperfection that gives

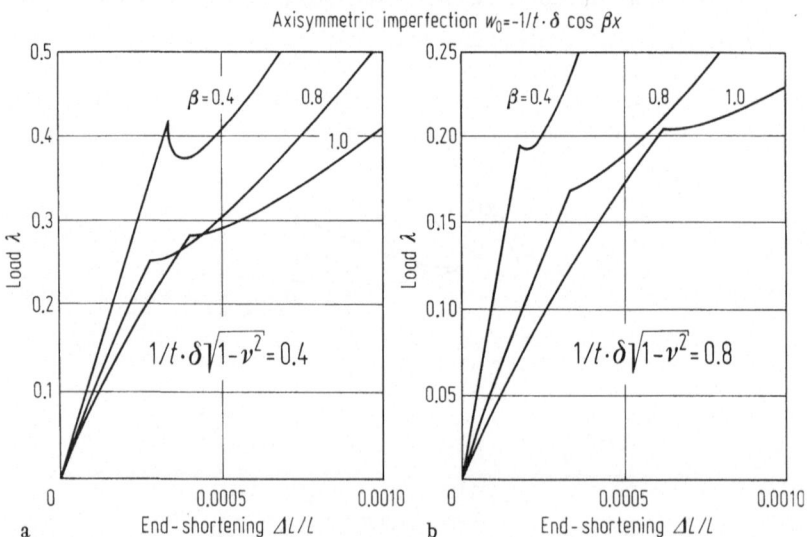

Fig. 5 a and b. Load-end-shortening equilibrium paths for medium size imperfections.

the smallest unstable bifurcation loads, we have here chosen to study the effect of imperfections with the wave numbers $\beta = 0.4$, 0.8 and 1.0.

The post-buckling curves corresponding to imperfections given by $\beta = 0.4$ and $(\delta/t)\sqrt{1 - \nu^2} = 0.4$ and 0.8 in Fig. 5a and b, respectively, show unstable bifurcations. The equilibrium curve corresponding to $\beta = 0.4$ in Fig. 5a shows a relatively large "end-shortening jump" from the bifurcation point to the equilibrium curve for constant load; therefore, the maximum support load is likely to be the smallest primary bifurcation load, which is about 0.38 times the classical buckling load. When the amplitude of the imperfection is increased to $(\delta/t)\sqrt{1 - \nu^2} = 0.8$, the theoretical results in Fig. 5b show that the bifurcation is still unstable. But now the bifurcation takes place at a load less than one-fifth of the classical load, i.e., a load comparable to the lowest reported experimental buckling loads for shells of this configuration. However, since the "end-shortening jump" from the bifurcation point to the equilibrium curve for constant load is relatively small, it may be of interest to study the effect of a small additional asymmetric imperfection.

The lowest bifurcation load associated with this imperfection corresponds to the axial wave parameter $k = \beta/5 = 0.08$. In Fig. 6a the dotted curve marked $\mu = 0$ shows the effect of a small additional asymmetric imperfection the shape of which is given by small amplitudes of the two bifurcation functions associated with the wave-parameter $k = 0.08$, that is, an imperfection of the form given by (5.2),

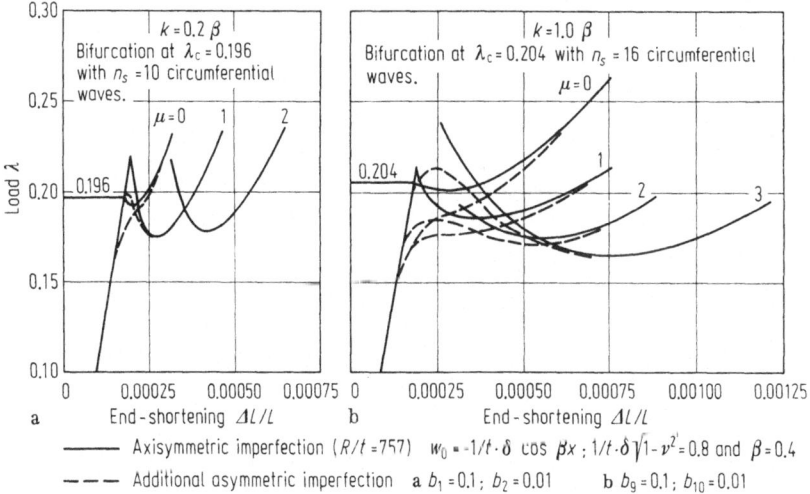

——— Axisymmetric imperfection $(R/t = 757)$ $w_0 = -1/t \cdot \delta \cos \beta x$; $1/t \cdot \delta \sqrt{1 - \nu^2} = 0.8$ and $\beta = 0.4$

— — — Additional asymmetric imperfection   a $b_1 = 0.1$; $b_2 = 0.01$ b $b_9 = 0.1$; $b_{10} = 0.01$

Fig. 6 a and b. Load-end-shortening equilibrium paths for different initial post-buckling modes (medium imperfection).

where $b_1 = 0.10$, $b_2 = 0.01$, and $b_3 = \cdots = b_{10} = 0$. The dotted equilibrium curve shows stable behaviour also in the region around the bifurcation load corresponding to the pure axisymmetric imperfection.

The full-line curves marked $\mu = 1$ and $\mu = 2$ represent the post-buckling behaviour of a shell with the same axisymmetric imperfection, but now determined from a set of coordinate functions given by (5.1) with $\mu = 1$ and $\mu = 2$. That is, instead of using coordinate functions in which the circumferential wave parameters s are chosen so that the associated bifurcation load attains a minimum, the functions are modified so that they represent buckling shapes with one or two waves less in the circumferential direction. The effect of using these coordinate functions is, of course, higher bifurcation loads but, as will be seen from the curves, also lower minimum loads of the load-end-deflection curves.

Fig. 6b shows a similar set of curves, except that here, the amplitude a_9 of the coordinate function corresponding the buckling mode with $k = \beta$ is prescribed.

From Fig. 6a we find that in this case—where the amplitude of the axisymmetric imperfection is relatively large—the addition of axisymmetric imperfections will most likely either raise the limit load above the lowest bifurcation load or result in the load-end-shortening curve increasing monotonously even here where the "perfect" axisymmetric imperfection causes unstable initial post-buckling behaviour. However, Fig. 6b shows that it is possible to construct specific imperfection shapes that will cause the shell to collapse at a smaller load than the lowest bifurcation load.

The smallest limit-point buckling loads found here agree well with the lowest experimental collapse loads for shells with $R/t \approx 750$ reported in [6] which are about 0.18.

Acknowledgement

The author wishes to thank Lektor Viggo Tvergaard for many valuable discussions.

References

1. Pedersen, P. Terndrup: Buckling of Unstiffened and Ring-Stiffened Cylindrical Shells under Axial Compression. Int. J. Solids Structures **9**, 671—691 (1973).
2. Budiansky, B.; Hutchinson, J. W.: Buckling of Circular Cylindrical Shells under Axial Compression. Contributions to the Theory of Aircraft Structures. Presented to A. van der Neut. Delft: University Press 1972, pp. 239—259.
3. Koiter, W. T.: The Effect of Axisymmetric Imperfections on the Buckling of Cylindrical Shells under Axial Compression. Proc. Koninkl. Nederl. Akademie van Wetenschappen. **66**, Series B, 265 (1963).

4. Hoff, N. J.; Madsen, W. A.; Mayers, J.: Postbuckling Equilibrium of Axial Compressed Circular Cylindrical Shells. AIAA J. **4**, No. 1, 126−133 (1966).
5. Thompson, J. M. T.; Walker, A. C.: A General Theory for the Branching Analysis of Discrete Structural Systems. Int. J. Solids Structures **5**, 281−288 (1969).
6. Harris, L. A et al.: The Stability of Thin-Walled Unstiffened Circular Cylinders Under Axial Compression Including the Effects of Internal Pressure. J. of the Aeronautical Sciences, 587−596 (1957).

Finite Element Representations for Thin Shell Instability Analysis

R. H. Gallagher

Cornell University, Ithaca, New York, U.S.A.

Summary

The development of finite element calculational procedures for thin shell instability analysis has involved the definition of appropriate element representations (i.e. the geometric form of the element and the approximation of displacement and/or stress), constitutive expressions, and computational algorithms. Among these, the status of thin shell finite element representations remains unsettled. This paper therefore emphasizes recent developments in the basic aspects of thin shell finite element analysis and discusses one simplified approach in more detail. Formulative and computational procedures for elastic instability analysis are then summarized and numerical results are shown for the simplified shell element formulation.

1. Introduction

In an earlier review [1] the writer outlined the component features of the finite element approach to thin shell instability analysis and observed that many serious developmental challenges remained in even these basic aspects. For example, the reliability in prediction of thin shell instability can be no better than in calculation of linear static shell analysis. Recent developments have shed new light on features of the latter which have heretofore been poorly understood, and have indicated more promising avenues of approach. One of these avenues, that of "generalized potential energy" [2], forms the basis of the triangular shell element described herein, which has been extended [3, 4] to deal with a variety of elastic instability phenomena.

This paper begins with a discussion of the linear static formulation and then progress from there to elastic instability analysis. The route followed represents a projection of the direct stiffness, assumed displacement procedures that have predominated in finite element analysis to date. We comment briefly on alternative routes at the close of the paper.

2. Finite Element Representations

Three distinct approaches to assumed-displacement finite element analysis of thin shell structures can be identified: (1) by means of three-dimensional (solid) elements with curved boundaries, (2) in "faceted" form with flat elements, and (3) via elements formulated on the basis of curved shell theory.

An appeal of solid elements is that they stem directly from three-dimensional theory of elasticity. The curved boundaries are easily defined through use of "isoparametric" curvilinear coordinates, which employ the same form of description for both geometry and displacement. Conditions of the Love-Kirchoff type are imposed, but this is insufficient to achieve economical solutions; it is necessary either to introduce certain supplementary displacement modes to describe bending action [5] or to de-emphasize the representation of shear deformation by approximation (reduced numerical integration) of the shear strain energy [6]. There has not been sufficient development of the non-linear analysis side of this approach to measure its effectiveness in thin shell stability analysis.

The use of flat plate elements is appealing in its simplicity. Among the objections to them, however, are the following: (a) they exclude coupling of stretching and bending within the element, (b) the restriction to triangular shapes when general shells are to be handled, (c) difficulty of treating junctions where all elements meeting at the point are co-planar, (d) "discontinuity" bending moments at element juncture lines, which do not appear in the continuously curved actual structure, and, for stability analysis, (e) the influence of the geometric approximation upon the solution for imperfection-sensitive structures. We might add the problems of definition of displacement fields which meet interelement continuity conditions, but these are even more serious for curved elements.

Straight finite elements represent the behaviour of curved structures in the limit and the errors due to (a) can be made small by use of a refined element network. Nearly all finite element programs suitable for stability analysis will contain triangles so that (b) is not a serious objection.

In item (c), co-planarity, a null stiffness corresponding to rotation about the axis normal to the plane will be present. One may define coordinate axes in the plane (which is generally at an angle with the global axes) and eliminate this rotational degree-of-freedom. This may be awkward in practical application since many different such planes may appear in the structure. Conversely, for small angles between the elements, dependence may be placed on this angle to maintain solution stability but now the stiffness eqnations approach singularity as the elements approach co-planarity. Note that related practical limitations appply to most existing curved shell element formulations, which admit only constant curvatures or restricted coordinate systems in the global idealization.

The formulation of the buckling problem does not involve the bending moments directly and often the influence of flexure on the distribution of in-surface forces is small. Thus, one might argue that objection (d) is not applicable to the analysis for instability. With respect to item (e), there is insufficient numerical experience in the related analysis situation to measure its significance.

In addressing the questions rasied by curved thin shell elements we assume that the strain-displacement equations of a "consistent" first-order deep shell theory will be adopted. We also confine attention to formulations in curvilinear coordinates, taking note of the alternative of formulations in projected rectangular coordinates. The principal concern is then the selection of displacement fields which meet the following conditions:

1. Zero strain energy under rigid body motion.
2. Inclusion of constant strain states.
3. Inter-element continuity to the degree required by the variational principle employed.

By using trigonometric as well as polynomials for coupled displacement fields it is possible to satisfy the rigid body motion condition. Such developments often fail to meet the constant strain condition, however. Ashwell and Sabir [8] satisfy both conditions by first selecting strain expansions which include the necessary constant terms and then integrating to produce the rigid body motion terms. This function fails to satisfy the interelement continuity condition because of the trigonometric terms introduced to meet the other two conditions.

The choice of conditions to be met exactly has therefore become a dilemma confronting those who formulate such elements and numerical evidence by Fonder and Clough [9] indicates that each alternative is unsatisfactory in certain cases. A logical alternative is therfore to use as many terms in a purely polynomial expansion—for which interelement continuity and the constant strain condition can generally be enforced—as are necessary to closely approximate the trigonometric terms associated with the rigid body motion condition.

To clarify this approach, we first note that the interelement continuity conditions on flexure demand a polynomial of such high order in the normal displacement that the approximation of the rigid body motion condition is usually satisfactory without additional terms. Under

certain conditions of coupling of the displacement components the in-
surface displacement components may be required to be of the same
order as the normal displacement to meet strict inter-element continuity conditions. Where these are not coupled, however, the in-
surface continuity can be achieved with lower order terms than for
flexure. In consequence, the polynomial terms which are employed
above those required for flat element interelement continuity can be
viewed as contributing principally to the satisfaction of the rigid body
motion condition.

There are two differing views regarding the degree of polynomial
representation of the inplane components in comparison with the degree
employed for normal displacement. In conjunction with the above
argument concerning approximate representation of the rigid body
motion condition, Morley [10] shows that when the in-surface displacements are quadratic polynomials a solution of acceptable accuracy
may require element planform dimensions that are of the same order
as the thickness. This has been confirmed by Morris [11, 12]. Thus, at
least a cubic polynomial should be specified and this is at the same
level as the simple normal displacement assumption to be described
subsequently.

Cowper, et al [13], however, believe that since the in-surface displacements appear in lower-order derivatives in the energy expression
they should be represented by lower order polynomials. The numerical
evidence of their work is not necessarily at variance with the view
above since they also use a cubic polynomial for in-surface components,
but a quintic polynomial for normal displacements.

To adapt these considerations to an element of minimal formulative
complexity we examine the element shown in Fig. 1. The shell middle
surface corresponds to the orthogonal curvilinear system $\alpha - \beta$. All
three displacement components are described by complete cubic polynomials in triangular coordinates

$$u = \lfloor N \rfloor \boldsymbol{u}, \quad v = \lfloor N \rfloor \boldsymbol{v}, \quad w = \lfloor N \rfloor \boldsymbol{w}, \tag{1}$$

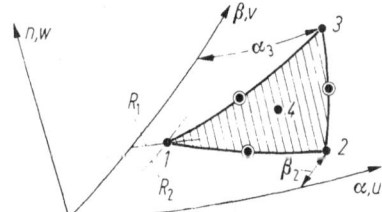

⊙ Points of application of constraint conditions

Fig. 1. Triangular shell element.

where

$$\boldsymbol{u}^T = \left\lfloor u_1 \left(\frac{\partial u}{\partial \alpha}\right)_1 \left(\frac{\partial u}{\partial \beta}\right)_1 \cdots \left(\frac{\partial u}{\partial \beta}\right)_3 u_4 \right\rfloor, \qquad (2)$$

and similarly for $\boldsymbol{v}$ and $\boldsymbol{w}$. $\lfloor\, N \,\rfloor$ contains the standard shape functions of a complete cubic corresponding to these degrees-of-freedom.

Equation (1) does not permit continuity of the angular displacement across element boundaries and leads to very poor results even for flat plate flexure. To resolve this, Harvey and Kelsey [2] introduce the notion of a constraint condition to "restore" continuity. That is, if A and B are neighbouring elements and the subscript n denotes the normal direction, the relative angular displacement (ϕ) of adjacent edges at their midpoint can be set equal to zero

$$\phi_n^{A-B} = \phi_n^A - \phi_n^B = 0. \qquad (3)$$

Differentiation of the displacement field [Eq. (1)] gives ϕ^A and ϕ^B in terms of the joint displacements $\boldsymbol{\Delta}$. Applying this to each boundary in each displacement component gives the set of algebraic constraint equations

$$\lfloor C \rfloor \Delta = 0. \qquad (4)$$

These are handled in the global analysis by use of Lagrange multipliers. Details of the element formulation are presented in [3].

It is noteworthy that the constraint condition can be directly incorporated in the variational principle and represented in the discretized system as a corrective element boundary stiffness matrix. Kikuchi and Ando [14] refer to this as the "simplified hybrid displacement method" and have recently [15] applied it to thin shell instability problems.

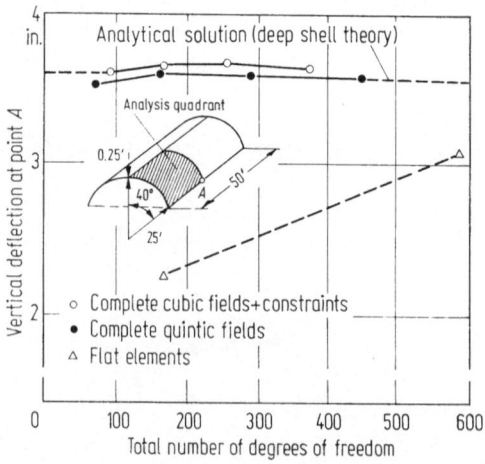

Fig. 2. Linear solution shell roof ($E = 3.0 \times 10^6$ lb/in^2, $\nu = 0$, weight $= 90$ lb/ft^2).

One can proceed from complete cubic to complete quintic displacement fields with improved satisfaction of both rigid body motion and interelement displacement continuity conditions at the expense of increased cost of element formulation. One must ask if the cost is justified by the increased accuracy. Complete quintic fields are employed for the highly accurate and sophisticated Sheba element [16], but this is not directly comparable with the above due to its use of a special coordinate system and exact satisfaction of all basic conditions. Recently, Dawe [17] presented a complete quintic element in curvilinear coordinates that is comparable with the above. One comparison is given for the case of the uniformly-loaded cylindrical shell [18, 19] in Fig. 2. Although these results suggest that higher sophistication does not have a corresponding payoff, extensive data on other numerical analyses in [17] demonstrate that the question is not clearly settled.

3. Element Relationships for Instability Analysis

The potential energy is a convenient starting point for the definition of element relationships for instability analysis. For simplicity we exclude initial strains and displacements, which are otherwise readily accounted for, and make no specification of the strain-displacement equations except to note that they are of the Lagrangian type.

The total strain ε can be written as the sum of components ε^{L} that are linear in the strain-displacement expressions and components ε^{N} that are non-linear functions of the displacements.

$$\varepsilon = \varepsilon^{L} + \varepsilon^{N}. \tag{5}$$

The potential energy is

$$\pi = (1/2) \int_{\mathrm{Vol}} \varepsilon E \varepsilon \, \mathrm{d}(Vol) - \int_{S_{\sigma}} \overline{T} \, \Delta^{b} \, \mathrm{d}S, \tag{6}$$

where E denotes the set of elastic constants, (Vol) is the volume of the structure, S_{σ} is the portion of the structure on which the tractions $\overline{T}$ are prescribed, and Δ^{b} are the surface displacements.

If the displacements are discretized by use of element shape functions, in the manner indicated previously, Eq. (1), we can write

$$\varepsilon^{L} = D^{L} \Delta^{e}, \quad \varepsilon^{N} = D^{N}(\Delta^{e}) \Delta^{e}, \tag{7, 8}$$

where D^{L} and $D^{N}(\Delta^{e})$ result from the application of the linear and non-linear operators of the strain displacement equations on the assumed displacement field, Δ lists the element discrete d.o.f. After substitution of (7, 8) into (5) and the result into (6) we have the element potential energy

$$\pi = k_{ij}\Delta_i\Delta_j/2 + n_{ijk}\Delta_i\Delta_j\Delta_k/6 +$$
$$n_{ijkl}\Delta_i\Delta_j\Delta_k\Delta_l/12 - F_i\Delta_i, \tag{9}$$

in which k_{ij}, n_{ijk}, and n_{ijkl} are second, third and fourth order tensors, and are fixed values, and F_i is the nodal or generalized force corresponding to Δ_i. After summation of element potential energies we have the global potential energy

$$\pi = K_{ij}\Delta_i\Delta_j/2 + N_{ijk}\Delta_i\Delta_j\Delta_k/6 +$$
$$+ N_{ijkl}\Delta_i\Delta_j\Delta_k\Delta_l/12 - P_i\Delta_i. \tag{10}$$

P_i denotes the external load in degree-of-freedom i.

The equilibrium state is represented by the solution to the equations resulting from application of the first necessary condition. Thus differentiation of (10) with respect to Δ_i gives

$$P_i = K_{ij}\Delta_j + N_{ijk}\Delta_j\Delta_k/2 + N_{ijkl}\Delta_j\Delta_k\Delta_l/3. \tag{11}$$

In the above we have implied the use of the same displacement function in all component terms, k_{ij}, n_{ijk}, and n_{ijkl}, i.e., a "consistent" formulation. Economic feasibility of analysis may depend on the use of simpler displacement fields for n_{ijk} and n_{ijkl}. Numerical evidence supports the view that very little accuracy is lost when linear approximations are made for these terms in the presence of cubic approximations for k_{ij} and recently Oden, et al. [20] proved theoretically the validity of this approach for beam elements.

Another factor in economic feasibility concerns the elimination of certain degrees-of-freedom prior to the start of non-linear analysis. Such "condensations" are facilitated when n_{ijk} and n_{ijkl} are formulated inconsistently and are also possible, if rather more complicated, in a consistent formulation. The usual practice is to eliminate angular degrees-of-freedom; the limits of condensation with respect to solution accuracy are not well understood, however.

4. Calculation of Load-Displacement Paths

The alternative methods of calculation of the load-displacement paths of finite element representations has recently been reviewed by the writer [21] for instability problems and by Stricklin and Haisler [22] for the prebuckling phase. In the following we present the outline only of the procedures employed in the numerical calculations to be described subsequently. Four features of the load-displacement path are treated: (1) general non-linear analysis, (2) limit point calculation, (3) bifurcation, and (4) post-buckling.

The widely-used combination of the tangent stiffness method with one-step Newton-Raphson (N-R) iteration has been adopted for the general non-linear analysis. In matrix form the N-R approach is

$$\Delta^{i+1} = \Delta^i_\bullet - [K^i_T]^{-1} f^i, \tag{12}$$

where Δ^{i+1} and Δ^i are the displacement vectors in successive cycles of iteration and f^i is the residual (imbalance of load) calculated from the equilibrium equation evaluated for Δ^i, i.e.

$$f^i = [[K] + (1/2)[N^i_1] + (1/3)[N^i_2]] \Delta^i - P. \tag{13}$$

In conventional N-R analysis one iterates on Eq. (12) until Δ^{i+1} is arbitrarily close to Δ^i. We see, however, that when the load path is followed incrementally $[K_T]^{-1}$ is available and one cycle of Eq. (12) can be applied at small additional cost.

[K_T] is singular at the limit point and poorly conditioned in its vicinity. It is therefore desirable to shift to a scheme of displacement, rather than load, incrementation. There are many ways of implementing this approach, the one adopted here being the superposition procedure described by Zienkiewicz [23].

The method of calculation of the bifurcation load is based on the zero-determinant condition. The determinant is evaluated at each load intensity and the intensity for zero determinant is obtained through Lagrange interpolation. The mode shape is calculated by solution of the stiffness equations after one degree-of-freedom has been assigned a reference value. Details are given in [4]. Few situations involving multiple or closely-spaced roots have yet been analyzed; it would appear that algorithms which exploit the Sturm-sequence property (e.g. [24 and 25] have the desired reliability in such cases.

Perturbation methods have a predominant role in analytical studies of postbuckling behaviour, and numerous studies have attempted to adapt these ideas to finite element analysis. This work is summarized in a report by Mau and the writer [26] and by Thompson and Hunt [27]. These demonstrate that the needed basic properties are given by the terms n_{ijk} and n_{ijkl}. The mode of combination entails so much computational effort, however, that perturbation analysis is less effective than pursuing the load-displacement path as a non-linear analysis problem, as described above and done in [4, 15, 28]. A different view of this situation is that finite element analysis has not exploited effectively the parametric analysis advantages of perturbation procedures.

5. Numerical Results

Numerical solutions for a number of beam, arch, plate and shell instability solutions, covering all of the phenomena described above, have

48   R. H. Gallagher

been given in [1, 3, 4, 26]. In each case the results are compared with
alternative analytical or numerical solutions. The problem shown in
Fig. 3 is chosen for presentation here since it represents the full range

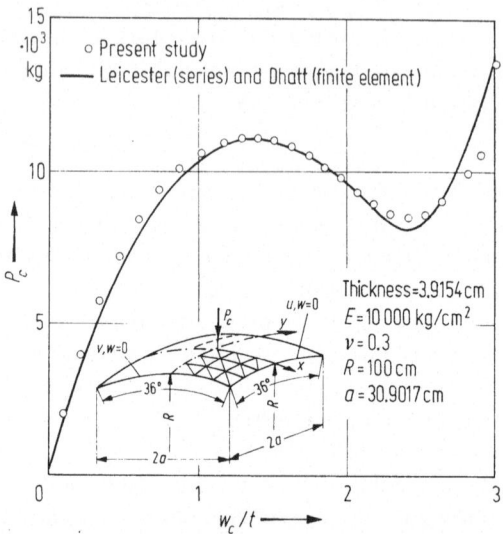

Fig. 3. Spherical cap under concentrated load at crown-nonlinear analysis.

of considerations described in prior sections, except for bifurcation and
non-zero twisting curvature. It involves a spherical cap with hinged
rectangular boundaries, subjected to a concentrated load at the crown.
The load-displacement response is of the snap-through type with recovery
of stiffness at higher displacements.

The results are shown for an 18-element (3×3 grid) representation
of the quadrant. The non-dimensionalized central deflection is plotted
versus the scaled intensity. An 8-term double series solution, based on
potential energy, by Leicester [29] and a "discrete Kirchoff" finite
element solution by Dhatt [30] are given for comparison. Agreement
among the three alternative solutions is very good, up to where a
stiffening behaviour is experienced.

6. Concluding Remarks

A principal motivating factor in the continuing development of the
finite element method is its suitability as a basis for the "general purpose" program, i.e. a program which can be used for the analysis of
many different structural forms and behaviour mechanisms. The framework of a general finite element computational procedure for elastic

instability phenomena is at hand but is too complex to permit implementation and efficient performance within many, if not all, of the existing general purpose programs.

As noted in the Introduction, the avenues outlined in this paper represent a natural projection of the direct stiffness, assumed displacement procedure that have grown up around linear, static finite element analysis and general purpose programs. It was earlier anticipated that strict adherence to potential energy concepts would enable "lower bound" solutions on strain energy to be realized. Due to factors described earlier, this goal has been difficult to reach in thin shell finite element analysis. Also, the strain energy is not often a parameter of design significance. The goal of a bounded solution is significant in bifurcation analysis, however, since it applies to the intensity of critical load.

Research into admissible approximations and more efficient procedures is therefore of key importance. The suggested alternatives include mixed variational principles (Reissner, Hu-Washizu, hybrid) and modes of algebraic formulation (Eulerian, convected coordinate). The formidable literature surrounding these attempts is beyond the scope of this paper. We observe, however, that these also confront the problem of integration in general purpose programs.

Comparisons with analytical solutions mainly serve to verify the basic adequacy of finite element thin shell formulations. These problems are usually best solved by analytical procedures however. The importance of finite element representations rests with stiffened shells of complicated geometry under arbitrary load. Careful, appropriate test data must be obtained to validate the method in these applications.

Acknowledgement

Work described in this paper was supported by NASA under Grant NGR 33-010-070. The author wishes to thank the former students who contributed to this work, but especially to Dr. Gareth Thomas who was responsible for the development of the triangular shell element and computational capabilities described herein.

References

1. Gallagher, R. H.: The Finite Element Method in Shell Stability Analysis. Computers and Structures **3**, 543—557 (1973).
2. Harvey, J. W.; Kelsey, S.: Triangular Plate Bending Element with Enforced Compatibility. AIAA J. **9**, No. 6, 1023—1026 (1971).
3. Thomas, G. R.; Gallagher, R. H.: A Triangular Thin Shell Finite Element: Linear Analysis. NASA CR 2482 (1975).
4. Thomas, G. R.; Gallagher, R. H.: A Triangular Thin Shell Finite Element: Nonlinear Analysis. NASA CR 2483 (1975).
5. Wilson, E. L. et al.: Incompatible Displacement Models. In: Numerical and Computer Methods in Structural Mechanics. New York: Academic Press 1973, pp. 43—57.

6. Zienkiewicz, O. C.; Taylor, R. L.; Too, J. M.: Reduced Integration Technique in General Analysis of Plates and Shells. Int. J. for Num. Meth. Engrg. **3**, 275—290 (1971).

7. Zienkiewicz, O. C.; Parekh, C.: Discussion of paper "Analysis of three-dimensional Thin-Walled Structures". Proc. ASCE, J. of the Struct. Div., No. ST8, Aug. 1970, pp. 1838—1846.

8. Ashwell, D.; Sabir, A.: A New Cylindrical Shell Finite Element Based on Simple Independent Strain Functions. Int. J. Mech. Sci. **14**, 171—183 (1972).

9. Fonder, G.; Clough, R.: Explicit Addition of Rigid-Body Motions in Curved Finite Elements. AIAA J. **11**, No. 3, 305—312 (1973).

10. Morley, L. S. D.: Polynomial Stress States in First Approximation Theory of Circular Cylindrical Shells. Quart. J. Mech. and Applied Math. **V. XXV**, Part I (1972).

11. Morris, A. J.: A Deficiency in Current Finite Elements for Thin Shell Applications. Int. J. Solids Struct. **9**, 331—346 (1973).

12. Morris, A. J.: A Summary of Appropriate Governing Equations and Functionals in the Finite Element Analysis of Thin Shells. Conf. in Finite Element Thin Shell Analysis. New York: J. Wiley (in Press).

13. Cowper, G. R.; Lindberg, G. M.; Olson, M. D.: Comparison of Two High-Precision Triangular Finite Elements for Arbitrary Deep Shells. Proc. of Third Air Force Conf. on Matrix Methods in Structural Mechanics. AFFDL TR—71—160, 277—304 (1971).

14. Kikuchi, F.; Ando, Y.: Some Finite Element Solutions for Plate Bending Problems by Simplified Hybrid Displacement Method. Nucl. Engrg. Des. **23**, 155—173 (1972).

15. Kikuchi, F.; Ando, Y.: Application of Simplified Hybrid Displacement Method to Plate and Shell Problems. Proc. of 2nd Int. Conf. on Struct. Mech. in Reactor Technology. Berlin, Sept. 1973, Vol. M, Paper M5/5.

16. Argyris, J. H.; Lochner, N.: On the Application of the SHEBA Shell Element. Comp. Methods in Applied Mechanics and Engrg. **1**, 317—347 (1972).

17. Dawe, D.: Some Higher-Order Elements for Arches and Shells. Conf. in Finite Element Thin Shell Analysis. New York: J. Wiley (in Press).

18. Scordelis, A. C.; Lo, K. S.: Computer Analysis of Cylindrcal Shells. ACI J. **61**, 539—561 (1964).

19. Forsberg, K.; Hartung, K.: An Evaluation of Finite Difference and Finite Element Techniques for Analysis of General Shells. Proc. IUTAM Symposium on High-Speed Computing of Elastic Structures. Tome 2, pp. 837—859.

20. Oden, J. T.; Akay, H. U.; Johnson, C. P.: Effect of Higher Order Terms in Certain Nonlinear Finite Element Models. AIAA J. **11**, 1589—1590 (1973).

21. Gallagher, R. H.: Finite Element Analysis of Geometrically Nonlinear Problems. In: Theory and Practice in Finite Element Structural Analysis. Y. Yamada and R. Gallagher, Eds., Univ. of Tokyo Press 1973, pp. 109—123.

22. Stricklin, J.; Haisler, W. E.: Survey of Solution Procedures for Nonlinear Static and Dynamic Analyses. Proc. of SAE Internat. Conf. on Vehicle Structural Mechanics. Detroit: Mich. Mar. 1974, pp. 1—17.

23. Zienkiewicz, O. C.: Incremental Displacement in Nonlinear Analysis. Int. J. Num. Method in Engrg. **3**, No. 4, 587—588 (1971).

24. Gupta, K. K.: Recent Advances in Numerical Analysis of Structural Eigenvalues. In: Theory and Practice in Finite Element Structural Analysis. Y. Yamada and R. Gallagher, Eds. Univ. of Tokyo Press, 1973, pp. 249—272.

25. Wittrick, W. H.; Williams, F. W.: A General Algorithm for Computing Natural Frequencies of Elastic Structures. Quart. J. of Mech. and Applied Math. **XXIV**, 263—284 (1964).
26. Mau, S. T.; Gallagher, R. H.: A Finite Element Procedure for Nonlinear Prebuckling and Initial Postbuckling Analysis. NASA CR-1936, Jan. 1972.
27. Thompson, J. M. T.; Hunt, G. W.: General Theory of Elastic Stability. London: J. Wiley 1973.
28. Batoz, J.; Dhatt, G.: Buckling of Deep Shells. Proc. 2nd Int. Conf. on Struct. Mech. in Reactor Technology. Berlin, Sept. 1973, Vol. M, Paper M5/7.
29. Leicester, R. H.: Finite Deformations of Shallow Shells. Proc. ASCE, J. of the Engrg. Mech. Div., **94**, No. EM6, 1409—1421 (1968).
30. Dhatt, G.: Instability of Thin Shells by the Finite Element Method. IASS Symp. for Folded Plates and Prismatic Structures. Vienna 1970.

Computer Solutions for Static and Dynamic Buckling of Shells

B. O. Almroth, E. Meller, F. A. Brogan

Lockheed Missiles & Space Company, Palo Alto, California, U.S.A.

Abstract

A computer program, STAGS, for analysis of the behavior of shells of general shape has been extended by the addition of a capability for transient response analysis. A brief discussion of the numerical integration procedures included as options in STAGS is presented here, together with some numerical results obtained by use of the program. The results are included to demonstrate how rapid development of computers and numerical analysis methods has enhanced our capability to solve complex shell stability problems. In addition, they illustrate some facets of shell behavior under transient and static loading.

The applications are concerned with an axially loaded cylindrical shell with unstiffened rectangular cutouts. The effect of the loading rate on the shell response is studied. The definition of a dynamic buckling criterion is discussed in view of the results obtained. Also considered is the possibility of determining a static collapse load (limit point) by use of a transient response analysis with the load applied slowly.

1. Introduction

The availability of high speed computers has made it possible to analyze structural models of increasing complexity. The analyst relies now in a lesser degree on handbooks and design charts. Instead he applies more frequently a computer analysis directly to a model of his particular structure. In view of these developments, it may not seem totally irrational to question the usefulness of the concept of stability in modern structural analysis. The criterion on structural adequacy is satisfied if the structure remains serviceable after it has been subjected to the loading environment that can be expected during its lifetime. A nonlinear transient response analysis of an adequate mathematical model of the structure will give all information that is needed, without reference to the concept of structural stability.

In the case of static loading, stability analysis over the years has been a very useful design tool. The primary reason for this is that this type of analysis yields somewhat limited but still useful information at a significant reduction in cost. The need for limited information at an acceptable price level remains and static stability considerations are still useful. The definition of a buckling load is useful, also because it makes possible the use of design charts for simple structural configurations. Such charts are particularly valuable in the preliminary design phase.

With static loading the event of buckling has generally been considered equivalent to the loss of stability of the equilibrium on the primary path. We can distinguish between two types of buckling: violent buckling, which means that the structure suddenly is set in motion when the critical load is reached, and gentle buckling, which means that a new deformation pattern begins to develop when the critical load is surpassed. Violent buckling occurs if the structural stability is lost at a limit point, or if it is lost at a bifurcation point and the equilibrium is unstable on the corresponding secondary path. Gentle buckling occurs if stability is lost at a bifurcation point and the equilibrium is stable on the secondary path.

The ultimate load carrying capability of a plate or shell structure can be determined only by a nonlinear analysis. However, such an analysis involves the solution of nonlinear partial differential equations and the corresponding numerical work is generally quite cumbersome. Therefore, the bifurcation approach often is used in lieu of a nonlinear analysis. For shells of revolution with axisymmetric loading, the deformation pattern on the secondary path in the immediate neighborhood of the bifurcation point is known to be harmonic in the circumferential direction. This knowledge makes it possible to express the solution to the bifurcation buckling problem in terms of ordinary differential equations. Even if prebuckling nonlinearity is included, the expenses involved in the solution of the bifurcation problem are small in comparison to those required for a two-dimensional nonlinear analysis. Consequently, the bifurcation approach is frequently the most practical for shells of revolution with axisymmetric loading. This approach yields the value of the load at the bifurcation point. Additional information needed in the design procedure is obtained from practical experience or possibly from a linearized analysis of the early post-buckling behavior.

It is possible to use the bifurcation theory as an approximation in some cases for which limit point buckling is critical [1]. However, a complete nonlinear analysis is generally needed. A transient analysis, including dynamic effects, would reveal what happens to the structure as the limit point is passed. Usually, with slow loading rate, structural behavior beyond the limit point is irrelevant and it is sufficient to find the load level at the limit point.

It is not obvious that the concept of buckling can be profitably extended into the domain of dynamic loading. Inclusion of time as an additional parameter complicates matters considerably and there seems to be no general consensus among the specialists on how the buckling and stability concepts should be extended. Dynamic buckling can, of course, be defined in any way that serves a useful purpose. Several reasonable definition have been proposed in the literature over the last few years. Clearly the structure can survive loads above the static buckling load if the loading is of sufficiently short duration. On the other hand, if the load is suddenly applied and then maintained at a constant level, the overshoot results in deformations that are larger than those caused by static loading. It appears that the dynamic buckling load, however it is defined, would be sometimes higher and sometimes lower than the static buckling load, depending on the loading history.

If a structure is subjected to a pulsating load, we encounter the special problem of resonance. For certain combinations of amplitude and frequency of the pulsating load, energy is continually added to the system and the structural response grows with time. When we refer to this phenomenon as "instability", we are not using the word in the same sense as in the discussion above of static stability.

The case of a pulsating load is not further discussed in this paper. Also excluded from the discussion is the case in which forces on the structure are caused by an interaction between the structural deformations and the surrounding medium (aeroelasticity).

The concept of bifurcation is less prominent in discussions of dynamic buckling than in discussions of buckling under static loading. Ho and Nash [2] discuss a perturbation method that is similar to the bifurcation approach in static stability analysis. Akkas [3] uses a similar approach for the asymmetric dynamic buckling of spherical caps. However, both these analyses disregard the effects of the time dependence of the precritical deformations. Simitses [4] regards this "quasi-static" buckling load as a lower bound to a dynamic buckling load.

The perturbation technique leads to an equation of the form

$$\ddot{\xi} + F(u_0, u_1)\, \xi = 0, \tag{1}$$

where $u_0 = u_0(t)$ is the unperturbed displacement solution, u_1 and $\xi = \xi(t)$ represent mode shape and amplitude of the perturbation. In the "quasi-static" approach, u_0 is at each instant considered independent of time, so that Eq. (1) can be treated as a linear differential equation with constant coefficients. If at some time t during the history of deformation F ceases to be positive definite, dynamic instability is said to occur.

The coupled equations for the axisymmetric precritical behavior and the asymmetric perturbation have been considered in a number of papers on the dynamic behavior of circular rings under an external pressure. This problem was first discussed in [5] by Goodier and McIvor.

In that paper, the behavior of a ring, or an infinitely long cylinder, was considered. A uniform external pressure impulse is applied, exciting vibration in the so-called breathing mode. If an imperfection exists in the form of one of the flexural modes, the nonlinear coupling term will act as a forcing function for this flexural mode. Then for some values of the size of the axisymmetric impulse, the parameters of the Mathieu type equation that governs the flexural mode will be in an "unstable region". In that case the amplitude of the response in the flexural mode grows with time. Since new energy is not added to the system, the amplitude of the breathing mode must decrease and the response in the flexural mode must be bounded. Energy is continously exchanged between the two modes. Additional results related to this problem have, for example, been presented by Lovell and McIvor [6], Lindberg [7] and Hubka [8].

In a series of publications ([9, 10], for example), Hsu and his co-workers discuss the topic of dynamic buckling. They define a dynamic system as stable if, with some damping present, the structure will eventually come to rest at a point on the primary path. A dynamic buckling load is defined then only if at least one secondary stable equilibrium configuration exists under the load at $t = \infty$. If the load vanishes for $t = \infty$, as in the case of impulsive loading, the motion of the system can be unstable only if equilibrium configurations exist under zero load that are distinct from the undeformed configuration. With this definition, the authors formulate a sufficient criterion for stability. This criterion is based on energy considerations and may be quite conservative.

A dynamic buckling load can be defined without reference to the concept of stability. In the static case the load corresponding to violent buckling may be defined as the load level at which a small load increment results in a disproportionately large response. A similar definition may be used in the dynamic case and was first suggested by Budiansky and Roth [11]. It was applied to the problem of axisymmetric snapping of spherical caps subjected to a uniform external pressure with a rectangular profile in time. In this investigation, the response (deformation at the pole) is determined as a function of load with the duration of loading being constant. It is noticed that at some load level the response increases dramatically with a small increase in load. This load level is defined as the dynamic buckling load. It appears that a similar criterion may be useful for all structures that are prone to violent buckling.

The buckling criteria of Budiansky and Roth has been applied in further studies of the dynamic behavior of spherical caps ([12], for example). These analyses indicate that if the applied load is not sufficient to cause snap-through on the first cycle of the response, large dis-

placements may occur much later. The delayed large response appears
to be a result of the phenomenon of exchange of energy between modes,
which was first discussed in [5].

In the preliminary design stage the analyst must still rely on design
charts for very simple "equivalent" structures. For complicated structures, the analyst needs a computer program for nonlinear transient
response analysis, and, in order to use this program intelligently, he
needs some basic understanding of how buckling-prone structures behave under dynamic loading. The results of analyses of simple structures
such as spherical caps, cylinders or rings under external pressure, cylinders under axial load [13], etc. are valuable in that they form the basis
of such understanding. It is hoped that the results presented in the
following on the dynamic behavior of cylinders with cutouts will be
helpful in a similar way.

2. The Computer Program

A computer program, STAGS, [14] for the analysis of shells of general
shape has been modified to include nonlinear transient response analysis. The analysis in STAGS is based on an energy formulation. Derivatives which appear in the energy expression are replaced by their
two-dimensional finite difference approximations. The energy is then
rendered stationary. This procedure results in a system of nonlinear
algebraic equations. These are solved by use of the modified Newton-
Raphson method.

For integration in time, the program includes as options an explicit
central difference scheme and an implicit stiffly stable method. The
explicit scheme is the central difference scheme discussed, for example,
in [15], and the implicit integration is based on the Gear method [16].
The program includes from second up to fifth order integration. In
general, the explicit scheme is the easier to use since the criterion for
stability is very simple (see [17]), and the experience indicates that if
the time-step remains within the bounds dictated by numerical stability, then the solution is also accurate. The numerical stability of the
stiffly stable methods is discussed in detail in [16]. Higher order methods
require larger time-steps in order that numerical instability be avoided.
The maximum size of the time-step is determined by required accuracy
in the solution. It is possible for the user of STAGS either to input a
fixed time-step or to use an option with automatic step size control.
Comparisons between the integration methods included in the program
indicate that in general the second or third order Gear method would
be most efficient in shell buckling analysis. The explicit method appears
to be the best choice if the loading history is such that it can only be
accurately determined by use of a very small time-step.

The equation of motion for a general structure can be written in the form

$$M\ddot{u} + D\dot{u} + Su = f, \tag{2}$$

where u is a displacement vector, M and D represent mass and damping matrices of the structure, and S is a nonlinear operator that represents the structural stiffness. The forcing function f can vary with time and with the shell coordinates. The mass matrix is diagonal and the damping matrix has the form

$$D = \alpha M + \beta S + \gamma g(x, y)\, \mathrm{d}A, \tag{3}$$

where α, β, γ are input constants, $\mathrm{d}A$ is the surface area element and g is a general user defined function of the surface coordinates x, y.

Some applications of STAGS are discussed in the following.

3. Effect of Loading Rate

A study was made of the effect of the loading rate on the response of a cylindrical shell with two diametrically opposite cutouts. The cylinder dimensions are shown in Fig. 1. The shell is loaded by a uniform axial

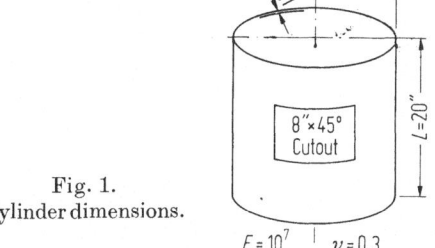

Fig. 1.
Cylinder dimensions.

load that gradually increases with time until it reaches approximately half the static collapse load. After that the load is held constant. With a relatively coarse grid (11 axial and 13 circumferential stations) the static collapse load is found to be about 800 lbs/in. For this model then,

$$P\,[\text{lbs/in}] = \begin{cases} 400\sin^2\,(\pi T/B) & \text{if } T < B/2, \\ 400 & \text{if } T \geq B/2. \end{cases} \tag{4}$$

The parameter B is twice the time it takes to apply full load. In the analysis it is assumed that symmetry conditions prevail at cylinder mid-length. This implies that the load is applied simultaneously at the two ends.

Figure 2 shows how the displacements on the cutout edge vary with time for a number of different values of the parameter B. There is practically no difference between the curves for $B = 25$ μsec and $B = 250$ μsec. Hence, $B = 250$ μsec results in the same response as does instantaneous application of the load.

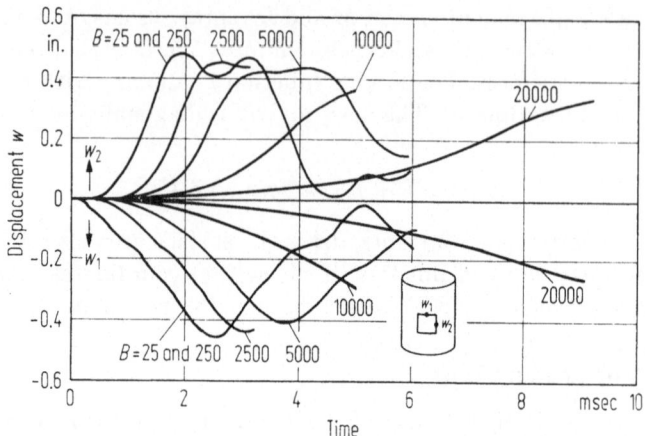

Fig. 2. Displacement versus time for different values of parameter B.

Figure 3 shows, for the same loading histories, the relationship between the cutout edge displacements and the applied axial load. With $B = 20000$ μsec the loading rate is such that it takes 0.01 sec to apply the full load, $P = 400$ lbs/in. With this loading rate there is still

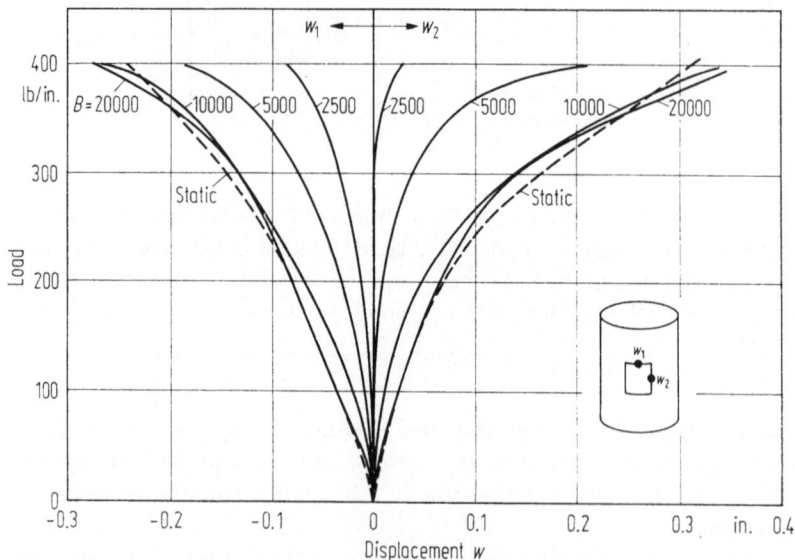

Fig. 3. Load versus displacement for different values of parameter B.

a significant difference between the static and the dynamic response. The lowest frequency for small vibrations in the unstressed state is 255 rad/sec. That is, a half period of vibration in the slowest mode corresponds to 0.0123 sec.

4. Static Limit Point Buckling

A limit point, i.e., a maximum in the load-displacement curve may occur if the primary equilibrium path is nonlinear. The process of determining the limit point usually consists of the solution of the governing nonlinear equations with a stepwise increasing load factor. The criterion for collapse then is that convergence cannot be obtained even if the load increment is very small. This is somewhat unsatisfactory since nonconvergence may be caused by mathematical difficulties. However, in most cases we can plot a suitable displacement versus the load factor and the shape of the curve will indicate whether a point of maximum is approached. Generally, if a sufficient number of values of the determinant of the stiffness matrix coefficients are avalable, a plot of the determinant versus the load factor will more clearly indicate the location of the limit point. This is illustrated in Fig. 4a for the static analysis (with STAGS) of the cylinder with two cutouts. Fig. 4b shows the part of the curve in the immediate neighborhood of the collapse load. In this figure the computed points on which the curve is based are indicated by small circles. Clearly, the critical load for the structural model can be read from this graph with greater accuracy than is needed.

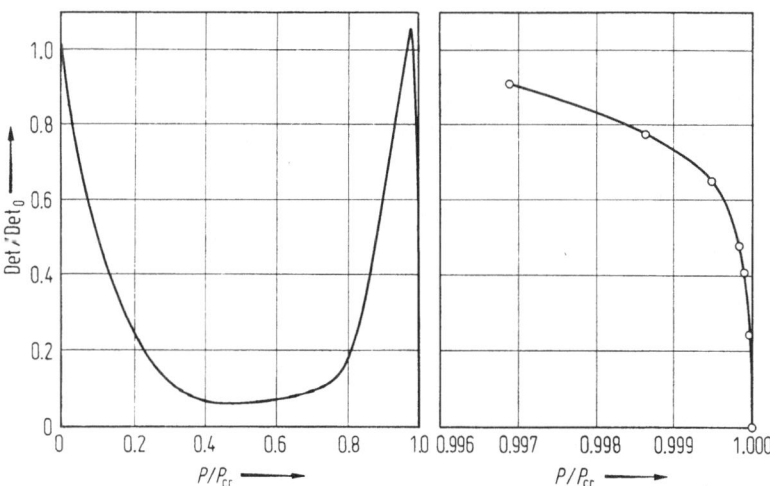

Fig. 4a. Normalized determinant versus normalized load (static).

Fig. 4b. Normalized determinant versus normalized load (static).

The presence of a horizontal tangent in the load-displacement curve does not necessarily mean that the structure will fail at that point. The horizontal tangent may indicate a point of inflection or a small kink in the load displacement curve as shown in Fig. 5. It may be

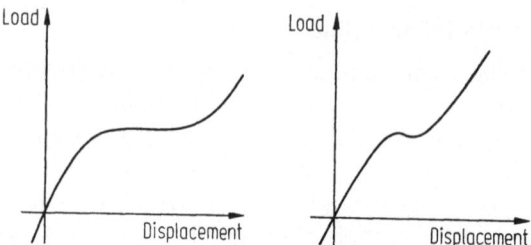

Fig. 5. Typical load versus displacement curves.

interesting to find out exactly what happens to the structure if loading is continued beyond the limit point. This can be done by use of transient response analysis. If the loading rate is sufficiently slow, the computed load displacement curve will be identical to the curve for static loading. At a limit point the shell will be set in motion. The analysis will reveal whether additional load can be applied without damage to the structure.

The transient response analysis capability of STAGS was used for a "static analysis" of the cylinder with two cutouts. The axial load was applied at a rate of 25 000 lbs/sec. The displacements at the edge of the cutout (w_1, w_2) and the displacement at mid-length and 90° from the center of the cutout (w_3) are shown in Fig. 6. The solid lines represent results from a static analysis. Up to a point just below the

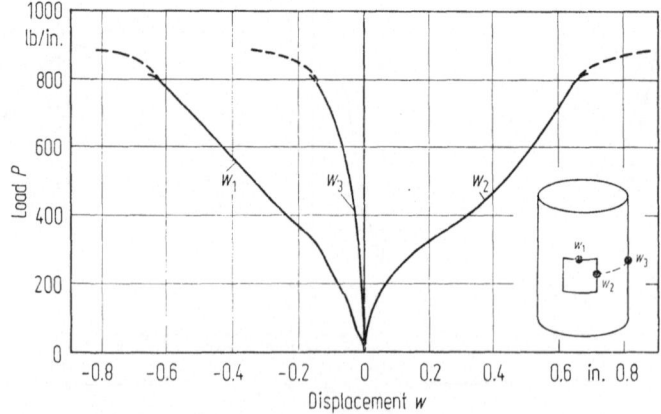

Fig. 6. Load displacement curves (static and transient).

maximum, results from the transient analysis and the static analysis are practically identical. The broken lines, representing results from the transient analysis, indicate a considerable increase in the axial load during the first phase of buckling. With a less rapid loading, the curves would break sharply at the critical load. The plot of the displacement acceleration, shown in Fig. 7, also fails (with a loading rate of 25 000 lbs/sec) to indicate distinctly the location of the limit point.

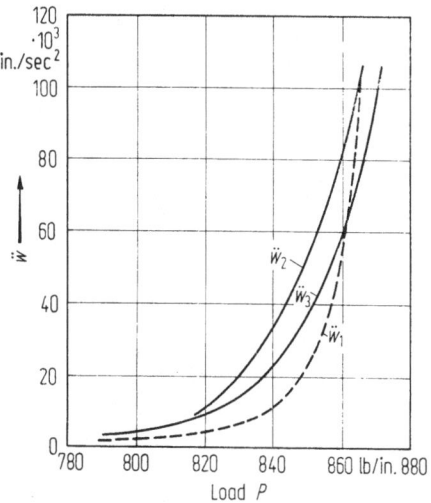

Fig. 7. Acceleration versus load.

The loading rate is low enough to reproduce the static behavior below the limit point. Much slower loading is needed if the behavior in the postbuckling range is to be determined for static loading. However, an analysis with a lower loading rate would be somewhat more expensive.

5. Dynamic Buckling

A study was undertaken of the response of the cylinder with cutouts (Fig. 1) to a load instantaneously applied at both ends and then held constant. Figure 8 shows the lateral displacements at three points on the cylinder surface as function of time for different values of the applied load. In the case with $P = 500$ lbs/in the integration was carried over a longer time interval (up to 0.0066 sec). During that time, the three displacement components did not exceed the maximum values shown in Fig. 8.

The displacement maxima at the three selected points are shown in Fig. 9 as a function of the applied load. The displacement (w_1) at mid length and at the cutout edge shows an almost linear growth with the applied load. Consequently, applied to this displacement, the dynamic buckling criterion by Budiansky and Roth [6] would be meaningless.

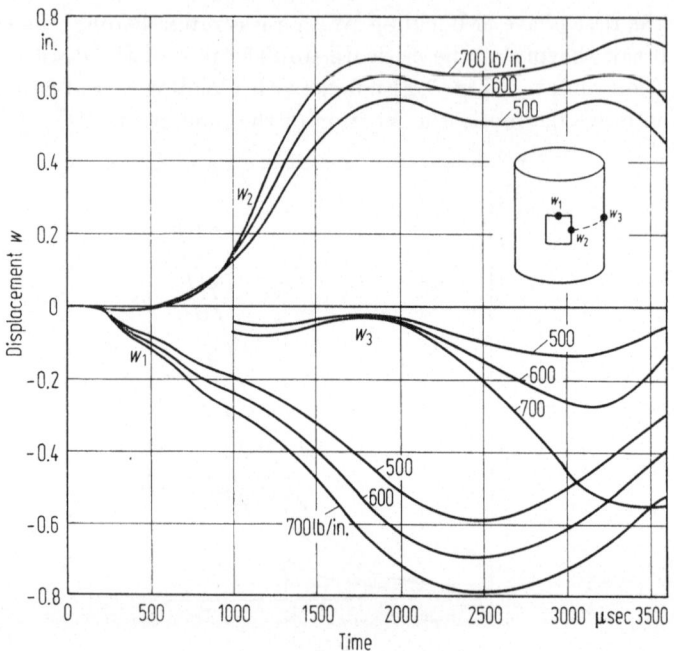

Fig. 8. Displacement versus time for different values of the applied load.

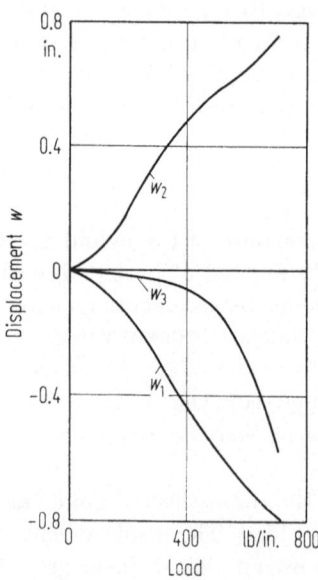

Fig. 9.
Displacement maxima versus load (static).

The displacement at the cutout edge is very large even at loads below the limit point. Collapse of the cylinder occurs when large displacements spread over the entire cylinder surface. Therefore, if we want to apply the Budiansky and Roth criterion, it would be more logical to plot the displacement at some point away from the cutout, like w_3. The curve for the displacement w_3 in the Fig. 9 indicates that the dynamic buckling load is approximately 700 lbs/in, i.e., somewhat below the static buckling load. The variation with time of the distribution of lateral displacements at mid-length is shown in Fig. 10.

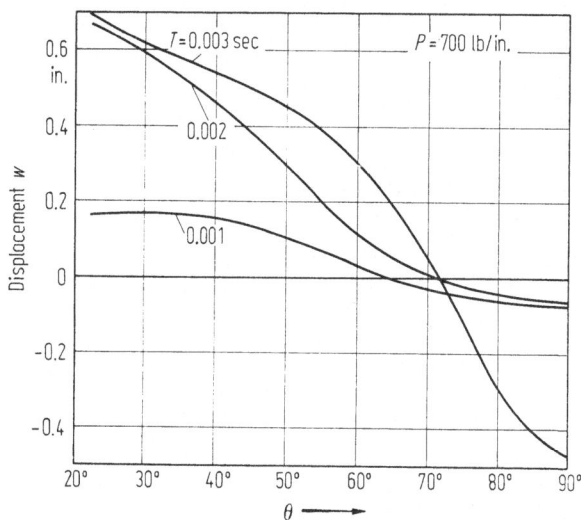

Fig. 10. Displacement versus circumferential coordinate (at cylinder midlength) for different values of time.

6. Conclusions

Numerical methods exist that, in principle, are suitable for the analysis of any structural shell problem. At least, this is true if we disregard the special problems that may arise from inclusion of inelastic deformation, unlimited strains and rotations, or the interaction with a surrounding medium. However, our capability to determine the structural response to the loading environment is bounded by the limitations imposed by computer run time and storage capacity.

The analysis of a cylinder with two cutouts, discussed here, is based on a model with a rectangular grid with 11 rows and 13 columns. This grid is somewhat too coarse, and the results obtained are consequently not very accurate. For the case in which an axial load of 500 lbs/in was instaneously applied the integration was carried over a time interval of 0.0066 sec. This is approximately the time from load application to the occurrence of the third maximum in the fundamental bending mode. The numerical analysis of this case required 11 minutes of computer time (CPU) on the CDC 6600.

In practical engineering we need to analyze more complicated structures, with a much larger number of degrees of freedom, over a longer time interval. The computer time, in a two-dimensional problem grows very rapidly with the size of the problem. Consequently, it is beyond our capability to obtain solutions of reasonable accuracy for many structural configurations. Presently this gap is filled by application of engineering skill in modeling of the structure and interpretation of the numerical results. Even at best this procedure introduces some uncertainty in the design process.

It is important, therefore, to pursue the possibilities of finding methods that can give results of acceptable accuracy with less effort. Let us consider first the case in which static buckling occurs through bifurcation. A representative case then is the shell of revolution with axisymmetric impulsive loading.

Taking the approach suggested by Akkas [3], we could obtain a lower bound by "freezing in time" the axisymmetric precritical displacement configuration. A more accurate solution is obtained if we extend the approach by Goodier and McIvor [5] to the general shell of revolution. In that case we would include small initial imperfections and integrate numerically the equations of motion for each of the harmonics separately. If it is assumed that there is no nonlinear coupling between different harmonics, the analysis will remain in the one-dimensional domain. This method will correctly determine the size of the impulse that will cause growth of the amplitude of the asymmetric mode. It may prove profitable to compare values of the critical impulse according to such analyses with the results of a two-dimensional analysis of a shell with a small imperfection. Only the two-dimensional analysis gives a complete picture of the shell behavior, but on the other hand its execution is so expensive that we frequently will be satisfied with the more limited information that can be obtained by use of any of the other two methods.

As pointed out above, it is possible in some cases with static loading to obtain an approximate solution by linearization of the precritical behavior so that the model loses stability at a bifurcation point rather than at a limit point. Presumably, this would be possible also in the dynamic case. However, in many cases this leads to intolerably inaccurate results. If the configuration is such that a limit point analysis is needed in the static case, we may define a dynamic buckling load either along the lines suggested by Hsu and his coworkers [10] or as suggested by Budiansky and Roth [11]. Hsu's definition is more restricted with respect to the cases to which it can be applied. So far it has not been shown that use of Hsu's sufficient criterion can be used as a conservative but time saving device for a reasonably large class of structural configurations. If the energy criterion is used, it appears that the major part of the work consists in determination of the strain energy in terms of the chosen degrees of freedom of the system. It is interesting to note that this part of the work need not be repeated if a number of different loading histories are considered.

The criterion by Budiansky and Roth [11] appears quite reasonable. However, to establish a dynamic buckling load according to this criterion will always require a greater effort than that involved in the determination of all details of the response to a given loading history. The critical load can only be determined after a number of transient response analyses have been performed. The advantage of this concept of a dynamic buckling load is that it makes it possible to construct design charts for some relatively simple structures.

While further pursuit of energy criteria and perturbation methods for approximate solutions seems warranted, it is also worthwhile to attempt to reduce the computer cost involved in straightforward numerical solution through improvement of the numerical procedures. For example, expansion of the displacements in vibration modes may prove to be profitable also in the nonlinear case.

References

1. Almroth, B. O.; Brogan, F. A.: Bifurcation Buckling as an Approximation of the Collapse Load for General Shells. AIAA J. 10, 463 (1972).
2. Ho, F. H.; Nash, W. A.: Dynamic Buckling of Shallow Spherical Caps. Proceedings of Symposium on Non-Classical Problems. International Association of Shell Structures, Warsaw 1963.
3. Akkas, N.: Asymmetric Buckling Behavior of Spherical Caps under Uniform Step Pressures. J. Appl. Mech. 39, 293 (1972).
4. Simitses, G. J.: On the Dynamic Buckling of Shallow Spherical Shells. J. Appl. Mech. 41, 299 (1974).
5. Goodier, J. N.; McIvor, I. K.: The Elastic Cylindrical Shell under Nearly Uniform Radial Impulse. J. Appl. Mech. 31, 259 (1964).
6. Lovell, E. G.; McIvor, I. K.: Nonlinear Response of a Cylindrical Shell to an Impulsive Pressure. J. Appl. Mech. 36, 277 (1969).
7. Lindberg, H. E.: Stress Amplification in a Ring Caused by Dynamic Instability. J. Appl. Mech. 41, 392 (1974).
8. Hubka, W. F.: Dynamic Buckling of the Elastic Cylindrical Shell Subjected to Impulsive Loading. J. Appl. Mech. 41, 401 (1974).
9. Hsu, C. S.: On Parametric Excitation of a Dynamic System Having Multiple Degrees of Freedom. J. Appl. Mech. 30, 367 (1973).
10. Hsu, C. S.; Kao, C. T.; Plant, R. M.: Dynamic Stability Criteria for Clamped Shallow Arches under Timewise Step Loads. AIAA J. 7, 1925 (1969).

11. Budiansky, B.; Roth, R. S.: Axisymmetric Dynamic Buckling of Clamped Shallow Spherical Shells. NASA TN-1510, 1962.
12. Stephens, W. B.; Fulton, R. E.: Axisymmetric Static and Dynamic Buckling of Spherical Caps due to Centrally Distributed Pressures. AIAA J. 7, 2120 (1969).
13. Lindberg, H. E.; Herbert, H. E.: Dynamic Buckling of a Thin Cylindrical Shell under Axial Impact. J. Appl. Mech. 33, 105 (1966).
14. Almroth, B. O.; Brogan, F. A.; Meller, E.; Zele, F.; Peterson, H. T.: Collapse Analysis for Shells of General Shape. II, User's Manual for the STAGS-A Computer Code, AFFDL-TR-71-8, March 1973.
15. Krieg, R. D.: Unconditional Stability in Numerical Time Integration Methods. J. Appl. Mech. 40, 417 (1973).
16. Jensen, P. S.: Transient Analysis of Structures by Stiffly Stable Methods. Computers and Structures 4, 615 (1974).
17. Geers, T. L.; Sobel, L. H.: Analysis of Transient Linear Wave Propagation in Shells by the Finite Difference Method, NASA CR-1885, Dec. 1971.

Theory and Experiment in the Creep Buckling of Plates and Shells [1]

N. J. Hoff

Stanford University, Stanford, California, U.S.A.

Abstract

It is shown that in the presence of creep the classical Eulerian concept of stability must be replaced by a more general one to obtain analytical results of usefulness to the engineer. Formulas derived earlier are presented in forms that are easy to apply to practical problems. Their predictions are compared with the results of two sets of tests. Satisfactory agreement is found between theory and experiment.

1. Elasticity, Plasticity and Creep

At the turn of the century engineering calculations were based mainly on the conditions of equilibrium, Hooke's law and empiricism. Gradually applications of the theory of elasticity began to find acceptance; this theory is based on Hooke's generalized law. Eventually the theory of plasticity was introduced and used to explain discrepancies between the failing loads calculated from Hooke's law and those observed in experiment. The accepted theories of plasticity have been found most useful even though they represent far-reaching simplifications of the real properties of materials; for instance, they disregard the well-known anisotropic nature of strain hardening. In the end, the particularly daring concept of a rigid-plastic material has become a tool of civil engineering practice.

If plasticity can be studied usefully on the basis of simplifying assumptions, the same procedure should be permissible for creep also. The idea that the creep phenomenon is too complicated to be reduced to mathematical formulas is simply an evasion of the issue. Many interesting phenomena can be understood and explained on the basis of the simplest possible formulation of a creep law according to which the rate of change of the strain is a function of the stress only when the temperature and the stress are constant. This so-called secondary creep law will be used here to study the creep buckling of plates and shells.

[1] The author acknowledges his indebtedness to the Mechanics Branch of the Office of Naval Research of the U.S. Navy for the support of the research whose results are presented here.

2. Nature of the Creep Buckling Phenomenon

It is important to realize that creep buckling is not a stability problem in the classical sense. This should be clear from the example of the simply supported column whose material deforms only in steady creep: When the initial straightline equilibrium state of such a column is disturbed by deflecting the column slightly sidewise, the disturbance will increase monotonically with time no matter how small is the axial compressive load. Thus in the classical sense all creeping columns are unstable and an investigation of their stability is superfluous. What the engineer needs to know is how much time it takes for the deflections to increase to such an extent that the column can no longer perform its structural function. The investigation of this question, and the determination whether the lifetime of the column is shorter or longer than the time for which the structure is intended to remain in use, are evidently not calculations of stability in the classical sense.

In a more general sense a structural element can be considered stable if a disturbance of a permissible kind and magnitude is followed by displacements that do not exceed permissible limits within the period of time for which the element is intended to be used [1]. This definition need not be changed if it is to be applied to columns, plates and shells whose material exhibits elastic deformations as well as primary and steady creep.

It is of some interest to note that investigations of this kind that have already been carried out reveal that the critical stress of pefectly elastic columns, plates and shells is replaced in the presence of creep by another critical quantity, namely the lifetime, or critical time, at which the analysis yields deflections that tend to infinity. This critical time is finite whenever the creep law is non-linear in the sense that a doubling of the stress leads to creep rates greater than twice the original ones [2]. Hence the purpose of most creep buckling investigations is to calculate the critical time.

On the other hand, in some cases creep can lead to states of deformation in which the structural element exhibits elastic, or elasto-plastic, instability in the classical sense. Two such cases have already been discussed in the literature.

In the first, the deflections due to steady creep of a column were calculated whose material deforms elasto-plastically under rapidly applied loads [3, 4]. The effective modulus of such a column decreases as the intensity of the stress increases. But as any initial deviation from the exact straightline shape of the column increases with time in consequence of creep caused by a constant load, the bending moment in the column increases with time and so does the maximum compressive stress in the column. Sooner or later the time is reached at which the reduction in the effective modulus reduces the buckling load of the column to the value of the constant applied load. Then the column, which has crept slowly for a long time, buckles suddenly. Snap-through of this kind has been observed in column tests [4].

A second kind of snap-through after creep is possible with axially compressed circular cylindrical shells. The buckling load of elastic shells is known to be sensitive to initial deviation from the exact cylindrical shape [5]. At the beginning of the test, axisymmetric deviations of such a small amplitude may be present in the circular cylindrical shell that it can easily carry the axial load imposed upon it. But these deviations increase with time because of creep and thus the initially rather accurate shell becomes increasingly inaccurate, or imperfect, as time passes. Finally the state of imperfection is reached for which the initially applied constant load is the snap-through load. At that time the shell suddenly develops large multilobed deformations after a long period of slowly growing axisymmetric deformations. The change-over from axisymmetric to multilobed deformations of shells that creep has already been described in the literature but the sudden jump has apparently not yet been observed [6].

3. Agreement between Theory and Experiment

In order to persuade practicing engineers that creep theories are of practical value, it is necessary to show that their conclusions can be verified by experiment. Unfortunately this is not an easy task because the experiments must be performed, as a rule, at high temperatures held rigorously constant over long periods of time, and because the scatter of all creep tests is very much higher than that of tests performed with perfectly elastic structural elements at room temperature. The large scatter is to some extent a consequence of the fact that small changes in stress and temperature give rise to large changes in the creep rate as a consequence of the non-linearity of the creep law.

Since the creep rate varies greatly in consequence of cold work, heat treatment and aging of the metals of construction, any attempt at evaluating the results of buckling tests on the basis of creep laws published in the literature is doomed to failure. The situation is similar to, but even worse than, that prevailing in the case of instantaneous buckling of short columns, and one should remember that the validity of Euler's concepts was proven for short columns only when von Kármán machined all his column test specimens and the specimens used for the establishment of the stress-strain curve out of a single block of steel.

In view of this situation an effort is made in this paper to correlate the results of some creep buckling tests with creep rate measurements made in the course of the same tests. The correlation is based on theoretical formulas derived earlier. They are presented here in a form that is likely to be most convenient for the designer.

4. Circular Cylindrical Shell Subjected to Axial Compression

When a compressive load P is applied to a cylindrical shell in the axial direction and is distributed uniformly around the circumference, the critical time is

$$t_{cr} = D_1 t_E \ln \left[(D_2 + e_0^{n-1})/e_0^{n-1} \right], \tag{1}$$

where

$$t_E = \varepsilon_E/\dot{\varepsilon}_{nom}, \quad \dot{\varepsilon}_{nom} = k\sigma^n, \quad e_0 = f_0/t^*. \tag{2}$$

$$\sigma = P/2\pi a t^*, \quad \varepsilon_E = 0.6(t^*/a), \tag{3}$$

and a is the mean radius and t^* the constant wall thickness of the shell. The material constants k and n are assumed to be known from creep tests. It is also assumed that axisymmetric initial deviations from the exact circular cylindrical shape exist in te radial direction, and they have the most dangerous wave length $\lambda \approx 1.6 \, (at^*)^{1/2}$ and an amplitude f_0. The values of the coefficients D_1 and D_2 depend on the creep exponent n [7, 8]. In the range

$$3 \leqq n \leqq 11 \tag{4}$$

they can be given by the approximate formulas

$$D_1 = 2.88/n^2, \quad D_2 = (0.55)^{n-1}. \tag{5}$$

When e_0 is small compared to 0.55, Eq. (1) simplifies to

$$t_{cr} = C_1 t_E \ln (C_2/e_0), \tag{6}$$

where

$$C_1 = 1.44/n^{0.732}, \quad C_2 = 0.55. \tag{7}$$

When the material of the shell is capable of simultaneous linearly elastic and nonlinear creep deformations [9], the critical time t_{cr} is shortened to t_{cr+el} with

$$t_{cr+el} = (\sigma_E - \sigma) \, t_{cr}/\sigma_E,$$

where

$$\sigma_E = \varepsilon_E E, \tag{8}$$

and E is Young's modulus of elasticity. At the same time e_0 must be replaced by $e_0 \sigma_E/(\sigma_E - \sigma)$.

As already mentioned, the slowly developing creep deformations can procdue a state from which the shell suddenly snaps through into a state of large deformations. The time at which this happens [10] will be denoted t_s; it can be calculated from

$$t_s = R t_{cr}. \tag{9}$$

Values of the redcution factor R are plotted in Fig. 1 in function of the load ratio

$$\varrho = \sigma/\sigma_E \tag{10}$$

with the initial deviation amplitude ratio e_0 as the parameter of the family of curves.

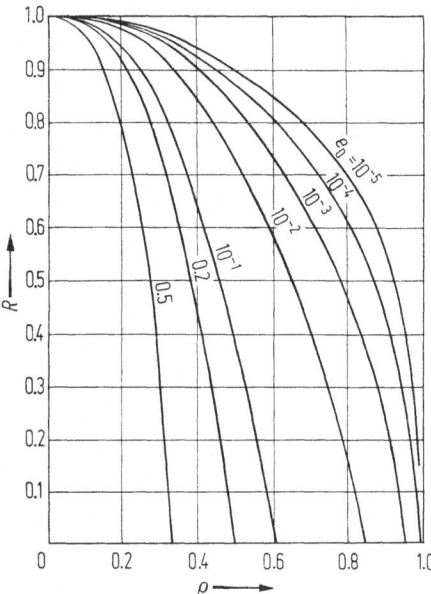

Fig. 1. Reduction factor for snap-through.

For a comparison with experiment, data are available on 31 specimens whose radius-to-thickness ratio varies from 30.6 to 96.4 [11]. They were manufactured of nickel by the electroplating method and were tested at 650°F. The modulus of elasticity of the material was found to be 24×10^6 psi. Since the loads applied to the specimens did not exceed 6 percent of the Euler load ($\varrho < 0.06$), the corrections for elasticity and snap-through were negligible. Substitutions, consideration of the relationship $e_0/0.55 \ll 1$ and solution for σ^n yield

$$\sigma^n = (0.6 C_1/k) \, [\ln(C_2/e_0)] \, (t^*/a t_{cr}). \tag{11}$$

Since t_{cr} varies slowly with e_0, and since the value of e_0 is not known accurately, it is reasonable to select an average value for e_0 and to consider it to be the same for all the test specimens. It follows then from (11) that

$$n \log \sigma = \log(t^*/a t_{cr}) + \text{const}. \tag{12}$$

A plot of σ as a function of t^*/at_{cr} is shown in Fig. 2. The test results can be represented by a straight line of slope $1:3$ although they show scatter of an amount usual with creep tests. With $n \doteq 3$ one has

$$C_1 = 0.645, \quad C_2 = 0.55.$$

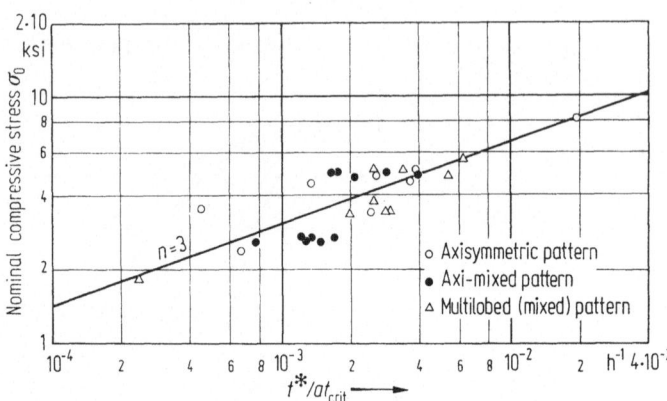

Fig. 2. Relationship between stress and critical time for cylindrical shell.

On the basis of buckling tests carried out with perfectly elastic shells it was found [12] that e_0 can be taken as $10^{-4}(a/h)$ if the specimens are prepared with great care. In this manner one obtains $e_0 = 0.00635$ for the average value 63.5 of a/h of the set of specimens. If these values are substituted in (11) and t^*/at_{cr} is taken as 4.5×10^{-3} h^{-1} when $\sigma = 5$ ksi, solution for k yields

$$k = 0.621 \times 10^{-4} \text{ h}^{-1} \text{ (ksi)}^{-3}. \tag{13}$$

To establish the sensitivity of the calculations to the choice of e_0, it will now be assumed tat e_0 has ten times the value assigned to it before (this corresponds to a very poorly manufactured specimen). The result is

$$k = 0.30 \times 10^{-4} \text{ h}^{-1} \text{ (ksi)}^{-3}. \tag{14}$$

At the beginning of the creep buckling tests the specimens are almost perfect and the deformations of the cylinder consist essentially of a shortening. If this divided by the initial length is plotted as a function of time, an essentially straight line is obtained after a transient stage immediately following load application and before the onset of a tertiary stage preceeding buckling. The slopes of these straight lines are the initial creep rates $\dot{\varepsilon}_{\mathrm{nom}}$; they are shown in Fig. 3. Although the data scatter greatly, it is possible to represent them by a straight line with a slope $1:3$. If this straight line is so placed as to make the mean square error a minimum, the equation representing the line is

$$\dot{\varepsilon}_{\mathrm{nom}} = 0.4 \times 10^{-4}\, \sigma^3\, \mathrm{h}^{-1}. \tag{15}$$

The satisfactory agreement between the creep laws derived from the critical times and from the secondary creep rates indicates that the theory is essentially correct.

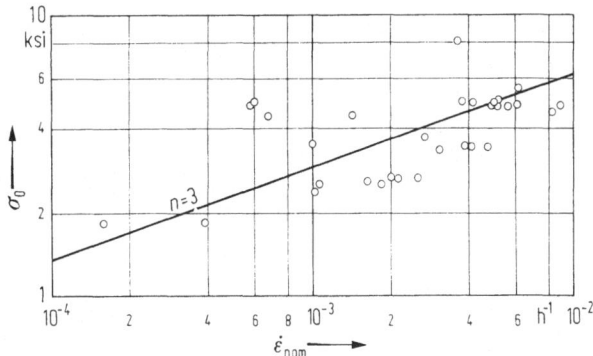

Fig. 3. Relationship between stress and creep strain rate for cylindrical shell.

5. Simply Supported Rectangular Plate Compressed Parallel to One Pair of Edges

When a compressive load P is applied parallel to one pair of the edges of a simply supported rectangular plate and is distributed uniformly along the loaded edge, the critical time is again given by (1) and (2). On the other hand (3) is replaced by

$$\sigma = P/bt^*, \quad \varepsilon_{\mathrm{E}} = 3.6(t^*/b)^2. \tag{16}$$

It is assumed that the initial deviations from the exact flat shape have the most dangerous wave length (approximately equal to b) and an amplitude f_0. The values of the coefficients D_1 and D_2 depend [13, 14] on the creep exponent n. In the range

$$3 \leqq n \leqq 7 \tag{17}$$

they can be represented by the approximate formulas

$$D_1 = 1.74/n^{1.42}, \quad D_2 = 3.08/e^{0.567n}. \tag{18}$$

When e_0 is small compared to $D_2^{1/(n-1)}$, (1) again simplifes to (6). The values of the coefficients C_1 and C_2 are for the flat plate

$$C_1 = 0.78 - 0.016n \ (\approx 0.7), \quad C_2 = 0.975/n^{0.239} \ (\approx 0.7). \tag{19}$$

When the material of the shell is capable of simultaneous linearly elastic and nonlinear creep deformations, the critical time t_{cr} is again replaced with $t_{\mathrm{cr+el}}$ in accordance with (8).

Calculations carried out with the aid of the digital computer have shown that the amplitude of the nondimensional initial deviations has a small secondary effect (in addition to the primary effect given by (1)) on the critical time. It is recommended therefore that the value of C_1 be retained when e_0 is less than 0.001, and that it be multiplied by factors of 0.85 and 0.55 when e_0 is 0.01 and 0.1 respectively, and $n > 4$.

When the displacements become large, the behaviour of the plate differs fundamentally from that of the shell. While the latter is likely to snap through into a different buckle pattern and lose much of its load-carrying capacity, the former continues to carry the load imposed upon it [15]. The physical significance of the critical time of a plate is simply that the rate of lateral deflections becomes large at approximately the critical time. Hence the plate cannot be used much beyond t_{cr} if large displacements are objectionable.

For a comparison with experiment, data are available on 33 test specimens [16] manifactured of 2024-T3 aluminum alloy. Sheets of this material having a thickness t^* of 0.020 in were bent into square boxes of a side width b of 1.5 in. and were tested in compression at a temperature of 600°F. Young's modulus E of the material was found to be 7.2×10^6 psi at this temperature.

The use of thin-walled square tubes was decided upon when two different sets of edge supports designed, built and tested earlier were found to develop undesirable restraints after the buckling displacement had reached large values. Ideally, buckles can form alternately in the inward and outward directions around the tube and the edges can act as frictionless hinges. Unfortunately, the ideal simple-support condition also requires that the initial deviations from flatness follow the same pattern.

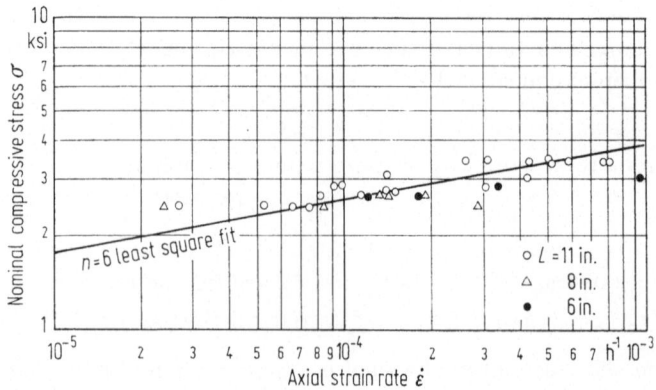

Fig. 4. Relationship between stress and creep strain rate for plate.

The shortening of each specimen was measured, and this quantity divided by the original length was plotted as a function of time. In the curves so obtained primary, secondary and tertiary phases can be discerned. The creep rates corresponding to the secondary phase are shown in Fig. 4 in function of the stress. From the steepest portion of the tertiary phase the location of the vertical asymptote was estimated and the corresponding time was taken as the critical time t_{cr} (see Fig. 5).

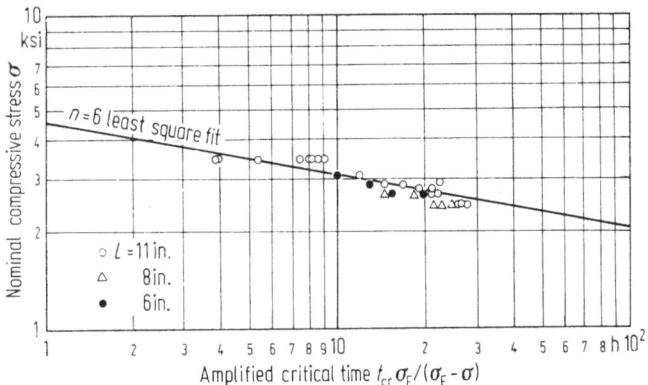

Fig. 5. Relationship between stress and critical time for plate.

As the loads applied to the flat plate specimens amounted to a much higher percentage (from 50 to 75 percent) of the classical critical loads than those applied to the circular cylindrical specimens, the correction for elasticity had to be used here although it was unnecessary in the case of the cylindrical shells. With this correction and after substitutions (8) can be written in the form

$$t_{cr}\sigma_E/(\sigma_E - \sigma) = (3.6C_1/k) \left[\ln (C_2/e_0)\right] (t^*/b)^2 (1/\sigma^n). \qquad (20)$$

Since t^*/b had the same value for all the specimens, all the symbols in this equation except σ and e_0 represent constants for all the tests made. In the case of flat plates no probable values are vailable for e_0. Nevertheless this quantity, which appears only in the argument of the logarithmic function, was considered as a constant (and, no doubt, its variation contributed to the scatter in the data). In a logarithmic form the equation is

$$\log \left[t_{cr}\sigma_E/(\sigma_E - \sigma)\right] = \overline{C} - n \log \sigma, \qquad (21)$$

where $\overline{C}$ is a constant for all the tests. At the same time the creep law (2) can be written in the logarithmic form

$$\log \dot{\varepsilon} = C^* + n \log \sigma, \qquad (22)$$

where C^* is a constant for all the tests.

It is evident that plots on double-logarithmic paper of $\dot{\varepsilon}$ and $t_{cr}\sigma_E/(\sigma_E - \sigma)$ against σ must appear as straight lines of opposite slope. This is indeed the case in Figs. 4 and 5 where straight lines of slopes $1:6$ and $-1:6$ satisfactorily represent the trends of the data. Again, the scatter is smaller for the critical time than for the creep rate, and both are smaller for he flat plate than for the cylindrical shell.

With $n = 6$ the value of the constant k in the creep law (2) was found from Fig. 4, to be

$$k = 3.4 \times 10^{-7} \text{ h}^{-1} \text{ (ksi)}^{-6}, \tag{23}$$

because $\dot{\varepsilon} = 2.48 \times 10^{-4}$ h^{-1} corresponds to a stress of 3 ksi. At this same stress the value of $t_{cr}\sigma_E/(\sigma_E - \sigma)$ is 11.8 h (see Fig. 5). One obtains successively

$$C_1 = 0.684, \quad C_2 = 0.636,$$

$$\varepsilon_E = 6.4 \times 10^{-4}, \quad \dot{\varepsilon}_{nom} = 2.48 \times 10^{-4} \text{ hr}^{-1}, \quad t_E = 2.58 \text{ h}. \tag{24}$$

From (6) and (8) one obtains

$$t_{cr}\sigma_E/(\sigma_E - \sigma) = 0.684 \times 2.58 \ln (0.636/e_0). \tag{25}$$

It follows then that

$$\ln (0.636/e_0) = 11.8/(0.684 \times 2.58) = 6.69. \tag{26}$$

Thus C_2/e_0 is 800 and

$$e_0 = 0.795 \times 10^{-3}. \tag{27}$$

This value is smaller by an order of magnitude than that assumed in the case of the cylindrical shell. It is certainly much smaller than the greatest deviation of the plate from a plane. However, e_0 is not the greatest deviation but the nondimensional amplitude of that deviation component which has the most dangerous wave length. Moreover, the existence of a relatively large amplitude on one of the four sides does not suffice to cause early buckling if similarly large amplitudes do not occur with the right signs on the other sides. In the absence of such deviations on he latter sides the simple support conditions are not satisfied for the former side.

On the basis of these considerations the value obtained for e_0 appears to be a reasonable one. This, and the equal and opposite slopes of the straight lines in Figs. 4 and 5, seem to indicate that the analysis here presented is a satisfactory one.

References

1. Hoff, N. J.: Dynamic Stability of Structures. Opening lecture of the conference on Dynamic Stability of Structures. Edited by George Herrmann. Oxford, New York: Pergamon Press 1967, p. 7.
2. Hoff, N. J.: Creep Buckling of Plates and Shells. Proceedings of the Thirteenth International Congress of Theoretical and Applied Mechanics. Edited

by E. Becker and G. K. Mikhailov. Berlin, Heidelberg, New York: Springer 1973, p. 124.

3. Fraeijs de Veubeke, B.: Creep Buckling. High Temperature Effects in Aircraft Structures. Edited by N. J. Hoff. Oxford: Pergamon Press 1958, p. 267.

4. Chapman, J. C.; Erickson, B.; Hoff, N. J.: A Theoretical and Experimental Investigation of Creep Buckling. Int. J. of the Mech. Sci. 1, No. 2/3, 145 (1960).

5. Koiter, W. T.: Over de Stabiliteit van het elastisch Evenwicht. Amsterdam: H. J. Paris 1945. — The Effect of Axisymmetric Imperfections on the Buckling of Cylindrical Shells under Axial Compression. Proceedings of the Royal Netherlands Academy of Sciences (Amsterdam) Series B, 66, No. 5 (1963).

6. Hoff, N. J.: On the Transition from Axisymmetric to Multilobed Creep Buckling. Contributions to the Theory of Aircraft Structures. A. van der Neut Anniversary Volume, 1972.

7. Hoff, N. J.: Axially Symmetric Creep Buckling of Circular Cylindrical Shells in Axial Compression. J. of Appl. Mech. 35, Series E, No. 3, 530 (1968).

8. Honikman, T. C.; Hoff, N. J.: The Effect of Variations in the Creep Exponent on the Buckling of Circular Cylindrical Shells. Int. J. of Solids and Structures 7, 1685 (1971).

9. Hoff, N. J.; Levi, I. M.: Short Cuts in Creep Buckling Analysis. Int. J. of Solids and Structures 8, No. 9, 1103 (1972).

10. Hoff, N. J.: The Effect of Geometric Nonlinearities on the Creep Buckling Time of Axially Compressed Circular Cylindrical Shells. Stanford University Dept. of Aeronautics and Astronautics Report SUDAAR No. 476, April 1974; also J. Appl. Mech. 42, 225—226 (1975).

11. Benoit, M.; Hoff, N. J.: Creep Buckling of Relatively Thick Circular Cylindrical Shells Compressed Axially. Stanford University Department of Aeronautics and Astronautics Report SUDAAR No. 461, July 1973.

12. Hoff, N. J.: Quelques nouveaux résultats de recherches sur le flambage des coques cylindriques. Technique et Science Aéronautiques et Spatiales. AFITAE, Paris, 1967, Nos. 5/6, p. 463.

13. Hoff, N. J.: Creep Buckling of Rectangular Plates under Uniaxial Compression. Engineering Plasticity. Sir John Baker Anniversary Volume. Edited by J. Heyman and F. A. Leckie, Cambridge/England: Cambridge University Press 1968, p. 257.

14. Hoff, N. J.: Berke, L.; Honikman, T. C.; Levi, I. M.; Creep Buckling of Flat, Rectangular Plates when the Exponent Ranges from 3 to 7. A. E. Johnson Memorial Volume, edited by A. I. Smith and A. M. Nicolson, Applied Sciences, London, 1971, p. 421.

15. Levi, I. M.; Hoff, N. J.: The Postcritical Behavior of Compressed Plates that Creep. Henry Görtler Memorial Issue. Ing.-Arch. 38, Nos. 4/5, 329 (1969).

16. Hoff, N. J.; Benoit, M.: An Experimental Check of the Theory of the Creep Buckling of Plates. Yu. N. Rabotnov Anniversary Volume, edited by Novozhilov, to be published.

Nonsymmetric Creep Buckling of Cylindrical Shells under Axial Compression and External Pressure

R. B. Rikards, G. A. Teters

Academy of Sciences of Latvian S.S.R., Riga, Latvian S.S.R., U.S.S.R.

Abstract

In the present paper the creep buckling problem of cylindrical shells made from polymer composite materials is investigated. It is assumed that all the deformations are due to nonlinear creep governed by Volterra's integral law. The investigation of problem is based on changes of the axisymmetric buckling mode to nonsymmetric one. The experimental results are presented and discussed.

1. Introduction

Polymer materials, in particular composites with a polymer matrix, have creep deformations under loads at normal temperature. Shells made from such material, if the loading is prolonged, buckle in time. The critical time of a shell depends on the material properties, lod level, geometry of the shell, and may vary from a few seconds to several months.

A real construction always has small initial geometrical imperfections of the middle surface of the shell. If we apply at $t = 0$ a load less than the elastic buckling load the shell will deform elastically. Due to creep the wall of a shell bends, and there is modification of the equilibrium state: displacements, stresses and strains slowly change. This process continues until $t = t^*$ when radial displacements begin to grow rapidly and the shell buckles and passes into a postbuckling state or the material of shell fails. So the buckling process is a process of development and change of the initial imperfection modes. The buckling modes completely determine the stress-strain state of the shell and from the analysis of the buckling modes one can determine the critical time of a construction. The buckling of polymer shells in general is similar to buckling of metal shells in high temperature fields. This problem was widely investigated by Hoff [1], Rabotnov [2], Grigoliuk and Lipovtsev [3], Samuelson [4] and other scientists. In all these works the authors assumed the incremental stress-strain law typical of metals. According to the incremental law strain rate depends only upon current strain. Polymers and composites with polymer matrixes are materials with memory and the strain rate in such materials depends on stresses in previous time moments.

2. Experimental Investigation

Buckling experiments on circular cylindrical shells under axial compression were carried out. The shells were made from polyethylene— nonlinear viscoelastic material. The test specimens had length $L =$ 200 mm, external diameter $D = 105$ mm, wall thickness $h = 2, 3, 4, 5$ and 6 mm. The total number of shells was 40. The applied load was $p^* = (0{,}60 \cdots 0{,}85) \, p^0$ where p^0 elastic critical load; the load p^* remains constant until buckling. The load level was chosen so that the critical time was in the time interval from 20 minutes to 2 hours. In Fig. 1 typical curves $\Delta l \sim t$ for the shells with $R/h = 25$ (1), 16,7 (2), 12,5 (3), 10 (4) are given. One can determine the critical time of the construction from these curves, but they do not give much information about the buckling process. For this purpose the radial deflections w were recorded [5] after load application. Registrations were done at regular intervals t_j.

In order to investigate the characteristic buckling mode the Fourier representation in complex form was used and approximation was done

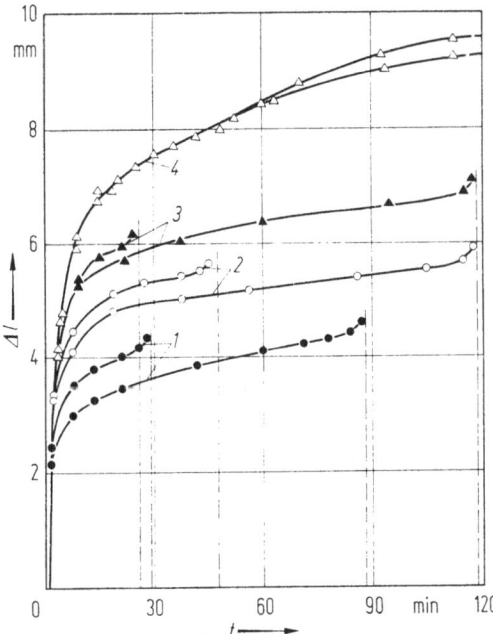

Fig. 1. The critical time of axially compressed shells with $R/h = 25(1)$, 16.7(2) 12.5(3), 10(4).

in discrete time moments t_j:

$$w(x, y, t_j) = \sum_{m=-M}^{M} \sum_{n=-N}^{N} \alpha_{mn}(t_j)\, e^{i[(m\pi x/L)+(ny/R)]} \qquad (1)$$

with m number of the half-waves in the axial direction, n number of the waves in the circumferential direction, R radius of a shell, x, y coordinates in the axial and circumferential directions, respectively. The determination of required complex coefficients from the registered radial deflection function $\tilde{w}(x, y, t_j)$ was carried out numerically using the Simpson's rule:

$$\alpha_{mn}(t_j) = (1/2\pi RL) \int_0^{2\pi R} \int_0^{L} \tilde{w}(x, y, t_j)\, e^{-i(m\pi x/L + ny/R)}\, \mathrm{d}y\, \mathrm{d}x. \qquad (2)$$

The complex Fourier representation as will be seen later is more convenient for solving the nonlinear buckling problem as compared with the trigonometrical form. But the trigonometrical form of Fourier representation is more convenient for the analysis of buckling modes and their development in time. The radial displacement field can be described by

$$\begin{aligned} w(x, y, t_j) = &\sum_{m=0}^{M} \sum_{n=0}^{N} a_{mn}(t_j) \cos(m\pi x/L) \cos(ny/R)\, + \\ &+ \sum_{m=1}^{M} \sum_{n=0}^{N} b_{mn}(t_j) \sin(m\pi x/L) \cos(ny/R)\, + \\ &+ \sum_{m=0}^{M} \sum_{n=1}^{N} c_{mn}(t_j) \cos(m\pi x/L) \sin(ny/R)\, + \\ &+ \sum_{m=1}^{M} \sum_{n=1}^{N} d_{mn}(t_j) \sin(m\pi x/L) \sin(ny/R). \end{aligned} \qquad (3)$$

The coefficients in the representation (3) are connected with the complex ones $\alpha_{mn}(t_j)$ by the following relations:

$$a_{00} = 4\alpha_{00}, \qquad a_{m0} = 2(\alpha_{m0} + \alpha_{-m,0}),$$

$$a_{0n} = 2(\alpha_{0n} + \alpha_{0,-n}), \qquad b_{m0} = 2i^{-1}(-\alpha_{m0} + \alpha_{-m,0}),$$

$$c_{0n} = 2i^{-1}(-\alpha_{0n} + \alpha_{0,-n}),$$

$$a_{mn} = (\alpha_{mn} + \alpha_{-m,n} + \alpha_{m,-n} + \alpha_{-m,-n}), \qquad (4)$$

$$b_{mn} = i^{-1}(-\alpha_{mn} + \alpha_{-m,-n} - \alpha_{m,-n} + \alpha_{-m,n}),$$

$$c_{mn} = i^{-1}(-\alpha_{mn} + \alpha_{-m,-n} + \alpha_{m,-n} - \alpha_{-m,n}),$$

$$d_{mn} = (-\alpha_{mn} - \alpha_{-m,-n} + \alpha_{m,-n} + \alpha_{-m,n}).$$

The coefficients (4) give the idea of initial imperfection and the predominant buckling modes and their development in time. Figure 2 shows the growth rate of axisymmetric (1, 2, 3) and nonsymmetric coefficients for shells with $R/h = 8.3$; 12.5 and 25.0, respectively. As can be

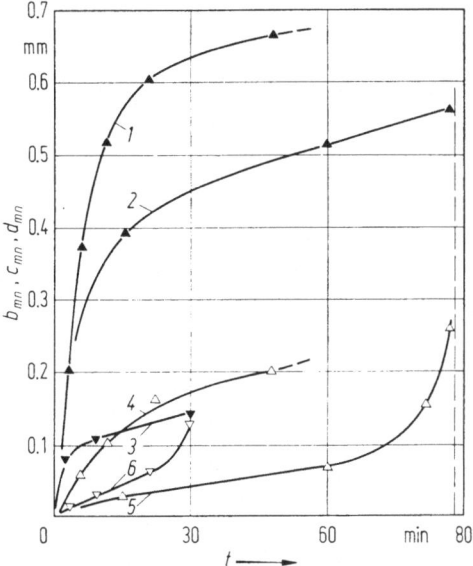

Fig. 2. The growth rate of axisymmetric (curves 1, 2, 3) and nonsymmetric (curves 4, 5, 6) coefficients for shells with $R/h = 8.3$, 12.5 and 25.0, respectively.

seen from these curves in the first period of time the axisymmetric mode predominates. Bending of the shell wall is local at the ends of the shell. The growth rate of the axisymmetric coefficients decreases in time, but increases for the nonsymmetric ones. The buckling mode at the critical time is nonsymmetric. The total buckling mode is determined by serveral predominant harmonics in the axial and circumferential directions. So the polymer shell buckles under prolonged axial load when the axisymmetric mode changes into a nonsymmetric one.

In order to determine the buckling time of composite shells an experimental investigation of glass-epoxy cylindrical shells under external pressure was carried out. Test specimens had lengths of $L = 300 \cdots 600$ mm, radius $R = 150$ mm, wall thicknesses $h = 3 \cdots 6$ mm. The number of shells was 15. The load level was $q^* = (0.80 \cdots 0.95) \, q^0$; q^0 is the elastic critical load. Depending upon the load level the critical time varied from 2 hours to 14.5 hours. Before the load application careful measurements of the initial imperfections were taken. After applying the load the radial displacement field was recorded in regular time intervals. The details of the experiment are described in [6]. Figure 3 shows the initial imperfection mode at the midlength of a shell (a dashed curve). The load level was $q^*/q^0 = 0.9$. Curve 1 shows the mode after the load application at $t = 0$, curve 2 at $t = 3$ h $30'$. The critical

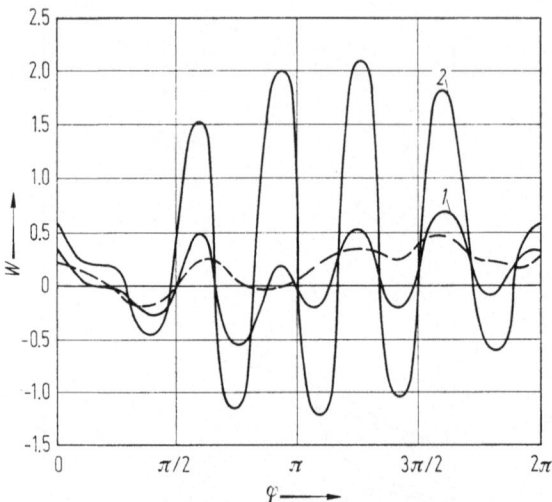

Fig. 3. The initial imperfections (dashed curve), buckling mode at $t = 0$ (curve 1) and at $t = 3$ h $30'$ (2).

time was $t^* = 3$ h $35'$. For this shell the predominant buckling mode consisted of 5 to 6 waves in the circumferential direction. It differs from the predominant initial imperfection mode consisting of 4 waves in the circumferential direction. The harmonic analysis of initial imperfections and buckling modes by Fourier representation (1) shows that the initial imperfections vary with time and at the critical time the buckling mode differs from the initial one. So we can conclude that the assumption of the similarity of the initial and buckling modes in most cases is not correct.

3. Theoretical Model

Let us consider a theoretical model of buckling of a cylinderical shell with initial imperfections, loaded in the axial direction and made from isotropic nonlinear viscoelastic incompressible material [7]. x is the axial coordinate of the shell, y the circumferential and z the normal coordinate. The stress-strain law will be assumed in the following form:

$$\varepsilon_{\alpha\beta}(t) = \frac{1}{2G} \left\{ s_{\alpha\beta}(t) + \int\limits_0^t K_1(t - \tau)\, s_{\alpha\beta}(\tau)\, \mathrm{d}\tau + \right.$$

$$\left. + \int\limits_0^t K_3(t - \tau)\, s_{\alpha\beta}(\tau)\, s_{\gamma\delta}(\tau)\, s_{\gamma\delta}(\tau)\, \mathrm{d}\tau \right\}, \tag{5}$$

$$s_{\alpha\beta}(t) = 2G \left\{ \varepsilon_{\alpha\beta}(t) - \int\limits_0^t \Gamma_1(t - \tau)\, \varepsilon_{\alpha\beta}(\tau)\, \mathrm{d}\tau - \right.$$

$$\left. - \int\limits_0^t \Gamma_3(t - \tau)\, \varepsilon_{\alpha\beta}(\tau)\, \varepsilon_{\gamma\delta}(\tau)\, \varepsilon_{\gamma\delta}(\tau)\, \mathrm{d}\tau \right\} \qquad (\alpha, \beta, \gamma, \delta = 1, 2),$$

$$\tag{6}$$

where $s_{\alpha\beta} = \sigma_{\alpha\beta} - b\, \delta_{\alpha\beta}$ is the stress deviator, and b is the hydrostatic stress.

The equilibrium equations of a shallow shell will be as follows:

$$N^{\alpha\beta}_{,\beta} + p^\alpha = 0,$$
$$M^{\alpha\beta}_{,\beta} - Q^\alpha = 0, \tag{7}$$
$$Q^\alpha_{,\alpha} + (b_{\alpha\beta} - K_{\alpha\beta})\, N^{\alpha\beta} + p^3 = 0,$$

where p^α, p^3 are the components of the resultant load vector, $N^{\alpha\beta}$, Q^α, $M^{\alpha\beta}$ the components of stress and couple resultant tensors, $b_{\alpha\beta}$ the second surface tensor, and $k_{\alpha\beta} = -w_{,\alpha\beta}$ is the bending deformation measure. Commas denote differentiation.

In the case of axial force ($p^\alpha = p^3 = 0$) the two first equilibrium equations are satisfied by introducing the stress function Φ:

$$N^{\alpha\beta} = e^{\alpha\gamma} e^{\beta\delta} \Phi_{,\gamma\delta}, \tag{8}$$

where $e^{\alpha\beta}$ is the permutation symbol. Eliminating Q^α from the last equation (7) and taking into account (8) we obtain

$$M^{\alpha\beta}_{,\beta\alpha} + (b_{\alpha\beta} + w_{,\alpha\beta})\, e^{\alpha\gamma}\, e^{\beta\delta} \Phi_{,\gamma\delta} = 0. \tag{9}$$

Using the stress-strain law (6) an the Kirchhoff-Love hypothesis let us express $M^{\alpha\beta}$ as a function of w. Finally we have the first equation of the problem

$$Z[w(x, y, t), \Phi(x, y, t)] = 0. \tag{10}$$

As the second equation let us use the nonlinear compatibility equation of shallow shells

$$e^{\alpha\delta}\, e^{\beta\gamma} [\varepsilon^*_{\alpha\beta,\gamma\delta} + (1/2)\, w_{,\alpha\beta} w_{,\gamma\delta} - w_{,\beta\delta} b_{\alpha\gamma}] = 0, \tag{11}$$

where $\varepsilon^*_{\alpha\beta}$ is the deformation measure of the middle surface of the shell. Substituting the stress-strain law (5) in the equation (11) and taking into account (8) and $h\sigma^{\alpha\beta} = N^{\alpha\beta}$ we obtain the second equation of the problem

$$V[w(x, y, t), \Phi(x, y, t)] = 0. \tag{12}$$

The system of integral-differential equations (10) and (12) allows us to investigate the equilibrium state of a shell in time. To solve the problem let us assume for the radial displacement w and stress function Φ the following representations:

$$w(x, y, t) = \sum_{m=-M}^{M} \sum_{n=-N}^{N} [\alpha_{mn}(t) - \alpha_{mn}^0] \, \psi_{mn}(x, y), \qquad (13)$$

$$\Phi(x, y, t) = \sum_{m=-M}^{M} \sum_{n=-N}^{N} \phi_{mn}(t) \, \psi_{mn}(x, y) - (py^2/2), \qquad (14)$$

where the Ψ_{mn} are functions which satisfy the boundary conditions:

$$\psi_{mn}(x, y) = e^{i[(m\pi x/L)+(ny/R)]}. \qquad (15)$$

Coefficients α_{mn}^0 in (13) correspond to the initial imperfection function.

By using Galerkin's method

$$\int_0^{2\pi R} \int_0^L V \bar{\psi}_{mn}(x, y) \, \mathrm{d}x \, \mathrm{d}y = 0,$$

$$\int_0^{2\pi R} \int_0^L Z \bar{\psi}_{mn}(x, y) \, \mathrm{d}x \, \mathrm{d}y = 0. \qquad (16)$$

($\bar{\Psi}_{mn}$ are the complex conjugate functions) instead of (10) and (12) we obtain a system of nonlinear integral equations, allowing us to calculate the coefficients $\alpha_{mn}(t)$ and $\phi_{mn}(t)$.

The system was solved numerically by digital computer using the iteration method. The kernels $K_i(t - \tau)$ and $\Gamma_i(t - \tau)$ were assumed to be exponential functions.

4. Numerical Results and Discussion

Let us consider the case when the load level applied to the shell is 90% of the elastic critical one. The geometrical characteristics of the shell were as follows: $L = 20$ cm, $R = 5$ cm, $h = 0.2$ cm. The coefficients α_{mn}^0 of the initial imperfection function were determined from the shell measurements. Figure 4 shows the numerical results of the problem. The curves show the growth rate of the coefficients of Fourier's representation as depending on dimensionless time $\bar{t} = t/r$ where r is the relaxation time of the material. Analysing these curves one can see satisfactory agreement with the experimental results. The growth rate of the axisymmetric coefficients a_{90} (curve 1), a_{80} (2), a_{10} (3) decreases in time in spite of the extreme growth in the initial period when the predominant mode was axisymmetric. The growth rate of the nonsymmetric coefficients has the opposite character. In the initial period they grow linearly until a definite moment, when the growth rate increases dramatically. The moment when the growth rate of the one of the harmonics tends to infinity $(\dot{a}_{mn} \to \infty)$ can be regarded as the critical time of the shell. For the present example the predominant post-buckling mode is determined by the coefficient a_{14} (4); this means that the shell buckles with 4 waves in the circumferential direction.

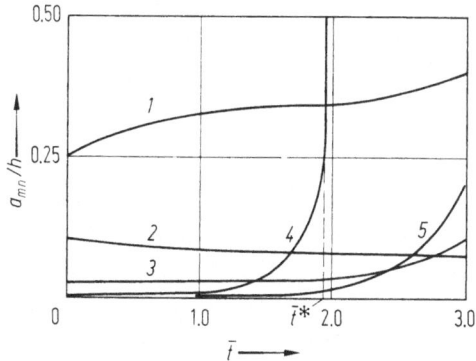

Fig. 4. The growth rate of axisymmetric coefficients a_{90} (curve 1), a_{80} (2), a_{10} (3) and nonsymmetric ones a_{14} (4), a_{13} (5) in the theoretical model.

A similar creep buckling process has been observed by Levi and Hoff [8] on metal shells at high temperatures. In our case the polymer shells never buckle in the axisymmetric mode, because the growth rate of the axisymmetric coefficients tends in time to zero in contrary to the results obtained in [8], where shells may still buckle axisymmetrically. This specific quality of buckling is due to the polymer material properties.

References

1. Hoff, N. J.: Axially Symmetric Creep Buckling of Circular Cylindrical Shells in Axial Compression. J. Appl. Mech. **35**, 530 (1968).
2. Rabotnov, Y. N.: Creep in Structures (in Russian). Moscow: Nauka 1966.
3. Grigoliuk, E. J.; Lipovtsev, Y. V.: On the Creep Buckling of Shells. Int. J. of Solids and Struct. **5**, No. 2, 1969.
4. Samuelson, L. A.: Creep Buckling of a Cylindrical Shell under Nonuniform External Loads. Int. J. of Solids and Struct. **6**, No. 1, 1970.
5. Rikards, R.; Brauns, J.: Investigation of Creep Buckling Pattern on Models of Cylindrical Shells. Mechanika Polimerov, No. 2, 1971.
6. Brauns, J.; Rikards, R.: Investigation of the Initial Imperfections and Buckling Modes of Fibreglass Cylindrical Shells under External Pressure (in Russian). Mechanika Polimerov, No. 6, 1971.
7. Rikards, R.; Teters, G.: Creep Buckling modes of Cylindrical Shells from Composite Materials (in Russian). Mechanika Polimerov, No. 4, 1971.
8. Levi, I. M.; Hoff, N. J.: Interaction between Axisymmetric and Nonsymmetric Creep Buckling of Circular Cylindrical Shells in Axial Compression. Creep in Structures, 1970, Symp. Gothenburg. Berlin, Heidelberg, New York: Springer 1972.

Creep Instability of Thick Shell Structures

F. A. Leckie, B. Hayman

University of Leicester, Leicester, England

1. Introduction

Recent studies by Hayman [1, 2] have shown that the time independent post-buckling characteristics of a structure can provide valuable information about the buckling behaviour of the structure when time dependent creep effects must be included. The effect of the creep deformation can generally be described in terms of the growth of an equivalent initial imperfection. An elastic structure whose load/deflection curve has the form shown in Fig. 1a generally experiences a lowering of its critical load on the introduction of successively greater imperfections. Under creep conditions such a structure undergoes instantaneous elastic buckling when the creep deformation has produced an equivalent imperfection sufficient to reduce the critical load to the leve of the applied load.

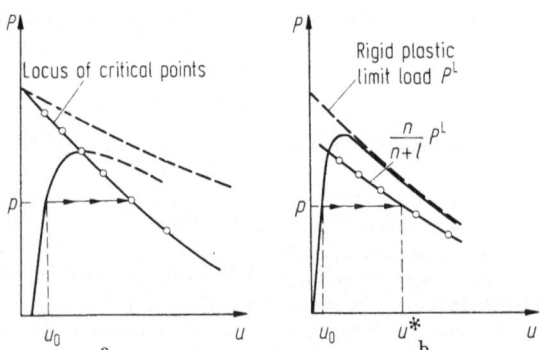

Fig. 1a and b. Collapse conditions. a) Elastic; b) Rigid plastic.

In relatively thick shell structures, the time-independent behaviour is likely to be influenced by plastic rather than elastic deformation. When in these circmustances time-dependent creep deformation must be taken into account, collapse occurs when the creep deformation has reduced he plastic collapse load to the level of the applied load. It has been suggested [2] that this happens when the applied load and the displacement of the structure are at a point on or near to the post-buckling equilibrium curve (Fig. 1b).

Important information about the creep buckling behaviour of certain structures can therefore be obtained from the form of the time independent post-buckling curve, which may, of course, be determined either theoretically or experimentally. In this paper consideration is given to the creep buckling performance of structures whose time independent post-buckling characteristics are determined by plastic strains. Estimates of the condition and time of buckling are found by applying some of the theorems of creep mechanics. Particular attention is paid to the behaviour of spherical shells subjected to a constant uniaxial radial force. This component has been selected for consideration since experimental results are available in the literature [3].

2. A Theorem of Creep Mechanics and Its Implications

Consider a body which is loaded over the area A_p with constant loads P_i while the displacements over the remaining area A_u are fixed to be zero. The body can suffer time independent plastic strains p_{ij} and time dependent creep strains v_{ij} so that the total strain ε_{ij} is

$$\varepsilon_{ij} = p_{ij} + v_{ij}. \tag{2.1}$$

The time independent plastic strain rate $\dot{p}_{ij}$ arises when the yield condition $f(\sigma_{ij}) = 0$ is satisfied, the strain rate being normal to the yield surface $f(\sigma_{ij}) = 0$ at σ_{ij}. The creep energy dissipation rate per unit volume is defined by

$$\dot{D}(\sigma_{ij}) = \sigma_{ij}\dot{v}_{ij}. \tag{2.2}$$

The creep strain rate is assumed to satisfy the time hardening form

$$\dot{v}_{ij}/\dot{v}_0 = \phi^n(\sigma_{ij}/\sigma_0)\, \partial\phi/\partial(\sigma_{ij}/\sigma_0)\, g(t). \tag{2.3}$$

In this relationship $\dot{v}_0$, σ_0 and n are material constants and ϕ is a linear homogeneous function of (σ_{ij}/σ_0). $\dot{D}(\sigma_{ij})$ is then given by

$$\dot{D}(\sigma_{ij})/\sigma_0\dot{v}_0 = \phi^{n+1}(\sigma_{ij}/\sigma_0)\, g(t). \tag{2.4}$$

For a uniaxial state of stress these equations take the form

$$\dot{D} = \sigma\dot{v}, \tag{2.2a}$$

$$\dot{v}/\dot{v}_0 = (\sigma/\sigma_0)^n\, g(t), \tag{2.3a}$$

and

$$\dot{D}/\sigma_0\dot{v}_0 = (\sigma/\sigma_0)^{n+1}\, g(t). \tag{2.4a}$$

An upper bound on the energy dissipation rate within the body given by the inquality [4]

$$\int_{A_p} P_i \dot{u}_i \, dA \leqq \int_V D(\sigma_{ij}^*) \, dV, \qquad (2.5)$$

where $\dot{u}_i$ are the displacement rates of the applied loads P_i and σ_{ij}^* is a statically admissible stress field in equilibrium with the applied loads subject to the restriction.

$$f\left(\frac{n+1}{n}\,\sigma_{ij}^*\right) \leqq 0. \qquad (2.6)$$

It is rather surprising that no time independent plastic strains $\dot{p}_{ij}$ need be included in the bound. However the restriction (2.5) means that the bound may not be used for loads in excess of $n/(n+1)$ times the limit load P_i^L. The implication is that for loads in excess of $n/(n+1)$ times the limit load P_i^L plastic strains can begin to be important. For minimum weight structures time independent plastic strains play no role until the limit load is reached, but for many structures with nonuniform states of stress it has been shown [5] that the deformation rate increases rapidly for loads in excess of $(n/n + 1) P_i^L$.

In the treatment given by Hayman [2] it was stated that instantaneous plastic collapse occurs when, as a result of creep deformations, the applied load coincides with the current value of the plastic collapse load. In fact, since the displacement rates are likely to increase rapidly for loads in excess of $(n/n + 1)$ times the current value of the rigid-plastic limit load P^L, it suggests that it is more practical to consider that buckling occurs when the applied load is equal to $(n/n + 1)$ times the current limit load. The effect of this criterion on the displacement for collapse is illustrated in Fig. 1b.

In addition to providing the condition for buckling, the inequality (2.5) can be used in conjunction with the post-buckling curve to determine a lower bound on the time for buckling to occur, and can also be used to give a conservative estimate of the displacement/time curve. The calculations necessary to determine the precise form of the post-buckling curve often prove to be very demanding and it is in these circumstances that the procedure developed by Batterman [6] provides a practical alternative. He represents the rigid plastic post-buckling curve in the form

$$P = \overline{P} - pu. \qquad (2.7)$$

$\overline{P}$ is the limit load of the body in its undeflected position and the constant p is given in terms of the stress and deformation fields required for the limit load calculations. In this way calculations involving finite deformations can be avoided.

3. Application to a Shell Structure

The problem considered is illustrated in Fig. 2 which shows a hemi-spherical shell subjected to an inward radial dead load P applied through a rigid boss. The experimental time independent post-buckling curve has the form shown in Fig. 4, and, for the geometry considered the loss of load carrying capacity is the result of plastic deformations.

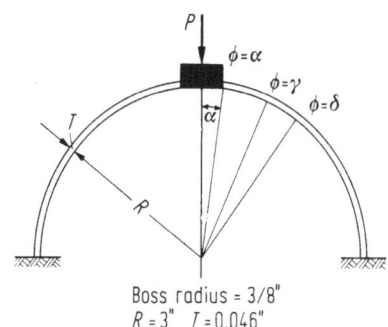

Fig. 2.
Shell geometry.

Boss radius = 3/8"
$R = 3"$ $T = 0.046"$

3.1. The Creep Energy Dissipation Rate Surface

In order to perform the calculations it is necessary to define sur-faces of constant creep energy dissipation rate in terms of the stress resultants N_θ, N_ϕ, M_θ and M_ϕ. Ponter and Leckie [5] have argued that the yield surfaces determined for plasticity theory are likely to provide a reasonable form for constant energy dissipation rate surfaces, and for convenience it is proposed to use the limited intersection surfaces shown in Fig. 3 which were first proposed by Hodge [7]. The stress resultants are normalised in terms of the constants N_0 and M_0 where

$$N_0 = \sigma_0 T \quad \text{and} \quad M_0 = \sigma_0 T^2/4.$$

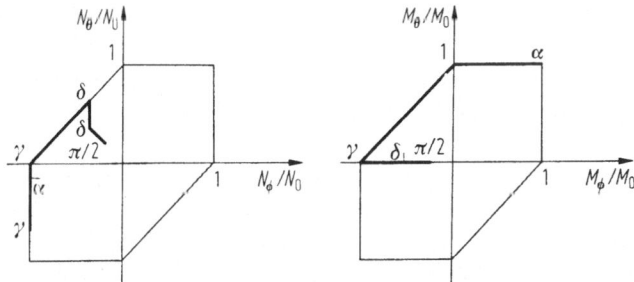

Fig. 3. Creep energy dissipation rate surface $\dot{D} = \sigma_0 \dot{V}_0$.

3.2. Stress Distributions and Upper Bound Energy Dissipation Calculation

A stress distribution used previously in a limit load calculation is available [8]. This stress field has four distinct regions which are illustrated in Figs. 2 and 3. For $\alpha < \phi < \delta$ the stress field everywhere lies on the yield surface, but for $\phi > \delta$ it lies within the yield surface.

If the applied load is $P = \lambda P^{\mathrm{L}}$ then a stress field suitable for use in the inequality (2.5) is obtained by applying the same factor λ to the limit load stress field. Since in the region $\alpha < \phi < \delta$ the limit load stress field is everywehre on the yield surface, the creep energy dissipation rate is everywhere constant in this region and is equal to $\lambda \sigma_y \dot{v}_\lambda$ where σ_y is the yield stress and $\dot{v}_\lambda$ the creep strain rate corresponding to an applied uniaxial stress $\lambda \sigma_y$. For $\phi > \delta$ the magnitudes of the stress decrease. Since the energy dissipation rate per unit volume depends on stress to the power $(n + 1)$ the contribution of this region to the total dissipation rate for the shell is likely to be small. Assuming that this contribution may be neglected, the bound on dissipation rate given by inequality (2.5) becomes

$$P\dot{u} \leq V\lambda \sigma_y \dot{v}_\lambda, \qquad (3.1)$$

where V is the volume of material in the region $\alpha < \phi < \delta$. Hence the displacement rate $\dot{u}$ is bounded according to the inequality

$$\dot{u} \leq V\lambda \sigma_y \dot{v}_\lambda / P. \qquad (3.2)$$

3.3. Estimates of Collapse Times

The inequality (3.2) may be integrated between appropriate limits to give bounds on the displacement-time relation and on the collapse time. For a given applied load P the initial deflection u_0 is readily found from the time-independent response curve of the shell (Fig. 1b). As the deflection of the shell increases under constant load P the limit load P^{L} decreases. The consequent increase in λ must therefore be taken into account in the integration. At any displacement u an estimate of P^{L} and hence $\dot{u}$ may be readily obtained from the post-buckling part of the time-independent response curve. It is assumed that the shell is in a state of collapse when $P = (n/n + 1)\,P^{\mathrm{L}}$, i.e. when $\lambda = \lambda^* = (n/n + 1)$. This condition is also readily identified in the time-independent response plot so that the corresponding deflection u^* may be found.

The integration itself may be performed by a step-by-step process using the above information from the time-independent response along with uniaxial creep data for the material of the shell. Alternatively an analytical solution may be found by using functions fitted to the time dependent creep data.

A simpler method is to use Eq. (2.7) in conjunction with the in-equality (3.2). From (3.2) and (3.2a)

$$\dot{u} \leq (\sigma_0 \dot{r}_0 V/P) (\lambda \sigma_y/\sigma_0)^{n+1} g(t). \tag{3.3}$$

Also

$$\lambda = P/P^{\mathrm{L}} = P/(\overline{P} - pu). \tag{3.4}$$

Substituting (3.4) in (3.3) and integrating between time 0 and t^* the time when $\lambda^* = n/n + 1$ gives

$$\dot{r}_0 \left(\frac{\sigma_y}{\sigma_0}\right)^n \int_0^{t^*} g(t) \, \mathrm{d}t \geq \frac{P^2}{(n+2) \, V\sigma_y p} \left\{\left(\frac{P_0}{P}\right)^{n+2} - \left(\frac{n+1}{n}\right)^{n+2}\right\},$$

where P_0 is the limit load corresponding to the initial displacement u_0. This expression can be written in a more suitable form if both sides are multiplied by a constant α^n. Then the left hand side becomes

$$\dot{r}_0 \left(\frac{\alpha\sigma_y}{\sigma_0}\right)^n \int_0^{t^*} g(t) \, \mathrm{d}t = \varepsilon_\alpha^*, \tag{3.5}$$

which is the creep strain observed at time t^* when a uniaxial specimen is subjected to a constant stress $\alpha\sigma_y$. The value of α can be chosen to correspond conveniently to available creep data, and thereby avoid complex curve fitting techniques. Hence

$$\varepsilon_\alpha^* \geq \frac{\alpha^n P^n}{(n+2) \, V\sigma_y P} \left\{\left(\frac{P_0}{P}\right)^{n+2} - \left(\frac{n+1}{n}\right)^{n+2}\right\}, \tag{3.6}$$

so that a conservative estimate for t^* is obtained by finding on the creep curve corresponding to stress $\alpha\sigma_y$ the time required to reach the strain given by the right hand side of (3.6).

4. Comparison with Experimental Results

The experimental results reported by Penny and Marriott [3] were obtained by applying constant loads at 200 °C to aluminium shells with the geometry shown in Fig. 2. The time-independent load/displacement curve is reproduced in Fig. 4 and the uniaxial creep data in Fig. 5. In Fig. 4 the post-buckling equilibrium curve has been extrapolated back to the load axis to give an estimate of the variation of P^{L} with u for the required range of displacements. For displacements up to about 0.060 inch it is seen that the variation of P^{L} with u is almost linear, and the post-buckling curve can be expressed in the form of Eq. (2.7) with $\overline{P} = 450$ lb and $p = 2650$ lb/in. Lines have also been drawn in Fig. 4 to show $(n/n + 1) P^{\mathrm{L}}$, thus defining the collapse configuration and enabling u^* to be estimated for any applied load P. It can be seen that the application of this criterion substantially de-

creases the displacement at collapse compared with the displacement
to cause plastic collapse.

Penny and Marriott suggest that the creep data (Fig. 5) fits a
time-hardening creep law of te form

$$\dot{v} = 5.76 \times 10^{-13} \sigma^2 t^{-0.6}. \tag{4.1}$$

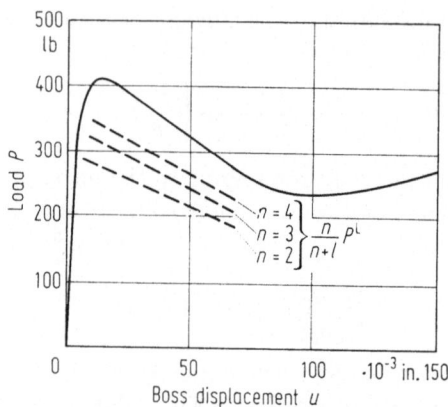

Fig. 4. Load displacement diagram at 200 °C.

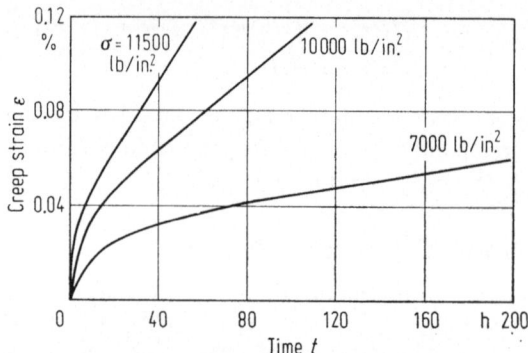

Fig. 5. Creep curves at 200 °C.

While this gives a reasonable fit at low strains it gives poor agreement
at the higher values of strain at which much of the integration is to
be carried out. For creep strains above about 0.05% it is found that
the data are better fitted by a steady state creep equation

$$\dot{v} = 1.30 \times 10^{-24} \sigma^{4.7}. \tag{4.2}$$

On this basis the value of n used in calculating λ^* has been taken as 4.7.
The yield stress σ_y of the aluminium is 16,200 lb/in^2.

For the particular shell geometry aused, $\delta = 0.48$ (Figs. 2, 3) and the volume of the region $\alpha < \phi < \delta$ is $V = 0.274$ in^3. Fig. 6 shows the displacement/time curves obtained experimentally for applied loads

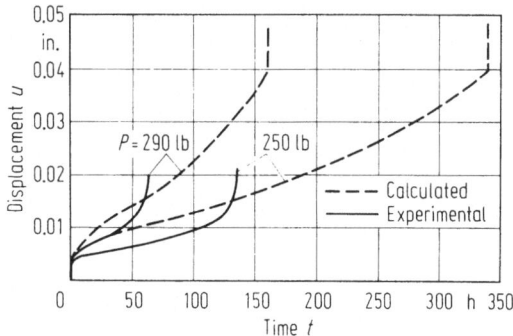

Fig. 6. Displacement/time curves.

$P = 250$ and 290 lb along with those obtained by integration of (3.2). (The integrations were actually performed analytically using (4.1) at creep strains below 0.05% and (4.2) at those above.) It is seen that the current theory gives quite good estimates of the response in the primary and steady-state parts of the creep response, but fails to show the rapid increase of deflection leading to collapse. The collapse time is overestimated by a substantial factor as a result of this. The reason for the failure of the analysis to describe the behaviour adequately and in a conservative manner may be either the neglect of the contribution of the region $\phi > \delta$ to the creep dissipation, or perhaps, the onset of large plastic strains at an earlier stage than that anticipated in the $(n/n + 1)\, P^L$ calculation, or an inability to estimate the value of the creep index n.

The importance of the value of n can be readily assessed by using the expression (3.6). The value of α is chosen to be $\alpha = 0.71$ so that the creep curve in Fig.5 corresponding to the uniaxial stress $\alpha\sigma_y = 11{,}500$ lb/in^2 may be used. The results of the calculations are shown in Table 1, and the pronounced effect of n can be readily appreciated. These results suggest that reasonable estimates of the collapse times are obtained if low values of n are assumed.

5. Conclusions

A simple method has been suggested for the estimation of displacement/time curves and collapse times of thick shell structures. Agreement of the displacement predictions and experimental results is satisfactory, but the collapse criterion while more conservative than that previously suggested still gives estimates of collapse times which are about double those observed experimentally.

Table 1. Comparison of experimental and theoretical collapse times

Load [lb]	250	290
Experimental time [h]	140	70
Theoretical time [h]		
$n = 2$	171	—
$n = 2.5$	241	60
$n = 3$	288	101
$n = 4$	355	130

References

1. Hayman, B.: Some Observations on Creep Buckling Analysis. Proc. of the Second International Conference on Structural Mechanics in Reactor Technology, Berlin 1973.
2. Hayman, B.: The Influence of Post-Buckling Characteristics on Creep Buckling. Leicester University, Engineering Dept., Report 74-8 (1974).
3. Penny, R. K.; Marriott, D. L.: Creep Buckling of Boss Loaded Spherical Shells. Report D/002/69, Dept. Mech. Engng. University of Liverpool 1969.
4. Leckie, F. A.; Ponter, A. R. S.: Deformation Bounds for Bodies which Creep in the Plastic Range. J. Appl. Mech. **37**, June 1970.
5. Ponter, A. R. S.; Leckie, F. A.: The Application of Energy Theorems to Bodies which Creep in the Plastic Range. J. Appl. Mech. **30**, 1970.
6. Batterman, S. C.: Load-Deformation Behavior of Shells of Revolution. Proc. ASCE Eng. Mech. Div. December 1964.
7. Hodge, P. G.: Limit Analysis of Rotationally Symmetric Plates and Shells. Prentice-Hall 1963.
8. Leckie, F. A.: Plastic Instability of a Spherical Shell. Second Symposium on Theory of Thin Shells. Berlin, Heidelberg, New York: Springer 1969.

A Plastic Flow Rule at a Yield Corner

M. J. Sewell

University of Reading, Reading England

Résumé

A simple macroscopic theory is constructed, and its consequences include the following results. At the uniaxial compressive stress point a yield vertex is supposed to have evolved with essentially 3 facets, and associated normal tensors $\boldsymbol{\nu}_x$ ($x = 1, 2, 3$). Tensors $\boldsymbol{\nu}_1$ and $\boldsymbol{\nu}_2$ are in the Tresca directions, and $\boldsymbol{\nu}_3$ is inclined to the (in-plane) shear stress axis by an amount proportional to a scalar k. The time-independent plastic flow rule associated with this vertex connects objective stress-rate $\dot{\boldsymbol{\tau}}$ and the plastic part $\gamma_x \boldsymbol{\nu}_x = \boldsymbol{\varepsilon}^{\mathrm{p}}$ of strain-rate by

$$\dot{\boldsymbol{\tau}}\boldsymbol{\nu}_\alpha - h_{\alpha\beta}\gamma_\beta \leqq 0, \quad \gamma_\alpha \geqq 0, \quad \gamma_\alpha(\dot{\boldsymbol{\tau}}\boldsymbol{\nu}_x - h_{\alpha\beta}\gamma_\beta) = 0.$$

Here the hardening matrix $h_{\alpha\beta}$ is postulated to be of the form

$$h_{\alpha\beta} = h_{11}\begin{bmatrix} 1 & \phi & \eta\psi \\ \phi & 1 & \eta\psi \\ \eta\psi & \eta\psi & \eta^2 \end{bmatrix}, \quad \eta\psi = (1 + \phi)/\sqrt{3 + k^2},$$

and thus implies coupled hardening between the facets. It is also assumed to be positive definite, so that $\dot{\boldsymbol{\tau}}\boldsymbol{\varepsilon}^{\mathrm{p}} = h_{\alpha\beta}\gamma_\alpha\gamma_\beta > 0$. Then the associated instantaneous shear modulus in fully active loading (all $\gamma_x > 0$) is

$$\mu^* \equiv \mu\Big/\left(1 + \frac{\mu k^2}{h_{11}[\eta^2(3 + k^2) - 2(1 + \phi)]}\right) < \mu,$$

where μ is the shear modulus.

This μ^*, and also the coefficients in the relations between the normal strain-rates and stress-rates under fully active loading, can be fitted if desired to the values furnished by J_2-deformation theory by choosing $h_{\alpha\beta}$ to be given by

$$\frac{E}{2h_{11}} = \frac{6[(E/E_s) - 1][(E/E_t) - 1]}{3[(E/E_s) - 1] + [(E/E_t) - 1]},$$

$$\phi = \frac{3[(E/E_s) - 1] - [(E/E_t) - 1]}{3[(E/E_s) - 1] + [(E/E_t) - 1]},$$

$$\eta^2 = \frac{4}{3 + k^2} \cdot \left[\frac{3[(E/E_s) - 1] + k^2[(E/E_t) - 1]}{3[(E/E_s) - 1] + [(E/E_t) - 1]}\right].$$

Here E, E_t and E_s are the Young's, tangent and secant moduli of an associated uniaxial compressive stress/strain curve. The hardening matrix and the flow rule can thereby be expressed entirely in terms of the uniaxial stress/strain curve and (via k^2) the shear angle of the considered compound vertex.

By this means a genuine flow theory is constructed whose local moduli, under fully active loading, have the same *values* as those of J_2-deformation theory, but which is not subject to the well-known objections to the latter theory. Such a flow theory could legitimize the use of these moduli, for example in certain "paradoxical" buckling problems whose resolution appears to require a reduced value of the shear modulus below the elastic value μ.

Introduction of a Ramberg-Osgood type of representation with exponent n, but with a definite yield stress σ_y, shows that

$$\phi = \frac{(3/n)\left[1 - (\sigma_y/\sigma)\right] - 1}{(3/n)\left[1 - (\sigma_y/\sigma)\right] + 1}.$$

This coupling parameter therefore increases from -1 as σ increases from σ_y. The present theory in fact generalizes one of Sewell (1973), where a 2×2 hardening matrix with possibly negative coupling ϕ was first considered.

The domain of stress-rate vectors enforcing fully active loading is wider than the prolongation of the yield vertex so long as the coupling remains negative. This domain is a calculable pyramid which, under the Ramberg-Osgood representation for example, contracts from the Mises half-space as the stress increases beyond yield. It may or may not ultimately become narrower than the prolongation of the yield vertex. In this respect the theory relaxes some assumptions of Budiansky (1959). He suggested that the plastic strain-rate predicted by J_2-deformation theory under a limited range of non-proportional loading within the prolongation could be consistent with a conical yield surface, although he did not propose an explicit hardening matrix.

The proofs of the present results will be given elsewhere (Sewell 1974). Justification of the generalized Shanley property that bifurcation will initially involve plastic loading everywhere, even in the presence of a vertex such as that considered here, is provided by an inequality of Sewell (1972, (3.65)).

Fundamental theoretical predictions of polycrystal aggregate behaviour based on the study of single crystals agree in general terms with some features of the present simple macroscopic theory. This applies to the contraction of the domain of loading stress-rates (Hill 1967), and to the reducing shear modulus (Hutchinson 1970).

The experimental evidence for the existence of vertices on yield surfaces has often been regarded as contradictory. Recently, however, Lin (1971) has discussed the matter with special reference to the problem of deciding upon the observable threshold of plastic strain. The burden of his remarks points to the existence of a vertex if infinitesimal plastic strain could be measured.

References

1. Budiansky, B.: A Reassessment of Deformation Theories of Plasticity. J. Appl. Mech. **26**, 259−264 (1959).
2. Hill, R.: The Essential Structure of Constitutive Laws for Metal Composites and Polycrystals. J. Mech. Phys. Solids **15**, 79−95 (1967).
3. Hutchinson, J. W.: Elastic/Plastic Behaviour of Polycrystalline Metals and Composites. Proc. Roy. Soc. London **A 319**, 247−272 (1970).
4. Lin, T. H.: Physical Theory of Plasticity. Advances in Applied Mechanics, Vol. 11. London: Academic Press 1971, pp. 255−311.
5. Sewell, M. J.: A Survey of Plastic Buckling. Ch. 5 of Stability, edited by H. Leipholz, University of Waterloo, Ontario, 1972.
6. Sewell, M. J.: A Yield Surface Corner lowers the Buckling Stress of an Elastic-Plastic Plate under Compression. J. Mech. Phys. Solids **21**, 19−45 (1973).
7. Sewell, M. J.: The present résumé is an extract from the full paper, "A Plastic Flow Rule at a Yield Vertex". J. of Mechanics and Physics of Solids **22** (1974), published by Pergamon Press.

Analytical and Numerical Study of the Effects of Initial Imperfections on the Inelastic Buckling of a Cruciform Column[1]

J. W. Hutchinson, B. Budiansky

Division of Engineering and Applied Physics, Harvard University, Cambridge, Massachusetts, U.S.A.

Abstract

The inelastic buckling of a cruciform column is investigated by a combination of analytical and numerical methods. An exact asymptotic analysis for the effect of small imperfections on the maximum load reveals clearly how it is possible for an exceedingly small imperfection to have a very large influence. As long as the strain hardening is sufficiently low, the numerical analysis confirms the Onat-Drucker conclusion that unavoidably small imperfections, together with the use of J_2 flow theory, give rise to a maximum load prediction which is approximated by the bifurcation load prediction based on a deformation theory of plasticity.

1. Introduction

The cruciform column is perhaps the best example for illustrating the discrepancy between the buckling predictions of the simplest flow and deformation theories of plasticity. According to *any* flow theory for an initially isotropic material in conjunction with a smooth yield surface, the critical compressive stress for twisting bifurcation is

$$\sigma_c = G(t/b)^2, \tag{1}$$

where G is the elastic shear modulus and t and b are shown in Fig. 1. According to *any* deformation theory of plasticity for an initially isotropic material the bifurcation stress is

$$\bar{\sigma}_c = \bar{G}(t/b)^2, \tag{2}$$

where $\bar{G}$ is a reduced effective shear modulus given by

$$\bar{G}/G = [1 + 3G(1/E_s - 1/E)]^{-1}, \tag{3}$$

where E is Young's modulus and $E_s \equiv \sigma/\varepsilon$ is the secant modulus of the compressive stress-strain curve at $\bar{\sigma}_c$.

---

[1] This work was supported in part by the Air Force Office of Scientific Research under Grant AFOSR-73-2476, and by the Division of Engineering and Applied Physics, Harvard University.

The experimental results for the cruciform column cited by Stowell (1951) are in good agreement with the deformation theory prediction as seen in Fig. 1. Arguments justifying the use of deformation theory

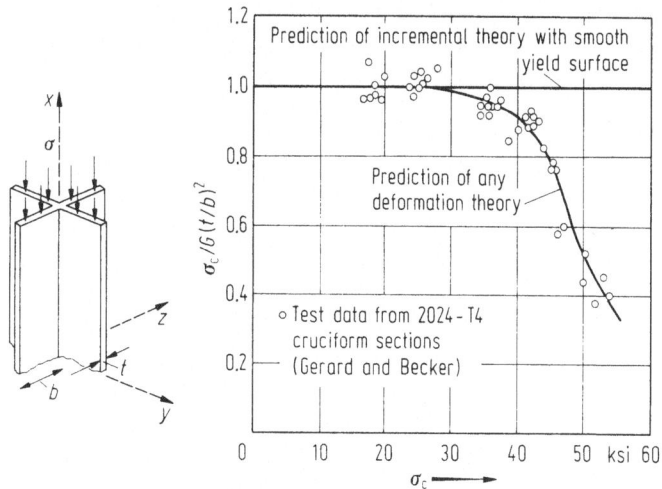

Fig. 1.
Comparison of tests and theory for torsional buckling of a cruciform column.

in bifurcation applications have appealed to the fact that the deformation theory predictions are identical to predictions of a more sophisticated flow theory which permits corners to develop on the yield surface. For example, in the case of the cruciform column, (2) is also the prediction of the slip theory of plasticity. Onat and Drucker (1953) have argued that the poor agreement between tests and the prediction (1) based on a smooth yield surface stems from the presence of unavoidably small imperfections which give rise to a maximum load which can be substantially below (1) and closely approximated by (2). The appropriateness of smooth or cornered yield surfaces in modeling elastic-plastic solids is still unresolved as discussed by Hutchinson (1974).

In this paper the imperfection-sensitivity associated with the simplest flow theory of plasticity is examined in somewhat more detail than has been previously reported. To set the stage, some accurate numerical results are given. Then an exact asymptotic analysis is carried out which reveals the source of the extreme imperfection-sensitivity associated with J_2 flow theory.

2. Governing Equations

The von Kármán nonlinear plate equations will be used together with the following geometric assumptions. Each cross section is assumed to rotate as a rigid body about the x-axis common to the four flange plates and to translate only in the x direction. The compressive strain ε of the axis and the rotation per unit length θ about the axis are taken to be independent of x. An initial imperfection is taken in the form of an x-independent initial rotation per unit length $\bar{\theta}$ so that the total twist per unit length is $\theta + \bar{\theta}$. The three middle surface displacements for the plate lying in the $x - y$ plane (see Fig. 1) are

$$U = -\varepsilon_0 x, \quad V = -(\theta^2 + 2\theta\bar{\theta}) x^2 y/2, \quad W = \theta x y. \qquad (4)$$

According to nonlinear plate theory the nonzero strains associated with the additional twist θ are given by

$$\varepsilon_{xx} = -\varepsilon_0 + y^2(\theta\bar{\theta} + \theta^2/2) \quad \text{and} \quad \gamma_{xy} = 2zK_{xy} = -2z\theta, \qquad (5)$$

where $K_{xy} = -W_{,xy}$ is the only nonzero bending strain.

The only stresses considered are σ_{xx} and σ_{xy} and these are necessarily even and odd functions of z, respectively. The two pertinent resultant stress and bending moment quantities are

$$N_{xx} = \int_{-t/2}^{t/2} \sigma_{xx} \, \mathrm{d}z \quad \text{and} \quad M_{xy} = \int_{-t/2}^{t/2} \sigma_{xy} z \, \mathrm{d}z. \qquad (6)$$

Equilibrium equations follow from the principle of virtual work which, in terms of work quantities per unit length in the x direction, is

$$\int_0^b \{2M_{xy} \, \delta K_{xy} + N_{xx} \, \delta\varepsilon_{xx}\} \, \mathrm{d}y - P \, \delta\varepsilon_0 = 0 \qquad (7)$$

for arbitrary $\delta\theta$ and $\delta\varepsilon_0$, where P is the compressive load carried by one flange plate. The two equilibrium equations resulting from (7) are

$$P = -\int_0^b N_{xx} \, \mathrm{d}y \qquad (8)$$

and

$$2 \int_0^b M_{xy} \, \mathrm{d}y - (\theta + \bar{\theta}) \int_0^b N_{xx} y^2 \, \mathrm{d}y = 0. \qquad (9)$$

The rate-constitutive relation is specialized assuming σ_{xx} and σ_{xy} are the only nonzero stress components. (Thus by the constitutive relation, ε_{yy} will not be zero, in general, in contradiction to the consequence of the geometric assumptions. This is an unimportant inconsistency since σ_{yy} is indeed expected to be very small.) The inverted form of the stress-strain relation according to J_2 flow theory is

$$\begin{aligned}
\dot{\sigma}_{xx} &= c_1 \dot{\varepsilon}_{xx} + c_2 \dot{\gamma}_{xy}, \\
\dot{\sigma}_{xy} &= c_2 \dot{\varepsilon}_{xx} + c_3 \dot{\gamma}_{xy},
\end{aligned} \qquad (10)$$

where

$$c_1 = E[1 - (\varrho - 1)\sigma_{xx}^2 D^{-1}], \quad c_2 = -3G(\varrho - 1)\sigma_{xx}\sigma_{xy} D^{-1},$$

$$c_3 = G[1 - 9(\varrho - 1)\sigma_{xy}^2\{2(1 + \nu) D\}^{-1}], \quad \varrho = E/E_t, \tag{11}$$

$$D = \varrho(\sigma_{xx}^2 + 3\sigma_{xy}^2) + \alpha(\varrho - 1)\sigma_{xy}^2, \quad \alpha = 3(1 - 2\nu)/[2(1 + \nu)].$$

Here, ν is Poisson's ratio and E_t is the tangent modulus of the uniaxial stress-strain curve which is regarded as a funtion of the effective stress $(\sigma_{xx}^2 + 3\sigma_{xy}^2)^{1/2}$. Equations (11) hold as long as the effective stress is increasing. The elastic constants pertain for unloading.

The bifurcation prediction based on Eqs. (4)—(11) (with $\bar{\theta} = 0$) is precisely (1). It can be noted that a bifurcation analysis based on the unapproximated von Kármán plate equations for a flange plate of length L which is simply supported on three sides and free on the fourth gives (1) plus smaller terms of relative order $(b/L)^2$. Thus the present theory is restricted to the torsional buckling of relatively long cruciforms.

3. Numerical Predictions

The above equations were solved numerically for a wide range of cases and imperfection levels. An incremental method was used. The region, $0 < y/b < 1$ and $|z/t| < 1/2$, was divided into subareas in which the stress was taken to be uniform. Integrals over the region were evaluated in such a way that when the moduli are actually uniform the integrals are evaluated exactly. Calculations were made using two, tensile stress-strain curves: The bilinear relation,

$$\dot{\sigma} = E\dot{\varepsilon} \quad \text{for} \quad \sigma < \sigma_y \quad \text{and} \quad \dot{\sigma} = E_t\dot{\varepsilon} \quad \text{for} \quad \sigma > \sigma_y, \tag{12}$$

where E_t is constant, and the Ramberg-Osgood relation

$$\varepsilon/\varepsilon_y = \sigma/\sigma_y + (3/7)(\sigma/\sigma_y)^n, \tag{13}$$

where $\varepsilon_y \equiv \sigma_y/E$.

Imperfection-sensitivity curves in terms of the maximum load of the imperfect column, P_{max}, normalized by the bifurcation load, $P_c = bt\sigma_c$, are plotted as a function of the imperfection parameter $\bar{\xi} = b^2\bar{\theta}/t$ in Fig. 2. The normalization $\bar{\xi} = b^2\bar{\theta}/t$ eliminates any explicit b/t dependence when results are plotted in the manner of Fig. 2. Note that $\bar{\xi}t$ is the initial relative displacement of any two points at the outer edge of the flange a distance b apart, i.e., $\bar{\xi}t = \overline{W}(x + b, b) - \overline{W}(x, b)$. Thus, for a typical cruciform an imperfection $\bar{\xi} = 0.1$ can be regarded as very large, while $\bar{\xi} = 0.01$ is small but not necessarily unavoidable. Imperfection levels less than about $\bar{\xi} = 0.001$ can probably be considered unavoidable.

The parameters for the examples in Fig. 2a were chosen so that the ratio of (1) to (2) is exactly 3/2. In Fig. 2b this ratio is 2; the deformation theory bifurcation prediction is indicated by a dashed line in both

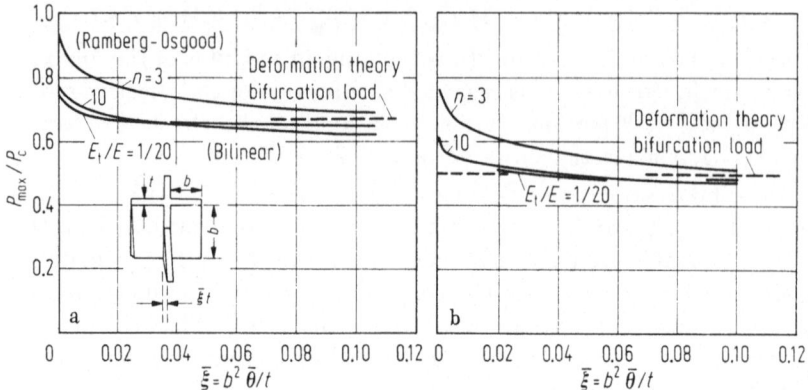

Fig. 2a and b. Numerical results for maximum load as a function of imperfection amplitude. a) $\sigma_c/\bar{\sigma}_c = 1.5$; b) $\sigma_c/\bar{\sigma}_c = 2$.

figures. The curves for the bilinear relation (12) with $E_t/E = 1/20$ apply to the example studied by Onat and Drucker. As they concluded, extremely small imperfections do indeed reduce the maximum load to about the deformation theory bifurcation load. The same is true for the case of a low strain hardening Ramberg-Osgood tensile relation ($n = 10$). However, when the strain hardening is high ($n = 3$) there is approximately a 20% variation in the maximum load over the range of imperfection levels which cannot necessarily be considered unavoidable.

These same calculations were repeated with J_2 deformation theory, and no elastic unloading. Over the imperfection range shown the maximum load differs very little from the deformation-theory bifurcation load of the perfect column.

4. Asymptotic Predictions

Torsional buckling of a cruciform column is one of the rare examples where bifurcation takes place in the plastic range under constant, as opposed to increasing, load to lowest order. In other words, the lowest bifurcation load is simultaneous with the so called reduced modulus load at which the straight configuration loses stability. For the *perfect* cruciform elastic unloading starts at bifurcation; but in most instances the numerical results for the *imperfect* cruciform showed that elastic unloading starts *after* the maximum load has been attained. We anticipate the load-twist behavior depicted for Case II in Fig. 3 and consider only the plastic loading relations given by (11). This does not mean that the analysis becomes the same as that for a nonlinear elastic material

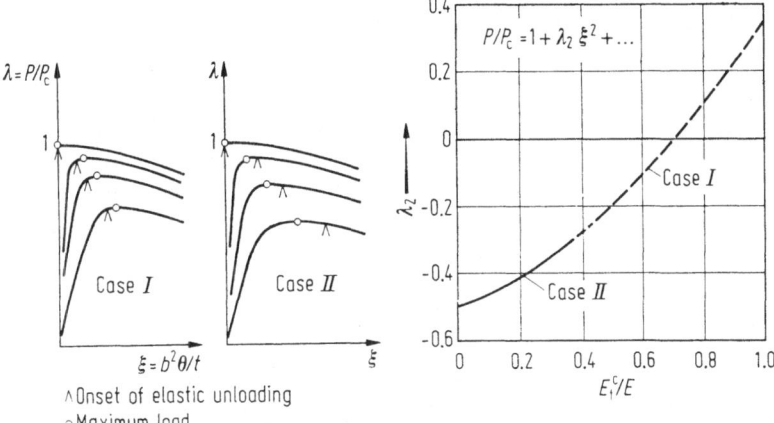

Fig. 3. Sketches of load-twist curves and initial post-bifurcation coefficient for perfect column calculated using J_2 theory with no elastic unloading.

Since the moduli in (11) cannot be derived from a strain energy density functional the solid characterized by J_2 flow theory with no unloading must be considered to be hypo-elastic. This distinction is essential as will be seen below. An a posteriori evaluation of the condition for attainment of the maximum load prior to the onset of unloading will be made.

In this section the results of an exact asymptotic analysis of the cruciform equations (4)—(11) are given. The details of the analysis are omitted. The analysis has three steps: a regular perturbation analysis giving the lowest order effect of $\bar{\xi}$ at loads below P_c, a singular perturbation analysis for the behavior in the vicinity of P_c, and, the final step, a matching of the regular and singular expansions.

Let $\lambda = P/P_c$ and define $f(\lambda)$ as

$$f(\lambda) = 3[E/E_t(\lambda) - 1][2(1 + \nu)]^{-1}, \qquad (14)$$

where $E_t(\lambda)$ denotes the value of the tangent modulus at the uniaxial stress $\sigma = \lambda \sigma_c$. The solution to the regular perturbation problem gives the following relation between $\xi = b^2\theta/t$, $\bar{\xi}$ and λ:

$$\xi = \bar{\xi}\{(1 - \lambda)^{-\varkappa} e^{-h(\lambda)} - 1\} + \mathcal{O}(\bar{\xi}^2), \qquad (15)$$

where

$$\varkappa = 1 + f(1) = 1 + 3[E/E_t^c - 1][2(1 + \nu)]^{-1}, \qquad (16)$$

and

$$h(\lambda) = \int\limits_0^\lambda [f(1) - f(\eta)]\,[1 - \eta]^{-1}\,\mathrm{d}\eta. \tag{17}$$

Equation (15) holds on the rising branch of the curve of λ as a function of ξ for $\lambda < 1$. Note that, through (17), ξ depends on the entire history of the loading up to λ. For linear or even nonlinear elastic solids $\varkappa$ in the exponent in (15) is always unity and there is no history dependence. However for a hypo-elastic solid that obeys (11), $\varkappa$ will be large compared to unity if E_t^c/E is small. For λ near 1, (15) can be rewritten as

$$(1 - \lambda) \approx [\mathrm{e}^{-\mathrm{h}(1)}\,\bar{\xi}/\xi]^{1/\varkappa}. \tag{18}$$

The singular perturbation analysis in the neighborhood of $\lambda = 1$ gives

$$\lambda = 1 + \lambda_2\xi^2 + c\bar{\xi}^\omega\xi^{-1/\varkappa} + \cdots, \tag{19}$$

where $\varkappa$ is the same as in (16). The exponent ω and the coefficient c are undetermined. The quantity λ_2 depends only on $\varrho_c = E/E_t^c$ and ν; it is

$$\lambda_2 = \varrho_c^{-1}\{4(1 + \nu)^2/15 - (\varrho_c - 1)^2/2 - 9(\varrho_c - 1)/10\}/\{\nu + \varrho_c\}. \tag{20}$$

For buckling in the elastic range, $\varrho_c = 1$ and $\lambda_2 = 4(1 + \nu)/15$, implying increasing load following bifurcation. For $\varrho_c \to \infty$, $\lambda_2 \to -1/2$. A plot of λ_2 as a function of E_t^c/E for $\nu = 1/3$ is shown in Fig. 3.

A matching of (15) and (19) determines ω and c. On the rising branch of the relation between λ and ξ, $\lambda_2\xi^2$ is negligible compared to $c\bar{\xi}^\omega\xi^{-1/\varkappa}$ for fixed $\lambda < 1$ in the limit as $\bar{\xi} \to 0$. Note that the exponents of ξ in (18) and (19) are the same, and to complete the match we choose

$$c = -\mathrm{e}^{-\mathrm{h}(1)/\varkappa} \quad \text{and} \quad \omega = 1/\varkappa. \tag{21}$$

A maximum value of λ occurs in the relation (19) between λ and ξ if λ_2 is negative. Using the condition $\mathrm{d}\lambda/\mathrm{d}\xi = 0$ in connection with (19) gives the asymptotic dependence of the maximum load on $\bar{\xi}$, i.e.,

$$\lambda_{\max} = P_{\max}/P_c = 1 - \mu\bar{\xi}^{2/(2\varkappa+1)}, \tag{22}$$

where

$$\mu = -\lambda_2(2\varkappa + 1)\,(-2\varkappa\lambda_2)^{-2\varkappa/(2\varkappa+1)}\,\mathrm{e}^{-2\mathrm{h}(1)/(2\varkappa+1)}.$$

The condition ensuring that elastic unloading does not occur prior to attainment of the maximum load for sufficiently small imperfections is found to be simply that

$$E_t^c/E \leqq 3/(7 + 4\nu). \tag{23}$$

Curves of μ and the exponent in (22), $2/(2\varkappa + 1)$, are shown in Fig. 4 as a function of the bifurcation strain ε_c normalized by ε_y for the Ramberg-Osgood tensile stress-strain relation (13). The dashed portions, on which

(23) is not satisfied, apply only to a hypo-elastic (rather than an elastic-plastic) cruciform. But (23) is satisfied on the solid line portions and thus these results are rigorously applicable to J_2 flow theory.

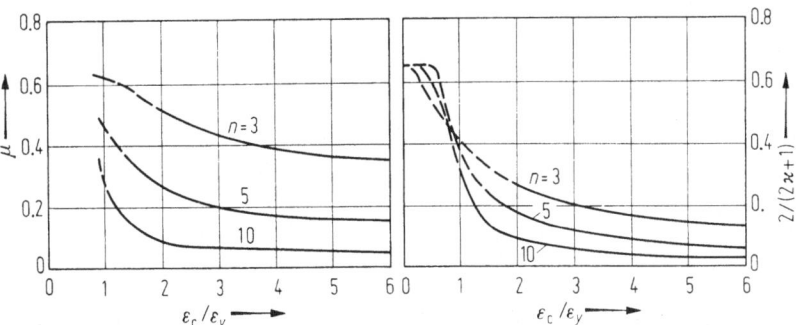

Fig. 4. Parameters for the asymptotic imperfection-sensitivity analysis calculated using the Ramberg-Osgood stress-strain relation and $\nu = 1/3$ (see Eq. (22)).

The essence of the extreme imperfection-sensitivity associated with the simple flow theory is the small exponent, $2/(2\varkappa + 1)$, in (22). As seen from (16) or in Fig. 4 this exponent may easily be as small as $1/10$ for buckling in the plastic range, implying that imperfection levels on the order of $\bar{\xi} = 10^{-10}$ will have a non-negligible effect. Essentially this same strong dependence on the imperfection amplitude was found for the cruciform model of Onat and Drucker (1953). In contrast, for elastic buckling $\varkappa = 1$ and the exponent in (22) is always $2/3$ for a symmetric bifurcation point (Koiter, 1945) (although (22) does not apply to the cruciform in the elastic range since $\lambda_2 > 0$ and there is no maximum in the vicinity of the bifurcation load). Even for nonlinear elastic solids (e.g., deformation theory) this exponent is $2/3$. Thus, the extreme imperfection-sensitivity typified by (22) is really a consequence of the hypo-elastic character of J_2 flow theory for loading.

References

Hutchinson, J. W.: Plastic Buckling. Advances in Applied Mechanics **14**, 67—144 (1974).
Koiter, W. T.: On the Stability of Elastic Equilibrium. Thesis (in Dutch), Delft, H. J. Paris, Amsterdam 1945. (English translations: (a) NASA TT-F10, 833 (1967), (b) AFFDL-TR-70-25 (1970)).
Onat, E. T.; Drucker, D. C.: Inelastic Instability and Incremental Theories of Plasticity. J. Aero. Sci. **20**, 181—186 (1953).
Stowell, E. Z.: Compressive Strength of Flanges. NACA Rep. 1029 (1951). (See also Gerard, G.; Becker, H.: Handbook of Structural Stability. NACA TN 3781 (1957)).

Stability Analysis of Rigid Plastic Structures at the Yield-Point Load

M. K. Duszek

Polish Academy of Sciences, Warsaw, Poland

Abstract

The behaviour of a rigid perfectly-plastic structure after reaching the yield point load is considered.

Criteria of material and geometric stability in Lagrangian description are formulated. The influence of assumed kind of definition of objective Cauchy stress rate on the material and geometric stability is discussed.

The theory is illustrated by examples of cylindrical shells subjected to a complex load. Stability in these cases depends on the ratio of the lateral tractions as well as on the geometry of the shell.

1. Introduction

An interest in the post-yield behaviour of a plastic structure is connected with the fact that some structures can support an applied load exceeding considerably the critical load predicted by the simple limit load theory. On the other hand, caution in utilizing the post-yield bearing capacity of the structure is also justified, because frequently the deflections increase quite rapidly and the structure collapses. The engineer's interest in the post-critical state of a structure is focused mainly on the question whether at the yield point load the structure is stable (safe) or not.

The problem of stability of a structure in the post-yield range depends on the stability of the material from which the structure is made as well as on geometry and boundary conditions.

According to Drucker's postulate, a stable plastic material is one whose additional deformation requires positive work done by the external agency. Because it is assumed that this postulate defines the behaviour of material at any time and any place, the variables of Euler-·ian description, namely Cauchy stress tensor σ and true strain rate tensor d are here suitable. The definition of material stability is closely connected with the definition of objective stress rate. Each definition of objective stress rate leads to a different criterion for material stability, that is to a different type of plastic material. While a limit load solution does not depend on the choice of this definition, the postcritical behaviour of the structure is very sensitive to the assumed type of plastic material.

Geometrical stability is associated with changes in geometry, so only nonlinear theory which takes into account non-linear kinematics can describe this phenomenon. It is convenient to apply then the variables of Lagrangian description (referred to the undeformed configuration), that is Kirchhoff stress tensor S and Green strain tensor E.

2. Basic Relations

In the subsequent analysis, capital letters refer to the material La-grangian) description $\{X^K\}$ and lower case letters to the spatial (Eulerian) description $\{x^k\}$.

If U is a displacement vector, the Green strain and strain rate tensors are specified by the following relations:

$$2E_{KL} = U_{K|L} + U_{L|K} + U_{M|K}U^M{}_{|L}, \qquad (2.1)$$

$$2\dot{E}_{KL} = \dot{U}_{K|L} + \dot{U}_{L|K} + \dot{U}_{M|K}U^M{}_{|L} + U_{M|K}\dot{U}^M{}_{|L}, \qquad (2.2)$$

where the vertical stroke denotes covariant differentiation in the original configuration.

For an incompressible material, the Kirchhoff stress tensor S^{KL} is related to the Cauchy (true) stress tensor σ^{ij} as follows [1]

$$\sigma^{ij} = x^i{}_{,K}x^j{}_{,L}S^{KL}. \qquad (2.3)$$

Equilibrium conditions, if referred to the undeformed configuration, take the form [1]

$$(S^{KL} + S^{KM}U^L{}_{|M})_{|K} = 0 \quad \text{in } V, \qquad (2.4)$$

$$(S^{KL} + S^{KM}U^L{}_{|M}) N_K = P^L \quad \text{on } S, \qquad (2.5)$$

where P^L stands for the surface tractions associated with an original area element dS having a unit outer normal N_K.

If p^k denotes the forces acting on a corresponding surface element ds of the deformed element, for an incompressible material, we have [1]:

$$P \, dS = p \, ds \qquad (2.6)$$

or

$$P^K \, dS = p^k g^K{}_k \, ds. \qquad (2.7)$$

where $g^K{}_k$ denotes the space shifter.

The equations governing the rates of stress can beobtained by taking the time derivative of Eqs. (2.4) and (2.5):

$$(\dot{S}^{KL} + \dot{S}^{KM}U^L{}_{|M} + S^{KM}\dot{U}^L{}_{|M})_{|K} = 0 \quad \text{in } V, \qquad (2.8)$$

$$(\dot{S}^{KL} + \dot{S}^{KM}U^L{}_{|M} + S^{KM}\dot{U}^L{}_{|M}) N_K = 0 \quad \text{on } S. \qquad (2.9)$$

Making use of (2.5) and Green's theorem it can be shown that at finite deformations the energy rate equation has the same form as that for infinitesimal deformations,

$$\int_S P^K \dot{U}_K \, dS = \int S^{KL} \dot{E}_{KL} \, dV. \tag{2.10}$$

Analogously, it can be shown (making use of (2.9) and Green's theorem) that a similar expression for the stress rates at finite deformations differs from that for infinitesimal deformations by nonlinear (with respect to velocity gradient) terms:

$$\int_S \dot{P}^K \dot{U}_K \, dS = \int_V (\dot{S}^{KL} \dot{E}_{KL} + S^{KL} \dot{U}_{M|K} \dot{U}^M{}_{|L}) \, dV. \tag{2.11}$$

3. Constitutive Relations and Material Stability Criterion

According to Drucker's postulate, a stable plastic material is one whose additional deformation requires positive work done by the external agency [2]. The following conclusion can be drawn from the above material stability postulate:

$$\overset{\triangledown}{\sigma}{}^{ij} d_{ij} \geqq 0. \tag{3.1}$$

$\overset{\triangledown}{\sigma}{}^{ij}$ denotes here an objective stress rate which vanishes under rigid-body motion and d_{ij} is the true strain rate.

A decision has to be now made which definition of stress rate is most appropriate for the postulate.

If $\overset{\triangledown}{\sigma}{}^{kl}$ denotes the Truesdell stress rate which is associated with axes rotating and deforming with the material element, then $\overset{\triangledown}{\sigma}{}^{kl}$ will be denoted by $\overset{\triangledown\text{T}}{\sigma}{}^{kl}$. The following relation can be written:

$$\overset{\triangledown\text{T}}{\sigma}{}^{kl} = \dot{\sigma}^{kl} + \sigma^{kl} v_p^p - \sigma^{kp} v_p^l - \sigma^{lp} v_p^k. \tag{3.2}$$

If $\overset{\triangledown}{\sigma}{}^{kl}$ denotes the Jaumann stress rate which is associated with axes rotating with material element but not deforming with it, then $\overset{\triangledown}{\sigma}{}^{kl}$ will be denoted by $\overset{\triangledown\text{J}}{\sigma}{}^{kl}$.

$$\overset{\triangledown\text{J}}{\sigma}{}^{kl} = \dot{\sigma}^{kl} + \sigma^{kp} \omega_p^l + \sigma^{lp} \omega_p^k. \tag{3.3}$$

Each of the above considered objective stress rates leads to a different type of plastic material. We may call it, respectively, plastic material in Truesdell's sense or plastic material in Jaumann's sense.

The theory of finite plastic deformations formulated in Lagrangian description, requires establishment of the constitutive relations using variables of this description, that is the Kirchhoff stress tensor S^{KL} and the Green strain rate tensor E_{KL}.

It may be shown that the following relation takes place:

$$d_{kl} \overset{\nabla T}{\sigma}{}^{kl} = \dot{E}_{KL} \dot{S}^{KL}. \tag{3.4}$$

From (3.1) and (3.4) it follows that for plastic material in Truesdell's sense the stability postulate may be written as:

$$\dot{E}_{KL} \dot{S}^{KL} \geqq 0. \tag{3.5}$$

Similarly, in view of the relation

$$d_{kl} \overset{\nabla J}{\sigma}{}^{kl} = \dot{E}_{KL} \dot{S}^{KL} + 2S^{KL} \dot{E}_{KM} \dot{E}^{M}_{L}, \tag{3.6}$$

the stability postulate for plastic material in Jaumann's sense may be written as

$$\dot{E}_{KL} \dot{S}^{KL} + 2S^{KL} \dot{E}_{KM} \dot{E}^{M}_{L} \geqq 0. \tag{3.7}$$

4. Geometrical Stability

An instability problem in the theory of plasticity arises when the instability associated with changes in geometry is great enough to overcome the stability of the material.

The question of geometrical stability depends on the geometry of the body and the boundary conditions which may change during the deformation process, so only nonlinear theory which takes into account geometry changes can describe this phenomenon.

According to the classical Dirichlet and Kelvin definition of stability, a sufficient condition for geometrical stability is that in any possible infinitesimal displacement from the position of equilibrium, the internal energy stored or dissipated should exceed the work done on the system by the applied external forces.

The criterion for geometrical stability may also be formulated in terms of the dead load intensity rate [3].

Let us consider a structure subjected to a system of dead surface loads P of monotonically increasing intensity

$$P(X, t) = \mu(t) \overset{0}{P}(X), \tag{4.1}$$

where $\mu(t)$ indicates the load intensity and $\overset{0}{P}(X)$ specifies the load distributions. So, at the yield point state $\mu(0) = 1$ and $P(X, 0) = \overset{0}{P}(X)$.

The rate of loading is therefore

$$\dot{P}(X, t) = \dot{\mu}(t) \overset{0}{P}(X). \tag{4.2}$$

If a quasi-static motion of a structure takes place only for increasing dead load, $\dot{\mu} > 0$, the structure is said to be stable. Geometrical instability is associated with $\dot{\mu} < 0$ and this means that a perfectly plastic structure continues to deform plastically even under decreasing dead loads.

Making use of Eq. (2.11) Eq. (4.2), the rate of load intensity $\dot{\mu}$ may be determined from the formula:

$$\dot{\mu}(t) = \int\limits_{V} (\dot{S}^{KL}\dot{E}_{KL} + S^{KL}\dot{U}_{M|K}\dot{U}^{M}{}_{|L})\, dV \Big/ \int\limits_{S} \overset{0}{P}{}^{K}\dot{U}_{K}\, dS. \qquad (4.3)$$

For $\mu \geqq 0$ it may be shown that

$$\int\limits_{S} \overset{0}{P}{}^{K}\dot{U}_{K}\, dS \geqq 0. \qquad (4.4)$$

From (4.3) and (4.4) it follows that the structure is stable, i.e. $\dot{\mu} > 0$ if

$$\int\limits_{V} (\dot{S}^{KL}\dot{E}_{KL} + S^{KL}\dot{U}_{M|K}\dot{U}^{M}{}_{|L})\, dV > 0. \qquad (4.5)$$

The foregoing stability condition is identical with that obtained by Hill [7] by employing the energy criterion.

For rigid perfectly plastic material in Truesdell's sense, Drucker's postulate of material stability (3.5) reduces to the following equality:

$$\dot{S}^{KL}\dot{E}_{KL} = 0. \qquad (4.6)$$

Then Eq. (4.3) may be written as

$$\dot{\mu}(t) = \int\limits_{V} S^{KL}\dot{U}_{M|K}\dot{U}^{M}{}_{|L}\, dV \Big/ \int\limits_{S} \overset{0}{P}{}^{K}\dot{U}_{K}\, dS \qquad (4.7)$$

and the geometrical stability condition (4.5) becomes

$$\int\limits_{V} S^{KL}\dot{U}_{M|K}\dot{U}^{M}{}_{|L} > 0. \qquad (4.8)$$

For rigid perfectly plastic material in Jaumann's sense, from (3.7) it follows that

$$\dot{E}_{KL}\dot{S}^{KL} + 2S^{KL}\dot{E}_{MK}\dot{E}^{M}_{L} = 0. \qquad (4.9)$$

Then Eq. (4.3) may be written as

$$\dot{\mu}(t) = \int\limits_{V} (S^{KL}\dot{U}_{M|K}\dot{U}^{M}{}_{|L} - 2S^{KL}\dot{E}_{KM}\dot{E}^{M}_{L})\, dV \Big/ \int\limits_{S} \overset{0}{P}{}^{K}\dot{U}_{K}\, dS \qquad (4.10)$$

and the geometrical stability condition (4.5) takes the form

$$\int\limits_{V} (S^{KL}\dot{U}_{M|K}\dot{U}^{M}{}_{|L} - 2S^{KL}\dot{E}_{KM}\dot{E}^{M}_{L})\, dV > 0. \qquad (4.11)$$

The above inequality, as shown by Hill, may be written in the following equivalent from [4]

$$- \int_V S^{KL} \dot{U}_{K|M} \dot{U}^M{}_{|L} \, dV > 0. \qquad (4.12)$$

Hill [4—7], employing the energy stability criterion for variables in Eulerian description expressed his result in terms of Cauchy stress tensor and velocity gradient in the form

$$- \int_V \sigma_{jk} v_{j,i} v_{i,k} \, dV > 0, \qquad (4.13)$$

which is analogous to (4.12) (describing the stability criterion for rigid perfectly plastic material in Jaumann's sense).

5. Simplified Analysis for Generalized Plastic Hinges

The above considered geometrical stability criteria apply to structures with deformation modes without any discontinuities or with discontinuities of the deflection slope solely. Whenever zones in which both concentrated extension and curvature changes appear, the second term in the numerator leads to some integration difficulties. This complication can be overcome by considering mechanisms with plastic hinges and accounting for large displacements and rotations.

Let us consider a hinged beam deforming so as the deflection W at the central hinge is comparable with the beam depth, Fig. 1. The rigid links rotate with respect to the supports and the axial extension Λ

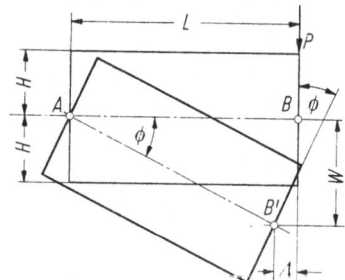

Fig. 1

appears at the plastic hinge, which eventually becomes a "generalized plastic hinge" in which both concentrated extension and curvature change occur. The geometrical relations yield [8]

$$\Lambda = W^2/2L, \quad \Phi = W/L, \quad \dot{\Lambda} = W\dot{W}/L, \quad \dot{\Phi} = \dot{W}/L. \qquad (5.1)$$

The energy rate equation results, for the case considered, in the relation

$$P\dot{W} = 2(N\dot{\Lambda} + M\dot{\Phi}), \qquad (5.2)$$

where bending moment M and axial force N are related by the inter-action curve $F(M, N) = 0$.

Differentiating (5.2) with respect to time and using material stability postulate (3.5) for rigid perfectly plastic material in Truesdell's sense (which in terms of generalized stresses and strains is $\dot{M}\dot{\varPhi} + \dot{N}\dot{\varLambda} = 0$), lead to the relation

$$\dot{P}W + P\dot{W} = 2(N\ddot{\varLambda} + M\ddot{\varPhi}). \tag{5.3}$$

By substituting in (5.3) the relation obtained from (5.2), namely

$$P\ddot{W} = 2(N\dot{\varLambda} + M\dot{\varPhi})\, \ddot{W}/\dot{W} \tag{5.4}$$

and using (5.1), we obtain that

$$\dot{P} = 2N\dot{W}/L. \tag{5.5}$$

This allows $\dot{\mu}$ to be evaluated from (4.2), yielding

$$\dot{\mu} = 2N\dot{W}/\overset{0}{P}L. \tag{5.6}$$

Since the rate of external work at the incipient point state is non-negative ($\overset{0}{P}\dot{W} \neq 0$), it follows that

$$\text{sign } \overset{0}{P} = \text{sign } \dot{W}. \tag{5.7}$$

It can be concluded from (5.6) and (5.7) that the sign of $\dot{\mu}$ is governed by the sign of axial forces occurring in the generalized hinges.

Moreover, from Eq. (5.5) the slope of the load-deflection curve is

$$dP/dW = 2N/L. \tag{5.8}$$

6. Application

As an illustration of application of the geometrical stability criterion for plastic material in Truesdell's and in Jaumann's sense let us consider a thin-walled rotationally symmetric cylindrical shell subjected to a rotationally symmetric deformation under action of uniformly distributed dead load P and the end load T, Fig. 2. The kinematic Kirch-hoff-Love assumptions are made[1].

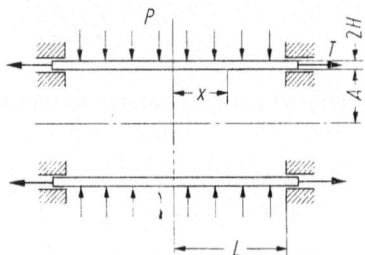

Fig. 2

[1] The above assumptions can be easily relaxed by allowing for the change in the shell thickness. This will not influence the solution much.

For a rigid perfectly plastic material in Truesdell's sense the expression (4.7) defining the rate of load intensity may be now written in cylindrical coordinates as

$$\dot{\mu} = \int_S \{n_x[(A^2/L^2)\,\dot{w}_{,x}^2 + \dot{v}_{,x}^2] + n_\theta \dot{w}^2 - (1/\alpha)\,(m_x \dot{w}_{,xx} + m_{x,x}\dot{w}_{,x})\,\dot{v}_{,x}\}\,\mathrm{d}S :$$

$$: \int_S (p\dot{w} + t\dot{v})\,\mathrm{d}S, \tag{6.1}$$

where the dimensionless quantities are defined as follows:

$$p = PA/2H\sigma_0, \quad t = T/4\pi AH\sigma_0, \tag{6.2}$$

$$w = U_R/A, \quad v = U_x/L, \quad x = X/L, \quad \alpha = L^2/AH, \tag{6.3}$$

$$n_x = (1/2\sigma_0 H) \int_{-H}^{+H} S_{xx}\,\mathrm{d}Z, \quad n_\theta = (1/2\sigma_0 H) \int_{-H}^{+H} S_{\theta\theta}\,\mathrm{d}Z,$$

$$m_x = (1/\sigma_0 H^2) \int_{-H}^{+H} S_{xx}Z\,\mathrm{d}Z. \tag{6.4}$$

Assuming the limited interaction yield condition, the "limit load" solution for a cylindrical shell with clamped edges was obtained by Hodge [9].

The curve $ABCDEF$ in Fig. 3 shows the yield-point loading curve for the particular case of $\alpha = \infty$ and the curve $A'B'C'D'E'F'$ for $\alpha = 2$.

Now, making use of (6.1) and (4.2) we shall supplement this solution by information about stability and load-deflection slope at the yield-point load.

(a) For the range of loads represented by line CD in Fig. 3:

$$0 \leq t < 1, \quad p = 1 - t + (2/\alpha),$$

$$m_x = 1 - 2x^2, \quad n_\theta = -1 + t, \quad n_x = t,$$

$$\dot{w} = \dot{w}_0(1 - x), \quad \dot{v} = \dot{w}_0[x - (x^2/2)],$$

$$\dot{\mu} = (2/3)\,[3\,(A^2/L^2)\,t + 2t - 1 + (1/\alpha)]\,\alpha\dot{w}_0/(\alpha + 2).$$

$$\dot{\mu} > 0 \quad \text{for} \quad t > \frac{\alpha - 1}{3(A/H) + 2\alpha} \quad \text{(shell is stable)},$$

$$\dot{\mu} < 0 \quad \text{for} \quad t < \frac{\alpha - 1}{3(A/H) + 2\alpha} \quad \text{(shell is unstable)}. \tag{6.5}$$

$$\frac{\mathrm{d}p}{\mathrm{d}w_0} = \frac{2}{3}\,\frac{[3(A^2/L^2)\,t + 2t - 1 + (1/\alpha)]\,(\alpha - \alpha t + 2)}{(\alpha + 2)},$$

$$\frac{\mathrm{d}t}{\mathrm{d}v_0} = \frac{4}{3}\,\frac{[3(A^2/L^2)\,t + 2t - 1 + (1/\alpha)]\,\alpha t}{(\alpha + 2)}.$$

(b) For the range of loads represented by line BC in Fig. 3:
$$-1 < t < 0, \quad p = 1 + (2/\alpha),$$
$$m_x = 1 - 2x^2, \quad n_\theta = -1, \quad n_x = t, \quad \dot{w} = \dot{w}_0(1 - x), \quad \dot{v} = 0.$$
$$\dot{\mu} = \frac{2}{3}\frac{[3(A^2/L^2)\, t - 1]\,\alpha\dot{w}_0}{(\alpha + 2)} < 0 \quad \text{(shell is unstable)}, \qquad (6.6)$$
$$dp/dw_0 = (2/3)\,[3\,(A^2/L^2)\,t - 1].$$

(c) For the range of loads represented by line AB in Fig. 3:
$$t = -1, \quad n_x = -1, \quad m_x = k(1 - 2x^2), \quad n_\theta = -p + 2k/\alpha, \quad -1 \leqq k \leqq 1,$$
$$\dot{w} = 0, \quad \dot{v} = -\dot{v}_0 x,$$
$$\dot{\mu} = -\dot{v}_0 < 0, \quad \text{(shell is unstable)} \qquad (6.7)$$
$$dt/dv_0 = -1.$$

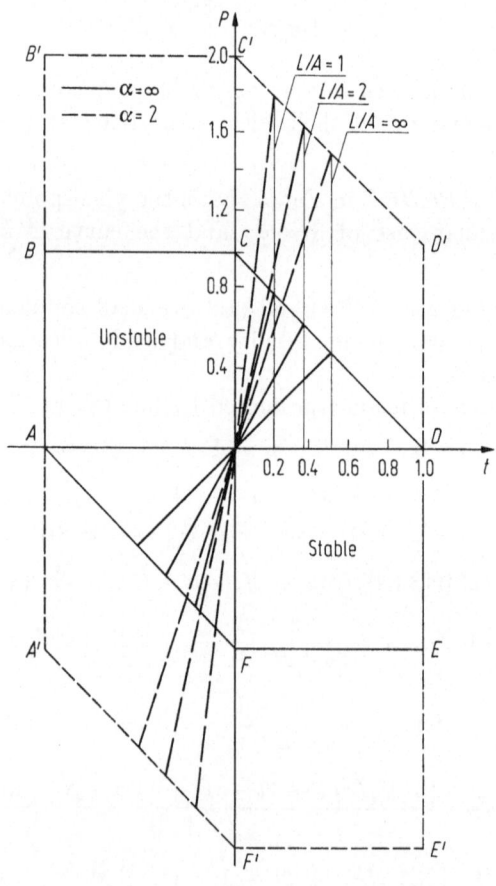

Fig. 3

Since the yield-point loading curve is evidently symmetric with respect to the origin, we consider only external dead load P, but the axial load may be positive or negative.

Stability in this case depends on the ratio of the surface tractions p to t as well as on the geometry of the shell described by the ratio A/L and the parameter α. The yield-point loading curve can be divided into two parts, Fig. 3. One of them, on the right-hand side of the heavy line, corresponds to a stable state at yield and the other, on the left-hand side, to an unstable state.

Putting $t = 0$ into Eqs. (6.5), we obtain a solution for a cylindrical shell under uniformly distributed dead load, as a particular case of the considered example.

Then, for $p = 1 + (2/\alpha)$ (external load)

$$\dot{\mu} = (2/3)\,(1 - \alpha)\,\dot{w}_0/(\alpha + 2).$$
$$\dot{\mu} < 0 \quad \text{for} \quad \alpha > 1, \tag{6.8}$$
$$\mathrm{d}p/\mathrm{d}w_0 = (2/3)\,(1 - \alpha)/\alpha.$$

For $p = -[1 + (2/\alpha)]$ (internal load)

$$\dot{\mu} = (2/3)\,(\alpha - 1)\,\dot{w}_0/(\alpha + 2).$$
$$\dot{\mu} > 0 \quad \text{for} \quad \alpha > 1, \tag{6.9}$$
$$\mathrm{d}p/\mathrm{d}w_0 = -(2/3)\,(1 - \alpha)/\alpha.$$

We now proceed to investigate the stability for the same cylindrical shell as considered above, but made from rigid perfectly plastic material in Jaumann's sense. Then from Eq. (4.10) we obtain that

$$\dot{\mu} = \int\limits_S \{n_x[(A^2/L^2)\,\dot{w}_{,x}^2 - \dot{v}_{,x}^2] - n_\theta \dot{w}^2 + (1/\alpha)\,(m_x \dot{w}_{,xx} - m_{x,x}\dot{w}_{,x})\,\dot{v}_{,x}\}\,\mathrm{d}S:$$
$$: \int\limits_S (p\dot{w} + t\dot{v})\,\mathrm{d}S. \tag{6.10}$$

In the particular case of $t = 0$, $p = 1 + (2/\alpha)$, an analysis similar to that leading to Eqs. (6.8) gives

$$\dot{\mu} = (2/3)\,(\alpha - 3)/(\alpha + 2).$$
$$\dot{\mu} > 0 \quad \text{for} \quad \alpha > 3, \tag{6.11}$$
$$\mathrm{d}p/\mathrm{d}w_0 = (2/3)\,(3 - \alpha)/\alpha.$$

From the comparison of these curves it follows that the post-yielding behaviour of perfectly plastic material in Truesdell's sense is quite different from that in Jaumann's sense.

Analytical solution of the problem of cylindrical shells at large displacements [11] indicates that rigid perfectly plastic cylindrical shell with clamped edges under internal uniformly distributed dead load behaves stable at the yield-point load. This result coincides with the

(6.9) obtained for the cylindrical shall made from a perfectly plastic material in Truesdell's sense.

On the other hand the stability criterion for plastic material in Jaumann's sense, applied to the necking problem, yields to the ordinary engineering criterion for the onset of necking [5].

So the question which definition of stress rate tensor should be used in the analysis of the stability problem is therefore essential and requires further investigations.

References

1. Fung, Y. C.: Foundations of Solid Mechanics. Englewood Cliffs, N.J.: Prentice-Hall: 1965.
2. Drucker, D. C.: A More Fundamental Approach to Plastic Stress-Strain Relations. Proceedings of the First U.S. National Congress of Applied Mechanics, ASME, 1951.
3. Onat, E. T.: The Influence of Geometry Changes on the Load-Deflection Behaviour of Plastic Solids in Plasticity. Ed. by L. S. Shu, P. S. Symonds, Pergamon Press 1961.
4. Hill, R.: Stability of Rigid-Plastic Solids. J. of Mech. and Physics of Solids **6** (1957).
5. Hill, R.: On the Problem of Uniqueness in the Theory of a Rigid-Plastic Solid. IV, ibid. **5** (1957).
6. Hill, R.: A General Theory of Uniqueness and Stability in Elastic-Plastic Solids. Ibid. **6** (1958).
7. Hill, R.: Some Basic Principles in the Mechanics of Solids without a Natural Time. Ibid. **7** (1959).
8. Sawczuk, A.: Simplyfied Methods of Analysis of Elastic-Plastic Structures (in Polish). Mech. Teoret. i Stos., **10** (1972).
9. Hodge, P. G.: Limit Analysis of Rotationally Symmetric Plates and Shells. Englewood Cliffs, N.J.: Prentice-Hall: 1963.
10. Augusti, G.; d'Agostino, S.: Test of Cylindrical Shells in the Plastic Range. J. of the Eng. Mech. Division **90**, EM1 (1964).
11. Duszek, M.; Sawczuk, A.: Load-Deflection Relations for Rigid-Plastic Cylindrical Shells beyond the Incipient Collapse Load. Int. J. Mech. Sci. **12** (1970).

Mode Interaction with Stiffened Panels

A. van der Neut

Technische Hogeschool, Delft, Netherlands

Summary

Previous work on mode interaction refers to a column model consisting of two equal flanges. This model is extremely sensitive to imperfection. Actual panel structures are unsymmetrical: the plate side has more cross section than the topside of the stiffeners; and the plate side is more affected by local buckling than the topside. This effect is exaggeratedly represented in a model where the stiffeners do not participate in local buckling. It appears that the sensitivity to imperfections of this model is very little and restricted to geometric parameters R in the near vicinity of $R = 1$. These two models represent extreme cases. The position of actual tophat stiffened panels between these extremes is being explored. It appears that mode interaction is less severe than with the two flange model though still significant.

Symbols

K	compressive force	β	amplitude of column imperfection: j
L	half of column length		
P	compressive force of a plate strip	ε	compressive strain at edge of strip
R	geometric parameter; K_E/K_l		
W	deflection of the axis of the structure	η	stiffness reduction factor
		λ	half wave length: width of strip
b	width of plate strip		
j	radius of gyration of the cross section	*Subscripts:*	
l	half wave length in local buckling	E	refers to Euler buckling load
w	deflection of plate strip	b	refers to overall buckling load
α	amplitude of plate imperfection: plate thickness	l	refers to local buckling load

1. Introduction Surveying the Problem

Thin-walled struts and panels are subject to local and to overall buckling. The usual conception that the structure where these buckling stresses are equal is optimal was attacked already in 1962 by Koiter and Skaloud [1]. They pointed at the possibility of unfavourable effect of mode interaction. This observation became confirmed by experimental evidence when panels failed explosively at stresses smaller than both the local and the Euler buckling stress. A first attempt to analyse this problem related to a simplified strut consisting of two equal load carrying flanges interconnected by webs without axial stiffness [2, 3, 4]. It demonstrates, that the equilibrium at the local buckling load K_1 is instable over a range of the geometrical parameter R close to unity, R being the ratio of Euler buckling load K_E and K_1; that consequently the structure is imperfection sensitive. Further the strength reductions due to the imperfections: initial waviness of the flanges and initial curvature of the strut were established. Using these results Thompson and Lewis [5] have shown that the effect of initial waviness moves the optimum from $R = 1$ to $R < 1$, eroding it more and more with increasing imperfection.

The two-flange model presents an extreme case of mode interaction, because the flanges are strongly and equally affected by local buckling and because the structure contains no components which temper the flange behaviour. The final object of this investigation is to establish the significance of mode interaction for heavily stiffened panels as used in aircraft wing structure. Here the cross section at the topside of the stringers is much smaller than the cross section at the plateside. Also the b/h-ratios of the strips at the two sides are quite different. This lack of symmetry of the two sides can have a major effect on the sensitivity to mode interaction.

A useful first reconnaissance in this area would be the overshoot of exaggerating the dissymmetry. A pronounced case of this type is the thin plate reinforced by stiffeners which do not participate in local buckling of the plate. It represents the lower extreme of the phenomen of mode interaction in contrast to the two-flange model which is representative for the upper extreme.

Tvergaard's work [6, 7] falls in this area. He investigated the integrally stiffened plate where the participation of the stiffeners in local buckling is confined to torsion. An important feature is that the compressive load is applied in the centroid of the cross section. However with unsymmetric structures, subject to local buckling the neutral plane in bending moves away from the side most affected by local buckling. The load in the centroid then causes bending. This exaggerates the strength reduction in comparison to the situation with a multi-bay panel.

So as to avoid bending below the buckling load the boundary conditions will be that the panel is rigidly restrained at its ends. Moreover these boundary conditions are realistic. In actual structures panels comprise a number of bays separated by supporting frames. The part of the structure between the centres of the first and the third bay corresponds to the panel clamped at its ends. In two consecutive bays the buckling mode yields increased compression at the plateside of one bay and decreased strain at the plate side of the next one. Due to the nonlinearity of the load-strain relation the bending stiffness of the first bay decreases but it increases in the adjacent one, thereby tempering the destabilisation of the combination of the bays. This concept leads to the conclusion that compression tests should be carried out on clamped structures when non-linear behaviour occurs.

It is to be expected that the model representing the lower extreme of mode interaction, will behave very mildly as to mode interaction.

Having considered these extreme cases the next problem is to locate between these extremes the position of panels where the stiffener is thin-walled and participates in local buckling.

The local buckling mode depends on quite a number of geometric parameters, such as the ratios of the widths of the strips composing the cross section and the ratios of the strip thicknesses. The coherence of the several strips requires equal longitudinal wave length of all strips, therefore different ratios of wave length and strip width. Equality of buckliug stresses means that the buckling coefficients of the strip will differ from each other. Equality of wave length and buckling stress is being achieved by bending moments occuring between the strips at their junction. These moments can have a stabilizing effect (positive elastic restraint) or a destabilizing effect (negative elastic restraint). The solution of the mode interaction problem requires the determination of the local buckling mode and of K_1. Methods are available.

Next comes the post-buckling behaviour of the perfect structure and the behaviour of imperfect structures. This is a very complex problem.

Let be assumed that the stiffnesses of edge restraints remain constant with increasing edge strain. Then the problem is to establish the stiffness in compression as a function of the edge strain of a strip for arbitrary wave length and edge restraint, whereas these restraints at the two edges can have any ratio, positive or negative. This is a tremendous problem though its solution does not even solve the problem of panel behaviour under compression without bending, where the strains at all edges are equal. The cause of this unsufficiency is that under the assumption of constant edge restraint the increase of edge rotations with increased compression will be different for adjoining strips. The more so with arbitrary imperfections of the individual strips. So in fact the stiffness of the edge restraints must vary during loading so as to yield compatible edge rotations.

The need to establish the bending stiffness of the panel introduces an additional problem. Plate strips not parallel to the panel plane obtain unequal edge strains due to bending. The position of the neutral plane is unknown and therefore the ratio of the edge strains. Consequently the behaviour of plates with arbitrary ratio of edge strains, arbitrary ratio of wave length to plate width, arbitrary ratio of edge restraints and arbitrary amount of edge restraint would have to be investigated. These ingredients are required to solve the mode interaction problem of this type of stiffened panel. If they were available the remaining interaction problem would not be too cumbersome. The capital problem is the one on strip behaviour just defined.

Since ignorance on the amount of the imperfections of actual structures excludes the possibility of theoretical prediction of the strength of these structures exact solution of the problem seems to have little practical value and would be too ambitious. On the other hand it serves a real need to explore the extent of the zone of R in which interaction presents reduction of strength and the order of magnitude of the reduction in the affected zone.

Therefore the logical step after having considered the two "extreme" models is to consider a model which is not identical to but representative of real panel structures. The problem of the behaviour of its composing plate strips should not be too complex.

2. Local Buckling Confined to One Side of the Panel

Panels with large stiffener spacing have local buckling stresses much lower than the overall buckling stress; R is far above unity. Then K_b is not affected by imperfections and can be established with the Euler formula thereby taking into account the effective modulus of the plate. Unfavourable mode interaction occurs only when R is close to unity. This means heavily reinforced panels, where stringer and plate sections do not differ much. As a representative structure has been taken the case where stringer and plate section are equal. The results obtained with ratios between 1/3 and 3 are only slightly different.

The method of solution is almost a special case of the analysis given in section 6.

Figure 1 shows the K_b/K_1, R — curve for the perfect plate and for one with the imperfection $\alpha = 0.05$. With the perfect structure the reduction factor of the bending stiffness for $K_b > K_1$ is 0.8. This reduces the transition zone at $K_b = K_1$ to $1 < R < 1.25$. The equilibrium at $K_b = K_1$ is instable over the narrow range $1.0 < R < 1.11$. These upper limits are very low in comparison to those of the two-flange model, where they are 2.45 and 1.725, respectively.

With the imperfection $\alpha = 0.05$ K_b is smaller than K_1 up to $R = 1.15$. The maximal reduction at $R = 1$ is 9%, whereas the two-flange model with $\alpha = 0.05$ has 17%.

Figure 2 depicts the load-strain curve together with the slopes of the load-shortening curves at the bifurcation $K = K_b$. The equilibrium at K_b is unstable only for $R < 1.01$. The steepest tangent at $R = 0.85$ is much milder than with the two-flange model.

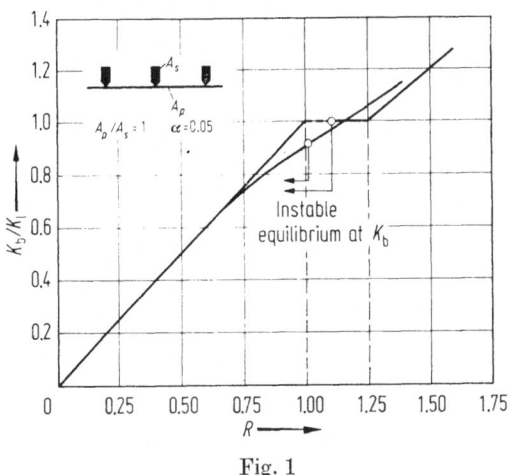

Fig. 1

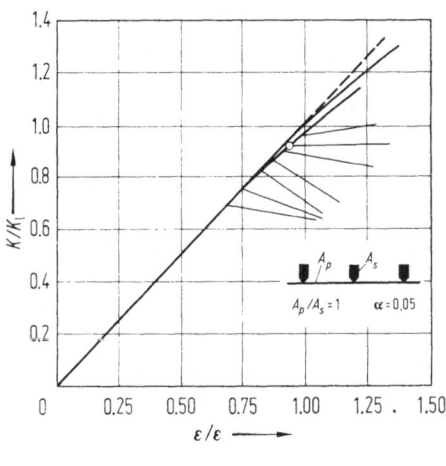

Fig. 2

Mode interaction appears to be confined to a narrow region in the immediate vicinity of $R = 1$. The strength there is slightly below K_l.

122   A. van der Neut

3. Simplified Panel Model

The selected structure is a panel stiffened by tophat stringers (Fig. 3).
The ratio of strip width and thickness is arbitrary. The cross sections
of plate and stringers have equal area. The local buckling strain has
been established by means of a straightforward method [8] as

$$\varepsilon_1 = 1.340 \frac{\pi^2}{12(1 - \nu^2)} \frac{h^2}{b^2}.$$

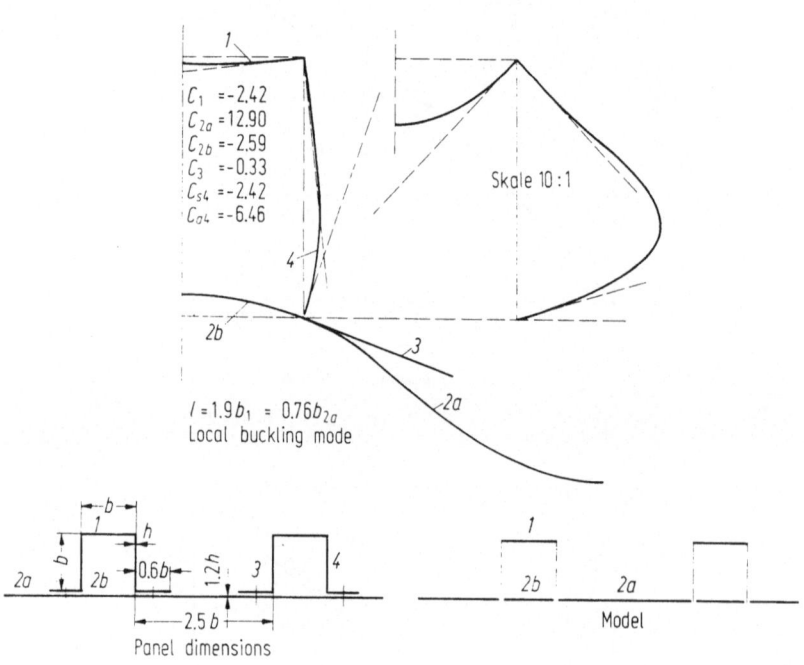

Fig. 3

The half wave length of the mode is $1.90 \cdot b$. The mode shape and the
edge rotations are shown in Fig. 3. The joints of plate and stringer have
been assumed in the intersection of the plate and the side strips of
the stringer, whereas in real structure the rivet joint will be somewhere
near the centre of the flanges. Chosing the joint along the rivet row
would involve that the edge deflection of the plate strips would be non-
zero, which would complicate the problem quite seriously. Another
difficulty would be to deal with the flange loaded along the joint.
However under these realistic conditions the deflections at the intersection of plate and side strips would be close to zero, as can be conjectured by inspection of Fig. 3. Further considering that the flanges

contain only 14% of the cross section a rude approximation of their behaviour seems acceptable.

The kind of restraint offered by the edges of the strips appears from the curvature of the mode. Plate strip 2a very clearly obtains support from plate-strip 2b and stringer strip 4; these latter then having negative restraint. The upper edge of strip 4 is being restrained by strip 1. The figure also mentions the coefficients of restraint C defined by:

$$\frac{\text{edge moment}}{\text{edge rotation}} = -C \frac{\text{bending stiffness of plate}}{\text{plate width}}.$$

With strip 4 C_s and C_a refer to the symmetrical and anti-symmetrical part of the mode. The figures indicate that the wide plate strip 2a, being the weakest part of the structure, gets very stiff edge restraint from the adjacent strips.

The model selected for the analysis of mode interaction is a simplified version of the basic structure because of various reasons. Only those strips are represented which have symmetrical buckling mode: strips nrs. 1, 2a and 2b. The flanges nr. 3 can be omitted since their contribution to the bending stiffness of the panel is very small compared to that of the plate. Also the axial load carried by the stringer sides no. 4 is being neglected so as to avoid serious complications arising from unequal restraint stiffness at upper and lower edge and unequal edge strains with panel bending. The function of the side strips to maintain the integrity of the assembly is being preserved. These simplifications affect the radius of gyration j only slightly, it increases from $0.377b$ to $0.394b$.

In this investigation the emphasis falls on the effect of stiffness reduction of the stringer top no 1, which is the essential difference with the structure where the stiffeners do not participate in local buckling. The model, now confined to strips nrs. 1 and 2, is again a two-flange model. The characteristic differences with the previous two-flange model are:

1. the areas of the cross sections of the flanges are quite different,
2. the ratio of half wave length to plate width $l/b \neq 1$,
3. the strips are elastically restrained at their edges.

A further simplification is that the coefficients of edge restraint C of the strips are assumed to be constant through the whole range of edge strains and equal to their value with local buckling given in Fig. 3. Finally it is being assumed that at the transition of strips 2a and 2b the in plane edge forces vanish. A better approximation would be to assume constant lateral displacement. But this would involve the necessity to solve another post buckling problem, whereas the present assumption permits to use the solution of the problem as occurring with respect to strip nr. *1*. To solve this latter problem had appeared to be a very laborious operation. When evaluating these various simplifications it should be kept in mind that exact quantitative knowledge of an actual structure is not being envisaged. The object being to establish to what extent mode interaction affects the panel strength. For this kind of orientation in the problem area a qualitatively correct representation of the behaviour of the structural elements seems acceptable.

Further assumptions have to be made on the imperfection of the several strips. For obvious reasons they have been taken similar to their local buckling mode shown in Fig. 3. The size of the imperfections however has no relation to the wave amplitudes in local buckling. It has been assumed that the ratios of the initial deflections in the centre of the strips to the strip widths is equal for the three strips. Three degrees of imperfection have been assumed (Table 1).

Table 1. Imperfections α = amplitude: strip thickness

Strip	I	II	III
1	0.012	0.024	0.036
2 a	0.025	0.050	0.075
2 b	0.010	0.020	0.030

4. The Elastically Restrained Strip

The plate strip is supported along its edges and has symmetric elastic restraint. With the perfect strip the post-buckling wave length can have any value. With the imperfect strip the wave length of the imperfection is also arbitrary; the lateral shape of the imperfection is similar to the buckling mode at that wave length; its amplitude is αh.

The governing parameters are: for the strip of width b the wave length parameter $\lambda = l/b$, the coefficient of restraint C, the imperfection α and the compressive strain parameter $\varepsilon/\varepsilon_1$. Instead of C can be taken the local buckling coefficient k defined by

$$\sigma_1 = k^2 \frac{\pi^2 E}{12(1 - \nu^2)} \left(\frac{h}{b} \right)^2. \tag{4.1}$$

The relation between C and k is

$$C = -2\pi \, (k/\lambda) \, [\beta \tanh (\pi/2) \, \beta + \gamma \tan (\pi/2) \, \gamma]^{-1},$$

where

$$\beta = (1/\lambda) \, (k\lambda + 1)^{1/2}, \quad \gamma = (1/\lambda) \, (k\lambda - 1)^{1/2}.$$

The local buckling mode is

$$w = \omega h \frac{\cosh{(\pi/2)}\beta \cos{\gamma\eta} - \cos{(\pi/2)}\gamma \cosh{\beta\eta}}{\cosh{(\pi/2)}\beta - \cos{(\pi/2)}\gamma} \sin{\xi}, \qquad (4.2)$$

where $\xi = \pi x/l$, $\eta = \pi y/b$ and ωh is the amplitude of the deflection.

The method by which the $P/P_1 - \varepsilon/\varepsilon_1$ relation has been established is as usual:

1. assume the deflection for any $\varepsilon/\varepsilon_1$ to be described by (4.2), where ω is unknown,

2. solve the Airy stress function from its linear differential equation,

3. solve ω from the equilibrium condition in the direction normal to the plate by means of the Ritz-Galerkin method.

The $P - \varepsilon$-relation so established is in parametric form for Poisson's ratio $\nu = 0.3$,

$$\varepsilon/\varepsilon_1 = \omega/(\omega + \alpha) + A\omega(\omega + 2\alpha)/2, \qquad (4.3)$$

$$P/P_1 = \omega/(\omega + \alpha) + B\omega(\omega + 2\alpha)/2, \qquad (4.4)$$

where A and B are highly complicated functions of λ and k, too lengthy to be reproduced here.

The non-linearity of the $P - \varepsilon$-relation is expressed by the functions η, η' and μ, occuring in the formulae describing the behaviour of the model.

$$\eta = \mathrm{d}(P/P_1)/\mathrm{d}(\varepsilon/\varepsilon_1) = (1 + BX)/(1 + AX), \qquad (4.5)$$

$$\eta' = \mathrm{d}\eta/\mathrm{d}(P/P_1) = -3(A - B) XY [\alpha(1 + AX)^2 (1 + BX)]^{-1}, \quad (4.6)$$

$$\mu = (2\eta'^2 - \eta\eta'')/6 = -(4 - 5BX) Y\eta'[6\alpha(1 + AX) (1 + BX)]^{-1},$$

$$(4.7)$$

where $X = Y^3/\alpha$ and $Y = \omega + \alpha$.

The values of λ and k pertaining to local buckling of the panel yield A and B given in Table 2 together with B/A. The post-buckling stiffness of the perfect plate strips being $\eta = B/A$. For comparison Table 2 gives these figures also for simply supported strips at the same values of λ and at $\lambda = 1$.

Table 2. Stiffness parameters A and B

Strip no.	λ	C	k^2	A	B	$\eta = B/A$
1	1.90	−2.42	1.340	2.1183	1.5067	0.7113
2a	0.76	12.9	5.82	1.1722	0.4463	0.3807
2b	1.90	−2.59	0.931	4.370	3.476	0.7954
—	0.76	0	4.31	1.7402	0.6432	0.3696
—	1.00	0	4	1.1536	0.4710	0.4083
—	1.90	0	5.89	0.2726	0.1441	0.5288

It appears that for $C = 0$ the post-buckling stiffness increases with λ. This is plausible because increased waviness (smaller λ) makes the strip more flexible. Comparing η for $C = 0$ and $C \neq 0$ at constant λ it appears that $C > 0$ yields a slight increase of η, whereas $C < 0$ increases the stiffness much.

5. Overall Buckling of the Perfect Model

The cross section of strip i is $A_i = F_i A$, where A is the total cross section of the model. The strips 2a and 2b can be combined to a flange 2 with the cross section $A_2 = A_{2a} + A_{2b}$ and with the effective cross section

$$(\eta F)_2 A = [(\eta F)_{2a} + (\eta F)_{2b}]\, A . \qquad (5.1)$$

Then the bending stiffness of the model is

$$\eta_b EI = \frac{(\eta F)_1\,(\eta F)_2}{(\eta F)_1 + (\eta F)_2}\, EAc^2 .$$

Since

$$EI = \frac{F_1 F_2}{F_1 + F_2}\, EAc^2$$

and by definition $F_1 + F_2 = 1$ follows

$$\eta_b = \eta_1 \eta_2 /[(\eta F)_1 + (\eta F)_2]. \qquad (5.2)$$

The dimensions given in Fig. 1 yield $\eta_2 = 0.4992$ and $\eta_b = 0.6576$. Figure 4 depicts

$$K_b/K_1 = \eta_b R . \qquad (5.3)$$

At $R > 1.52\ K_b$ exceeds K_1.

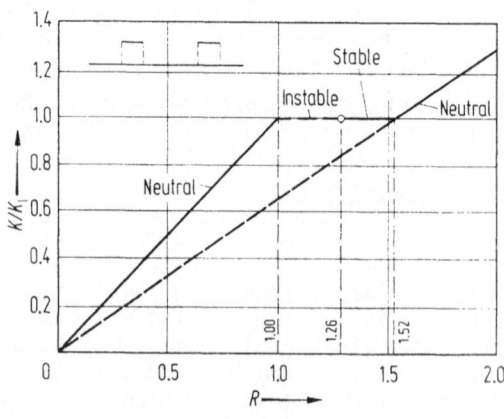

Fig. 4

The panel of length $2L_0$ (parameter R_0) is at $K = K_1$ in neutral equilibrium. Its overall buckling mode consists of 2 parts with different stiffness to infinitesimal deflection. In part I flange nr. 1 is at the concave side of the mode therefore in the post-buckling range and flange nr. 2 in the unbuckled condition. Then its bending stiffness following from (5.2). is

$$\eta_I = \eta_1/[(\eta F)_1 + F_2].$$

In part II the condition of the 2 flanges is reversed yielding

$$\eta_{II} = \eta_2/[F_1 + (\eta F)_2].$$

Having equal buckling loads the lengths L_I and L_{II} are not equal, but their sum is $2L_0$. It follows

$$R_0 = 4(\eta_I^{1/2} + \eta_{II}^{1/2})^{-2} = 1.258.$$

Panels of greater length than $2L_0$ ($1 < R < R_0$) are at $K = K$ instable to infinitesimal deflection. This instable range of R is much smaller than with the two-flange model ($1 < R < 1.725$) but wider than with the panel where the stiffeners do not participate in local buckling ($1 < R < 1.11$).

6. The Buckling Load of the Imperfect Model

With respect to infinitesimal deflection the imperfect model is in neutral equilibrium at K_b following from (5.1), (5.2) and (5.3) after establishing η of the several strips by means of (4.5).

For a given $\varepsilon/\varepsilon_1$, equal for the 3 strips, their ω follows from (4.3). Then (4.4) yields $(P/P_1)_i$, whereupon the compressive force can be established

$$K = \sum P_i = K_1 \sum (P/P_1)_i F_i. \tag{6.1}$$

The reduction factor of the bending stiffness with respect to infinitesimal deflection derives from (4.5) and (5.1), (5.2) whereupon the geometric parameter R at which K is the buckling load can be established by means of (5.3).

The $K_b/K_1 - R$-curves pertaining to the three cases of imperfection (Table 1) are shown in Fig. 5. The maximal reductions due to initial waviness occurring at $R = 1$ are resp. 9, 12.5 and 14%. For $R > 1.37 \, K_b$ exceeds K_1. With R close to unity the strength reduction by strip imperfection appears to be not negligible.

So as to answer the question about the character of the equilibrium at K_b, whether it is stable or instable, the behaviour at small finite deflections has to be established. The possibility exists that $(dK/dW_0^2)_b \neq 0$ due to the non-linearity of the $K - \varepsilon$-relation. Then at the deflection W

$$K = K_b + K_1 t(W_0/2j)^2,\qquad (6.2)$$

where

$$t = \mathrm{d}(K/K_1)/\mathrm{d}(W_0/2j)^2.$$

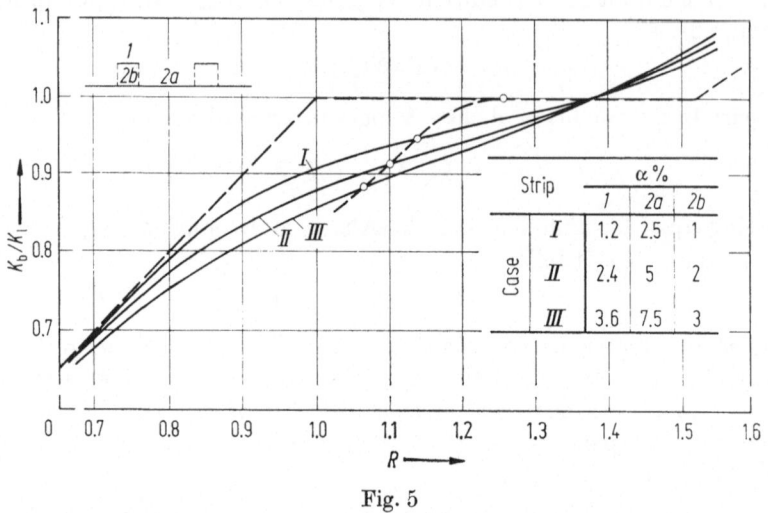

Fig. 5

The condition of equilibrium in the deflected state is

$$(\mathrm{d}^2/\mathrm{d}x^2)\left[M + K_b\left(1 + \frac{1}{\eta_b R}t\left(\frac{W_0}{2j}\right)^2\right)W\right] = 0.\qquad (6.3)$$

The last term of this equation is of third degree in W; then M has to be established as well up to terms of the third degree. This can be done by expressing the strip loads P_i by the truncated Taylor expansion

$$P = P_b + \frac{\mathrm{d}P}{\mathrm{d}\varepsilon}(\varepsilon - \varepsilon_b) + \frac{1}{2}\frac{\mathrm{d}^2P}{\mathrm{d}\varepsilon^2}(\varepsilon - \varepsilon_b)^2 + \frac{1}{6}\frac{\mathrm{d}^3P}{\mathrm{d}\varepsilon^3}(\varepsilon - \varepsilon_b)^3.\qquad (6.4)$$

The flange strains have the relation

$$\varepsilon_1 - \varepsilon_2 = c(\mathrm{d}^2W/\mathrm{d}x^2).\qquad (6.5)$$

The compressive force K and the bending moment M are resp.

$$K = P_1 + P_2,\qquad (6.6)$$

$$M = [P_1\gamma_2/(\gamma_1 + \gamma_2) - P_2\gamma_1/(\gamma_1 + \gamma_2)]\,c,\qquad (6.7)$$

where $\gamma_i = (\eta F)_i$.

Eliminating $\varepsilon_1, \varepsilon_2, P_1, P_2$ from the equations (6.2) (6.4), (6.5), (6.6), (6.7), thereby neglecting terms of higher order than the third, the relation between M, W and t is obtained

$$M/K_b c = \left(1 - A_3 \frac{1}{4\varrho} tV_0^2\right) V^{\cdot\cdot} + A_1 \frac{R}{\varrho} V^{\cdot\cdot 2} -$$

$$- (A_2 - 2A_1^2)\frac{R^2}{\varrho^2} V^{\cdot\cdot 3} + M_h/K_b c, \qquad (6.8)$$

where

$$d^2 V/d\xi^2 = V^{\cdot\cdot} = \frac{L^2}{\pi^2}\frac{d^2 W/c}{dx^2},$$

$$\varrho = j^2/c^2 = F_1 F_2$$

and A_1, A_2, A_3 are functions of η and its derivatives

$$A_1 = (-\gamma_2^2\phi_1 + \gamma_1^2\phi_2)/(\gamma_1 + \gamma_2)^2,$$

$$A_2 = (\gamma_2^3\mu_1 + \gamma_1^3\mu_2)/(\gamma_1 + \gamma_2)^3,$$

$$A_3 = 2(\gamma_2\phi_1 + \gamma_1\phi_2)/(\gamma_1 + \gamma_2)^2,$$

$$\gamma_1 = \eta_1 F_1, \quad \gamma_2 = \gamma_{2a} + \gamma_{2b} = (\eta F)_{2a} + (\eta F)_{2b},$$

$$\phi_1 = -\eta_1'/2, \quad \phi_2 = -[(\eta'\gamma)_{2a} + (\eta'\gamma)_{2b}]/2\gamma_2,$$

$$\mu_2 = [(\mu\gamma)_{2a} + (\mu\gamma)_{2b}]/\gamma_2 - \gamma_{2a}\gamma_{2b}'(\eta_{2a}' - \eta_{2b}')^2/2\gamma_2^2,$$

Substitution of M into (6.3) yields the differential equation

$$\{V^{\cdot\cdot}[1 + A_1 RV^{\cdot\cdot}/\varrho - (A_2^\blacksquare - 2A_1^2) R^2 V^{\cdot\cdot 2}/\varrho^2 -$$

$$- A_3 tV_0^2/4\varrho] + V(1 + tV_0^2/4\varrho\eta_b R)\}^{\cdot\cdot} = 0. \qquad (6.9)$$

The expression between square brackets represents the modification of the bending stiffness with small finite deflections.

The solution of this equation is not unique; the amplitude V_0 is undetermined within the range of small finite deflections. Substitution of the solution into (6.9) yields a left hand side which is the sum of terms of the orders V_0, V_0^2, and V_0^3. Since V_0 is undetermined each of these terms must vanish which yields 3 linear differential equations for the 3 components of V. $V = V_1 + V_2 + V_3$, where V_1, V_2, V_3 are resp. of the order V_0, V_0^2, V_0^3.

Their solution for the panel of length $2L$ clamped at its ends is

$$V = V_{10}(\cos\xi + 1) + A_1^\blacksquare RV_{10}^2(\cos 2\xi - 1)/6\varrho +$$

$$+ [A_2 + (2A_1^2/3)] R^2 V_{10}^3 (\cos 3\xi + 1)/32\varrho^2 \qquad (6.10)$$

and

$$t = \frac{d(K/K_1)}{d(W_0/2j)^2} = -\frac{3R^2[A_2 - (10A_1^2/9)]}{4\varrho[(1/\eta_b R) + A_3]}. \qquad (6.11)$$

The slope of the load shortening curve at the bifurcation K_b is

$$\left[\frac{d(K/K_1)}{d(-\Delta L/L\varepsilon_1)}\right]_b = \{(\gamma_1 + \gamma_2)\times[1 - \gamma_1\gamma_2((1/\eta_b R) + A_3)^2 :$$

$$: 3(A_2 - (10A_1^2/9)]^{-1}\}_b. \qquad (5.11)$$

The load shortening curves in the three cases of imperfection (Table 1), together with the tangents at the bifurcation are given in Fig. 6. The equilibrium at K_b appears to be instable for $R < 1.1$.

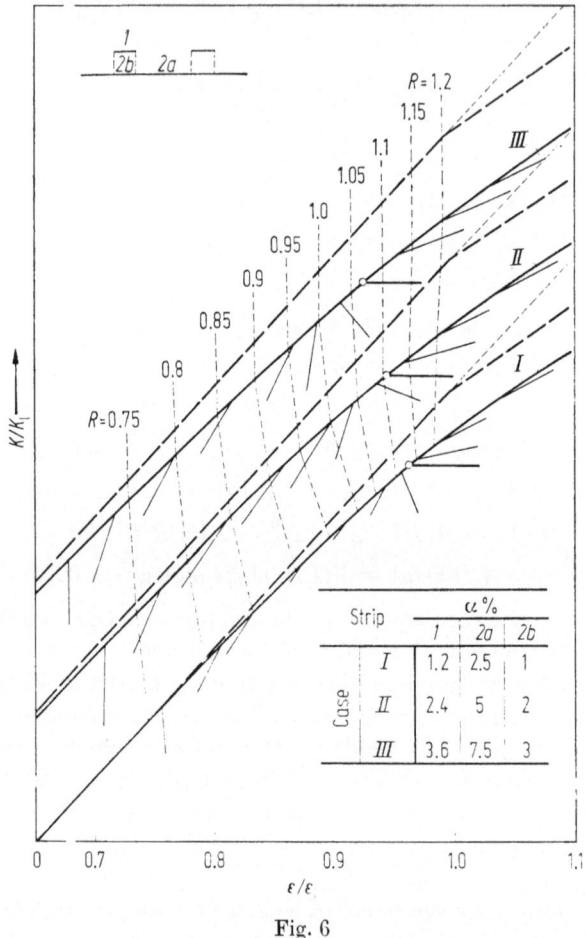

Fig. 6

In comparison to the "extreme" cases, [2] and Fig. 2, the peaks of the loadshortening curves at the bifurcation are much sharper with the "simplified panel". This implies that a panel without imperfection of the panel axis would fail explosively at K_b. Presumably it does not imply that the sensitivity to axis imperfection is very great. More significant for this effect is the negative slope $d(K/K_1)/d(W_0/2j)^2$. Figure 7 compares these slopes in case II ($x_{2a} = 0.05$) with the slopes in the "extreme" cases, both at $\alpha = 0.05$. It appears that the lower extreme (section 2) is insensitive to axis imperfection and that the maximal slope with the "simplified" panel is half of the maximal slope with the two-flange model. The slope is of the order of the slope with the two-flange model at $\alpha = 0.1$, where initial curvature adds little to the reduction of strength. The effect of axis imperfection is a matter of further investigation.

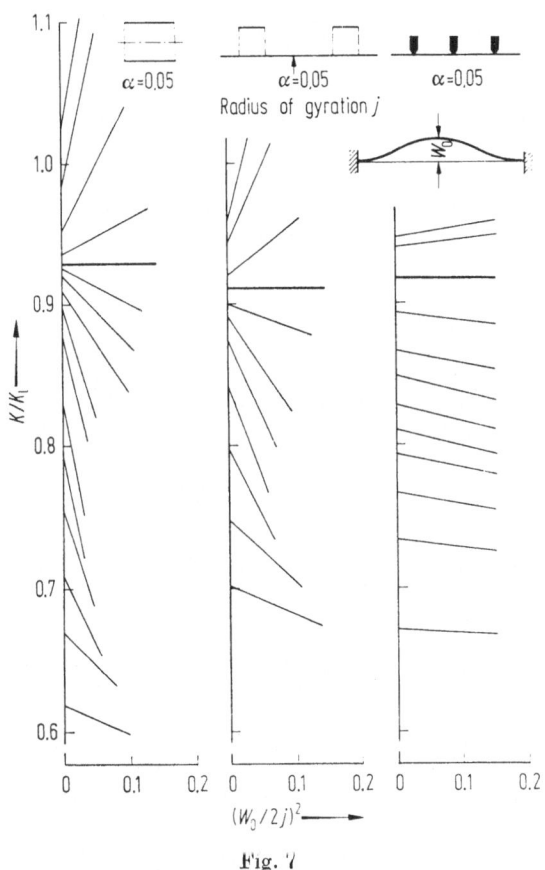

Fig. 7

7. Conclusions

Tophat stiffened panels are subject to mode interaction. With the geometric parameter $R > 1.35$ the imperfections: initial waviness of the plate strips and initial curvature of the panel axis have negligible effect. However the overal buckling load K_b depends on the post-buckling characteristics of the structure. This means that theoretical determination of K_b requires knowledge of the reduction of the bending stiffness of the panel in the post-buckling state. This problem is as yet unsolved.

When $R \gg 1$ the stiffeners are negligibly little affected by local buckling. Then the effective bending stiffness can be established with available methods. With $R < 1.35$ K_b is below both Euler and local buckling load; the reduction depending on the amount of initial waviness and to some, presumably less, extent on initial curvature. The maximal reduction occurring at $R = 1$ is in the order of 10%. Scatter of imperfections yields scatter of strength. Therefore the strength of structures with R close to unity cannot be predicted with great precision. A number of tests on identical specimen has te be carried out.

The strength of pin-ended panels is not representative for the strength of multibay panels. The position of the neutral plane in bending shifts with increasing compressive strain away from the side of the plate, causing bending before buckling. Tests should be carried out on clamped specimen.

References

1. Koiter, W. T.; Skaloud, M.: Interventions. In: Comportement Postcritique des plaques utilisées en construction métallique. Colloque intern. à l'Université de Liège, 1962, Mémoires de la Société Royale des Sciences de Liège, 5me série, tome VIII, fasc. 5, pp. 64—68, 103, 104.
2. van der Neut, A.: The Interaction of Local Buckling and Column Failure of Thin-Walled Compression Members. Delft University of Technology, Dept. of Aer. Eng., Rep. VTH 149 (1968). In: Proc. 12th Int. Congr. Appl. Mech. Berlin, Heidelberg, New York: Springer 1969, pp. 389—399.
3. Meyer, J. J.; van der Neut, A.: The Interaction of Local Buckling and Column Failure of Imperfect Thin-Walled Compression Members. Rep, VTH 160 (1970).
4. van der Neut, A.: The Sensitivity of Thin-Walled Compression Members to Column Axis Imperfection. Rep. VTH 172 (1972). In: Int. J. Solids and Structures 9, 999—1011, (1973).
5. Thompson, J. M. T.; Lewis, G. M.: On the Optimum Design of Thin-Walled Compression Members. J. Mech. Phys. Solids 20, 101—109 (1972).
6. Tvergaard, V.: Imperfection Sensitivity of a Wide Integrally Stiffened Panel under Compression. The Danish center for Appl. Math. and Mech., Techn. Univ. of Denmark, Rep. no. 19 (1971).
7. Tvergaard, V.: Influence of Post-Buckling on Optimum Design of Stiffened Panels. Rep. no. 35 (1972).
8. van der Neut, A.: The Local Instability of Compression Members Built up from Flat Plates. Rep. VTH 47(1952). In: C. B. Biezeno Anniversary Volume on Appl. Mech. 1953, pp. 174 to 197. Techn. Uitgeverij H. Stam.

An Alternative Approach to the Interaction between Local and Overall Buckling in Stiffened Panels

W. T. Koiter

Technische Hogeschool, Delft, Netherlands

M. Pignataro

Università di Roma, Rome, Italy

Abstract

The present analysis deals with stiffened panels in which plate buckling, possibly involving torsion of the stiffeners, dominates the local buckling mode. Previous work on the post-buckling behaviour of long flat plates is employed to write reasonably simple and still sufficiently accurate expressions for the displacements in the initial post-buckling range. The concept of slowly varying functions is employed in the derivation of a relatively simple approximate energy expression governing combined local and overall buckling of the panel. In addition to the ratio of the critical stresses for overall and local buckling $\lambda_E = \sigma_E/\sigma_{loc}$ the energy expression depends essentially on only one additional nondimensional parameter. Moreover, the latter parameter depends only on properties of the cross-section, and it does not appear to vary widely for the cross-sections likely to be encountered in practice.

The analysis is applied both to single-bay panels and to the more important multi-bay panels, continuous over a (large) number of equidistant transverse ribs. The nonlinear interaction proves to be far less severe than for the idealized model of a column with two heavy plate flanges, considered earlier by van der Neut and others. The high sensitivity to initial geometric imperfections, noted by Tvergaard, is confirmed for extremely small imperfections, but it is far less serious for imperfections of a magnitude likely to occur in practice.

1. Introduction

The explosive buckling phenomenon exhibited by many shell structures, and the associated extreme imperfection-sensitivity of these structures, are mostly due to a nonlinear interaction between linearly independent simultaneous or nearly simultaneous buckling modes [1—5]. This experience has led us to caution against a "naive" approach to the optimum design of structures liable to failure by buckling in several (linearly) independent modes, in which it is attempted to equalize the critical stresses for the individual buckling modes [6].

The first detailed investigations of the nonlinear interaction between overall buckling of a built-up column and local buckling of the plate flanges are due to van der Neut [7] and to Graves Smith [8]. Several authors have contributed further developments in recent years, both for built-up columns and stiffened panels [9—16], and current interest in this topic is also reflected in a number of papers at the present IUTAM Symposium [19—22]. Our attention has also been drawn to a much earlier experimental investigation on possible interactions between Euler buckling and torsional buckling of columns with an angle cross-section [17].[1]

It is the purpose of the present paper to give a simplified (approximate) analysis, in the elastic range, of the interaction problem for stiffened panels of the type investigated by Tvergaard [12, 14], where plate buckling of the skin between the stiffeners, possibly involving torsion of the stiffeners, dominates the local buckling mode. Previous work on the post-buckling behaviour of flat plates [18] is employed to write reasonably simple, and still sufficiently accurate expressions for the displacements in the initial post-buckling range. The concept of slowly varying functions, introduced in our simplified analysis in [10] of van der Neut's idealized column [7], is again employed in the derivation of a relatively simple approximate energy expression which governs combined local and overall buckling of the panel as well as the influence of initial imperfections. It appears that the behaviour of the panel depends on a single geometric parameter of the cross-section, in addition to the ratio $\lambda_E = \sigma_E/\sigma_{loc}$ of the Euler critical stress to the critical stress for local buckling.

2. Assumptions for Overall Buckling Mode

We shall consider both a single-bay panel, simply supported at its ends $x = 0$ and $x = L$, and a continuous multi-bay panel, supported on transverse ribs at regular longitudinal distances L. In both cases we shall assume the normal deflection in overall buckling to be described by a function $W(x)$, independent of the transverse coordinate y, thus ignoring transverse bending of the plate between adjacent stiffeners. For the axial displacement component we shall employ the Bernoulli assumption that plane cross-sections remain plane, thus ignoring the effect of shear lag. Finally, we shall assume that transverse direct stresses may also be ignored (except in plate bending), and the resulting critical stress is given by the famous Euler formula $\sigma_E = \pi^2 E i^2/L^2$, where the radius of gyration i of the cross-section is given by (Fig. 1).

$$(bh + A_s)\, i^2 = bhe^2 + A_s(e - e^*)^2 + I_s + [bh^3/12(1 - \nu^2)]. \tag{2.1}$$

---

[1] We are indebted to Dr. Lloyd H. Donnell for this early reference.

The neglection of transverse normal stresses introduces only a very small error in so-called "moderately wide" panels with a total width B not exceeding the bay length L, less than 1 percent. The ignoration of

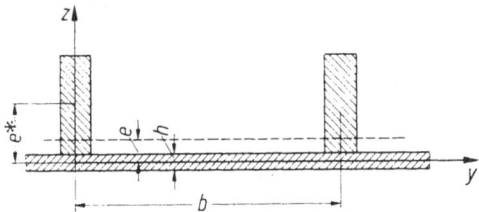

Fig. 1. Data of panel cross-section.

A_s Area of cross-section of stiffener; EI_s Flexural rigidity of stiffener with respect to horizontal axis; S_t Torsional rigidity of stiffener; EI_z^* "Effective" flexural rigidity of stiffener with respect to vertical axis; e^* Eccentricity of stiffener c.g.; e Eccentricity of c.g. of entire cross-section.

transverse plate bending and of shear lag may be more significant. It would be easily possible, by a simple refinement of our assumptions, to allow for the effects of transverse plate bending and shear lag, but such a refinement would be hardly relevant to our present purposes. Even in the rather extreme case of the configuration considered by Tvergaard [12] the error of the simple Euler formula is less than 5 percent.

3. Assumptions for Local Buckling Mode

The local buckling mode of a plate bay $0 < \eta < 1$, where $\eta = y/b$, may always be written in the form

$$w(x, y) = fg(\eta) \sin (\pi x/a), \tag{3.1}$$

where a is the axial half-wave length, $g(\eta)$ describes the transverse wave profile, and f is a constant amplitude factor. Except in the case of simply-supported longitudinal plate edges, the case of stiffeners with a negligible torsional rigidity, the exact profile $g(\eta)$ is somewhat inconvenient, involving both trigonometric and hyperbolic functions. We prefer a simple, suffiently accurate approximation [18]

$$g(\eta) = q \sin \pi\eta + (1 - q)(1 - \cos 3\pi\eta)/3 \quad \text{for} \quad 0 < \eta < (1/6),$$

$$g(\eta) = q \sin \pi\eta + (1 - q)[1 + 2\sin(3\pi\eta/2 - \pi/4)]/3$$

$$\text{for} \ (1/6) < \eta < (5/6), \tag{3.2}$$

$$g(\eta) = q \sin \pi\eta + (1 - q)(1 + \cos 3\pi\eta)/3 \quad \text{for} \quad (5/6) < \eta < 1,$$

symmetric with respect to the mid-point $\eta = 1/2$. The number q $(0 \leqq q \leqq 1)$ and the half-wave length a depend on the edge restraint, i.e. on he torsional rigidity of the stiffeners. The values $q = 1$ and $a = b$ correspond to simply-supported plate edges (stiffeners with negligible torsional rigidity), and the values $q = 0$, $a = 0.657b$ to clamped plate edges (stiffeners with a very large torsional rigidity). The accuracy of our approximate profile is evident from the largest error in the associated critical stress, which occur in the case of clamped edges ($q = 0$, $a = 0.657b$) and amounts to only 0.7 percent.

The inconvenience of the exact solutions is even more apparent for the in-plane displacements in the initial post-buckling range. Here again we employ the approximate expressions given in [18]

$$\left. \begin{aligned} u(x, y) &= -\varepsilon x - (\pi f^2/8a)\, g^2(\eta) \sin (2\pi x/a), \\ v(x, y) &= v_0(\eta) + (1/8b)\, f^2 g(\eta)\, \dot{g}(\eta) \cos (2\pi x/a), \end{aligned} \right\} \tag{3.3}$$

where a dot denotes differentiation with respect to the nondimensional coordinate $\eta = y/b$. The constant ε represents the additional effective end shortening due to buckling, and the function $v_0(\eta)$ is obtained by minimizing the total potential energy with respect to this function. The result is

$$v_0(\eta) = - (f^2/4b) \int_0^\eta \dot{g}^2(\eta)\, \mathrm{d}\eta - \nu\, (\pi^2 b/4a^2)\, f^2 \int_0^\eta g^2(\eta)\, \mathrm{d}\eta. \tag{3.4}$$

The accuracy of the expressions (3.3), (3.4) may be assessed from the effective stiffness of the plate against (additional) axial compression immediately after buckling. From a comparison with exact results it follows that the largest error occurs in the case of simply-supported edges ($q = 1$, $a = b$), and it is an overestimate by no more than 2.4 percent.

4. Assumptions for Combined Buckling Modes

As a consequence of the nonlinear interaction between the Euler buckling mode $W(x) = F \sin (\pi x/L)$ and the local buckling mode $w(x, y) = fg(\eta) \sin (\pi x/a)$, te amplitude f of the local mode will become a function $f(x)$ of the axial coordinate. This modulating factor $f(x)$, however, may be expected to be a *slowly varying* function in the sense of [10], with a half-wave length of order L, in contradistinction to the rapidly varying trigonometric factor $\sin (\pi x/a)$, with a half-wave length a which does not exceed the stiffener spacing b. Hence we assume the normal deflection of the plate bay $0 < y < b$ in the form

$$w(x, y) = W(x) + f(x)\, g(\eta) \sin (\pi x/a), \tag{4.1}$$

where both $W(x)$ and $f(x)$ are slowly varying functions.

The post-buckling in-plane displacement components (3.3), associated with the local buckling mode, are modified in correspondence with the normal deflection by replacing the constant factor f^2 by the square of the modulating function $f(x)$. The in-plane displacements associated with overall buckling are assumed conforming to the Bernoulli hypothesis and the neglection of transverse normal stresses. For reasons explained in [10], which reasons are also apparent from our energy evaluation in Appendix A, we need a further term in the axial displacement component which represents a coupling between overall and local buckling. Our final assumptions for the in-plane displacements are

$$u(x, y) = U(x) + eW'(x) - (\pi/8a) f^2(x) g^2(\eta) \sin (2\pi x/a) -$$
$$- W'(x) f(x) g(\eta) \sin (\pi x/a), \qquad (4.2)$$

$$v(x, y) = -\nu y [U'(x) + eW''(x) + (1/2) W'^2(x)] -$$
$$- (f^2(x)/4b) \int_0^\eta \dot{g}^2(\eta) \, d\eta - \nu (\pi^2 b/4a^2) f^2(x) \int_0^\eta g^2(\eta) \, d\eta$$
$$+ (f^2(x)/8b) g(\eta) \dot{g}(\eta) \cos (2\pi x/a), \qquad (4.3)$$

where $U(x)$ is the axial displacement of the centroid of the cross-section, and primes denote differentiations with respect to the axial coordinate x.

The normal deflection of the stiffeners is specified by $W(x)$, and the axial displacement of its centroid is $U(x) + (e - e^*) W'(x)$. We ignore deformations of the cross-section, and the rotation around the x-axis is $(1/b) f(x) \dot{g}(0) \sin (\pi x/a)$ for the stiffener at $y = 0$. Lateral bending of the stiffener is taken into account, if necessary.

5. The Energy Functional

The energy functional is defined as the increment in potential energy in a deflected configuration, in comparison with the unbuckled fundamental state at the same load. This energy functional has been evaluated in Appendix A for a single plate bay $0 < y < b$ and for a single stiffener, and this functional is representative for the entire panel.

It is convenient to reduce the functional obtained in the appendix to a functional in terms of dimensionless variables. We write

$$x = \xi L, \ W(x) = (C_3/C_5) L^2 \psi(\xi), f(x) = (C_3/C_6)^{1/2} \phi(\xi),$$
$$\sigma = \lambda \sigma_{\text{loc}} = \lambda (C_3/C_4), \lambda_E = \sigma_E/\sigma_{\text{loc}} = \pi^2 C_1 C_4/L^2 C_2 C_3, \qquad (5.1)$$

where the constants C_1 to C_6 are defined by (A 18)—(A 23). Henceforward *primes will denote differentiations with respect to the non-dimensional coordinate* $\xi = x/L$. The energy functional (A 17) now takes the form

$$P[\psi(\xi), \phi(\xi); \lambda] = K \int [A\{\psi''^2(\xi) - \pi^2 (\lambda/\lambda_E) \psi'^2(\xi)\} +$$
$$+ (1 - \lambda) \phi^2(\xi) + \psi''(\xi) \phi^2(\xi) + \phi^4(\xi)] \, d\xi, \qquad (5.2)$$

where the constants K and A are defined by

$$K = L(C_3^2/C_6), \quad A = C_1 C_6/C_5^2. \tag{5.3}$$

The most significant simplification achieved in the energy functional (5.2) is that it depends on only two nondimensional parameters of the panel, A and λ_E, in addition of course to the load factor λ. Moreover, the parameter A depends only on *properties of the cross-section*, and the bay length L appears only (implicitly) in λ_E.

In order to assess the range of numerical values of the parameter A we write

$$A = \frac{C_1 C_6}{C_5^2} = \frac{1}{4}\left[1 + \frac{bh}{A_s} + \frac{I_s}{bhe^2} + \frac{h^2}{12(1-\nu^2)\,e^2}\right] \times$$

$$\times \left[-\frac{bh}{bh+A_s} + \frac{\int_0^1 \{g^4(\eta) + (a^4/8(1-\nu^2)\,\pi^4 b^4)[-\dot{g}^2(\eta) + g(\eta)\ddot{g}(\eta)]^2\}\,d\eta}{\left\{\int_0^1 g^2(\eta)\,d\eta\right\}^2}\right]. \tag{5.4}$$

The *first factor* is 1.195 for the panel considered by Tvergaard [12]. It is smaller for a panel with heavier rectangular stiffeners, e.g. for $A_s = bh$ this factor is $5/6 = 0.833$. In the second factor the second fraction depends on the edge conditions of the plate. It ranges from a value 1.949 for clamped edges, obtained from [18], to a value 2.049 for simply-supported edges (where its exact value would actually be 2). For the Tvergaard panel with virtually clamped edges we obtain $A = 1.51$, for a similar panel with the heavier stiffeners $A_s = bh$ we have $A = 1.21$. *The practical range of values for A seems to be from 1.1 to 1.6.*

The dimensional factors for the overall and local buckling deflections in (5.1) are evaluated most easily for *clamped* plate edges. In this case we have

$$\frac{C_3}{C_5}\,L^2 = \frac{C_3}{C_4}\frac{C_4}{C_5}\,L^2 = \frac{\pi^2}{\lambda_E}\frac{i^2}{e},$$

$$(C_3/C_6) = 1.648h/[1 - 0.513bh/(bh + A_s)]^{1/2}. \tag{5.5}$$

In the presence of *overall imperfections* $W_0(x)$ and *local imperfections* $w_0(x, y) = f_0(x)\,g(\eta)\sin(\pi x/a)$ in the panel we write as in (5.1)

$$W_0(x) = (C_3/C_5)\,L^2\psi_0(\xi), \quad f_0(x) = (C_3/C_6)^{1/2}\,\phi_0(\xi). \tag{5.6}$$

The energy functional (5.2) for the perfect panel is now replaced by

$$P^*[\psi(\xi), \phi(\xi); \psi_0(\xi), \phi_0(\xi); \lambda] =$$

$$= K\int[A\{\psi''^2(\xi) - \pi^2\,(\lambda/\lambda_E)\,\psi'^2(\xi) - 2\pi^2\,(\lambda/\lambda_E)\,\psi_0'(\xi)\,\psi'(\xi)\} +$$

$$+ (1 - \lambda)\,\phi^2(\xi) - 2\lambda\phi_0(\xi)\,\phi(\xi) + \psi''(\xi)\,\phi^2(\xi) + \phi^4(\xi)]\,d\xi. \tag{5.7}$$

6. Stability of Perfect Panel

The unbuckled fundamental state is stable, if and only if the energy functional (5.2) is positive definite. The stability limit is $\lambda = \lambda_E$ for $\lambda_E \leqq 1$, and $\lambda = 1$ for $\lambda_E \geqq 1$. By the same argument as in [10] it is easily seen that equilibrium at the stability limit is strictly neutral for $\lambda_E < 1$, but that it is highly unstable at this limit in the case $\lambda_E = 1$, in view of the nonvanishing cubic term $\psi''(\xi)\,\phi^2(\xi)$ in (5.2) in the case of simultaneous buckling modes.

For $\lambda_E > 1$ the discussion of the stability of equilibrium at the stability limit $\lambda = 1$ requires some further analysis, and the results are different for a single-bay panel and for multi-bay panels. Minimizing (5.2) at $\lambda = 1$ with respect to $\phi(\xi)$, we obtain $\phi^2(\xi) = -(1/2)\,\psi''(\xi)$ whenever the right-hand member is positive, and $\phi(\xi) \equiv 0$ wherever $\psi''(\xi)$ is positive. The integrand of (5.2) reduces to $[A - (1/4)]\,\psi''^2(\xi) - (\pi^2 A/\lambda_E)\,\psi'^2(\xi)$ when $\psi''(\xi) < 0$, and to $A\{\psi''^2(\xi) - (\pi^2/\lambda_E)\,\psi'^2(\xi)\}$ where $\psi''(\xi) > 0$.

In a *single-bay* panel we may have $\psi''(\xi) < 0$ over the entire length, and the integral is minimized for $\psi(\xi) = \alpha \sin \pi\xi$ with $\alpha > 0$. The integral is positive for $\lambda_E > 4A/(4A - 1)$ and negative for $1 < \lambda_E < 4A/(4A - 1)$, and equilibrium at the stability limit $\lambda = 1$ is therefore stable or unstable accordingly as λ_E is larger or smaller than the critical value $4A/(4A - 1)$.

In *multi-bay* panels intervals with negative values of $\psi''(\xi)$ must alternate with intervals where $\psi''(\xi)$ is positive. Taking again $\psi(\xi) = \alpha \sin \pi\xi$, now as an approximation which is entirely adequate for our purposes, we find that the equilibrium at the stability limit $\lambda = 1$ is stable or unstable accordingly as λ_E is larger or smaller than the critical value $8A/(8A - 1)$.

Local plate buckling is in itself no limitation on the load carrying capacity of a structure. Stable post-buckling equilibrium configurations may therefore conceivably exist for stiffened panels at $\lambda > 1$, provided λ_E has at least the value which ensures stability at $\lambda = 1$. It is easily seen, both by inspection of the panel and from the functional (5.2) for simply-supported edges at $\xi = 0$ and $\xi = 1$, that such a post-buckling solution must involve the overall mode $\psi(\xi)$ in the case of a single-bay panel. In the more important case of *multi-bay* panels, however, zero overall deflections $\psi(\xi) \equiv 0$ represent a solution of the equilibrium equations, if the modulating function $\phi(\xi)$ for the local buckling mode has a constant value, independent of ξ. This constant value $\pm \sqrt{(\lambda - 1)/2}$ is obtained by minimizing (5.2) with respect to $\phi(\xi)$ at a zero value $\psi(\xi)$.

In order to investigate the stability of this post-buckling solution for multi-bay panels we consider the associated second variation of the energy functional (5.2)

$$\delta^2 P = K \int [A\{\delta\psi''^2(\xi) - \pi^2 (\lambda/\lambda_E) \delta\psi'^2(\xi)\} + 2(\lambda - 1) \delta\phi^2(\xi) \pm$$

$$\pm 2\sqrt{(\lambda - 1)/2} \, \delta\psi''(\xi) \, \delta\phi(\xi)] \, d\xi. \tag{6.1}$$

The secondary buckling mode is described by $\delta\psi(\xi) = \delta\alpha \sin \pi\xi$, $\delta\phi(\xi) = \delta\beta \sin \pi\xi$, the critical load factor for secondary buckling is $(1 - 1/4A) \lambda_E$, and the post-buckling solution $\phi(\xi) = \pm\sqrt{(\lambda - 1)/2}$ is stable in the range $1 < \lambda < (1 - 1/4A) \lambda_E$ which is evidently only significant for $\lambda_E > 4A/(4A - 1)$.

The results discussed above are summarized in Fig. 2. The most important conclusion to be drawn at this stage is that the *unstable local buckling range* $(1 < \lambda_E < 1.2$ for a single-bay panel and $1 < \lambda_E < 1.091$ for a multi-bay panel, both for $A = 1.5$) is *much smaller* than for van der Neut's idealized column [1, 10]. Hence we should also expect a reduced sensitivity to initial imperfections.

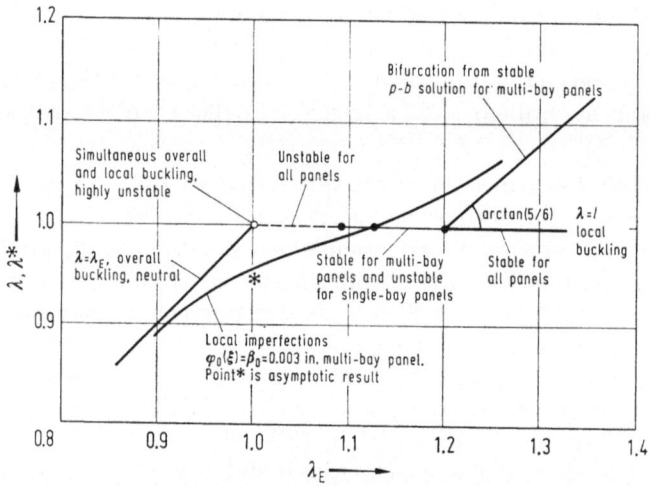

Fig. 2. Results for panel with $A = 1.5$.

7. The Influence of Local Imperfections in a Multi-Bay Panel

We shall consider only the simplest case of a constant value $\phi_0(\xi) = \beta_0$ of the local imperfections in a multi-bay panel. It follows immediately from the functional (5.7), where $\psi_0(\xi) \equiv 0$ and $\phi_0(\xi) = \beta_0$, that a solution of the equations of equilibrium is $\psi(\xi) \equiv 0$. Moreover, the solution for $\phi(\xi)$ is a constant β, to be solved for from the equation

$$2\beta^3 + (1 - \lambda) \beta - \lambda\beta_0 = 0. \tag{7.1}$$

The stability of this solution is governed by the second variation of (5.7) associated with the solution of the equations of equilibrium

$$\delta^2 P^* = K \int [A\{\delta\psi''^2(\xi) - \pi^2 (\lambda/\lambda_{\mathrm{E}}) \,\delta\psi'^2(\xi)\} +$$

$$+ (1 - \lambda + 6\beta^2) \,\delta\phi^2(\xi) + 2\beta \,\delta\psi''(\xi) \,\delta\phi(\xi)] \,d\xi. \quad (7.2)$$

The critical mode is again $\delta\psi(\xi) = \delta\alpha \sin \pi\xi$, $\delta\phi(\xi) = \delta\beta \sin \pi\xi$, and the second variation (7.2) reduces to

$$\delta^2 P^* = (Kl/2) [A\pi^4[1 - (\lambda/\lambda_{\mathrm{E}})] \,\delta\alpha^2 + (1 - \lambda + 6\beta^2) \,\delta\beta^2 - 2\pi^2\beta \,\delta\alpha \,\delta\beta], \quad (7.3)$$

where l is the nondimensional length of the panel. The critical load factor is attained when the determinant of the quadratic form (7.3) vanishes. Combining this condition with the equation of equilibrium (7.1), we arrive at the equation for the critical load factor λ^*

$$(1 - \lambda^*) \frac{1 - 4A(1 - \lambda^*/\lambda_{\mathrm{E}})}{1 - 6A(1 - \lambda^*/\lambda_{\mathrm{E}})} \left\{ \frac{A(1 - \lambda^*) \,(1 - \lambda^*/\lambda_{\mathrm{E}})}{1 - 6A(1 - \lambda^*/\lambda_{\mathrm{E}})} \right\}^{1/2} = \lambda^*\beta_0. \quad (7.4)$$

For the Tvergaard panel [12] with $A = 1.5$ we have plotted in Fig. 2 the curve of λ^* as a function of λ_{E} for a value $\beta_0 = 0.003$ of the imperfection parameter which corresponds to an amplitude equal to 0.6 percent of the plate thickness h for this panel. We have also indicated the value of λ^* according to the asymptotic formula at $\lambda_{\mathrm{E}} = 1$

$$\lambda^* = 1 - A^{-1/4}\beta_0^{1/2}, \quad (7.5)$$

holding for sufficiently small values of β_0, and we note that this formula is already considerably in error for the very small value of β_0 under consideration.

We note two further significant properties of the curves of λ^* as a function of λ_{E} with β_0 as a parameter. All these curves have the point $\lambda_{\mathrm{E}} = 6A/(6A - 1)$, $\lambda^* = 1$ in common. Even more important is the fact that for $\lambda_{\mathrm{E}} < 6A/(6A - 1)$ the critical load factor λ^* is *bounded below* by the inequality

$$\lambda^* \geq [1 - (6A)^{-1}] \lambda_{\mathrm{E}} = 0.89\lambda_{\mathrm{E}} \quad \text{for} \quad A = 1.5. \quad (7.6)$$

This confirms our earlier conjecture, at least in the case of local imperfections, that a stiffened panel is considerably less sensitive to imperfections than van der Neut's idealized built-up column[1].

---

[1] It is perhaps of interest to note that the lower bound for λ^* for van der Neut's column is 0.508 at $\lambda_{\mathrm{E}} = 1$.

8. The Influence of Overall Imperfections

In the absence of local imperfections it is evident by inspection of (5.7) that $\phi(\xi) \equiv 0$ is a solution of the equations of equilibrium for sufficiently small values of the load factor λ. For overall imperfections $\psi_0(\xi) = \alpha_0 \sin \pi\xi$ the solution of the equation of equilibrium is $\psi(\xi) = \alpha \sin \pi\xi$ with $\alpha = \lambda\alpha_0/(\lambda_E - \lambda)$. The stability condition for this solution at $\lambda < \lambda_E$ reads

$$\int [1 - \lambda - \pi^2\alpha \sin \pi\xi] \, \delta\phi^2(\xi) \, d\xi > 0, \tag{8.1}$$

and the stability limit is reached when the integrand vanishes at $\xi = 1/2$. The critical load factor λ^* is thus obtained from the equation

$$(1 - \lambda^*) \, (\lambda_E - \lambda^*) = \pi^2\alpha_0\lambda^*. \tag{8.2}$$

The imperfection sensitivity at $\lambda_E = 1$ is somewhat more severe than in Tvergaard's analysis [12] because he requires $\delta\phi(\xi)$ to be a constant, and this requirement implies a replacement of the factor π^2 in (8.2) by 2π in his case of a single-bay panel. We note that (8.2) holds for both single-bay and multi-bay panels.

We have not confirmed Tvergaard's observation that the post-buckling path emanating from the bifurcation point λ^* obtained from (8.2) would imply a drop in load. On the contrary, we obtain in most cases a *rising post-buckling path*, either monotonically up to an asymptotic value or rising to a maximum load and subsequently descending to the asymptotic value.

The analysis is quite similar for single-bay and multi-bay panels. We minimize the functional (5.7), in which $\phi_0(\xi) \equiv 0$, first with respect to $\phi(\xi)$, and we obtain

$$\phi^2(\xi) = [\lambda - 1 - \psi''(\xi)]/2 \tag{8.3}$$

if the right-hand member is positive, and $\phi(\xi)$ is zero elsewhere. Substituting from the minimizing value into (5.7), we obtain the reduced energy functional

$$K \int [A\{\psi''^2(\xi) \quad - \pi^2 \, (\lambda/\lambda_E) \, \psi'^2(\xi) - 2\pi^2 \, (\lambda/\lambda_E) \, \psi_0'(\xi) \, \psi'(\xi)\} - \underline{(1/4) \, \{\lambda - 1 - \psi''(\xi)\}^2}] \, d\xi, \tag{8.4}$$

where the dotted underlining of the last term indicates that it should be included, if and only if the right-hand member of (8.3) is positive at the value of ξ under consideration.

We omit details of a complete analysis of the functional (8.4) because it is rather laborious, and the most significant results may already be obtained from the asymptotic values of the load factor. In a *single-bay panel* the asymptotic case occurs when the dotted term in

(8.4) is present over (nearly) the entire nondimensional length, equal to unity, and it is easily seen that the asymptotic value of he load factor is

$$\lambda_{as} = \lambda_E (4A - 1)/4A = 5\lambda_E/6 \tag{8.5}$$

for the value $A = 1.5$ of Tvergaard's panel.

In the case of a *multi-bay panel* we have to make a distinction between intervals of different sign of $\psi''(\xi)$. For our purposes it is adequate to make the approximation $\psi(\xi) = \alpha \sin \pi\xi$, and the asymptotic value of the load factor is

$$\lambda_{as} = \lambda_E (8A - 1)/8A = 11\lambda_E/12 \tag{8.6}$$

for the value $A = 1.5$ of Tvergaard's panel.

9. Conclusions

The main conclusion to be drawn from our investigation of stiffened panels of the type depicted in Fig. 1 tends to be reassuring to the designer. We have confirmed he extreme sensitivity of such panels to both local and overall imperfections, but we have also found that the classical asymptotic formulae for imperfection-sensitivity grossly overestimate this effect, except for extremely small imperfections.

In particular we have established a *lower bound* for the actual critical load factor λ^* of a multi-bay panel in the presence of *local imperfections* in the form (7.6) which is also readily verified from the shape of the curve in Fig. 2 for $\lambda^* < 1$. In the case of *overall imperfections* our formula (8.2) is "exact", but the post-buckling behaviour is similar to the behaviour of a normal solid Euler-type column with imperfections, and the load factor approaches the asymptotic value (8.5) or (8.6) for a single-bay and a multi-bay panel respectively. Time and space do not permit a detailed discussion in this paper of the influence of combined overall and local imperfections, but some calculations, to be reported elsewhere, have already confirmed our expectation that they will not lead to any spectacular change in the findings reported here.

Little information is available about the magnitude of the imperfections in actual stiffened panels. In the absence of such information we suggest for design purposes to employ the broken line in Fig. 2 with a *cut-off* of the corner near $\lambda_E = 1$. A cut-off which is likely to be conservative is a straight line from the poine $\lambda_E = \lambda = 0.8$ to the point $\lambda_E = 6A/(6A - 1)$, $\lambda = 1$ in the case of a multi-bay panel, and a straight line from the point $\lambda_E = \lambda = 0.8$ to the point $\lambda_E = 4A/(4A-1)$, $\lambda = 1$ in the case of a single-bay panel. The maximum penalty to be paid as a consequence of our ignorance occurs at $\lambda_E = 1$, and it amounts to 8 percent for a multi-bay panel and 10 percent for a single bay panel, both for $A = 1.5$.

Appendix A. Evaluation of Energy Functional

The *membrane strains* in the middle plane of the plate are evaluated from the displacements (4.1), (4.2) and (4.3)

$$\boxed{\gamma_{xx} = U'(x) + eW''(x) + (1/2)\, W'^2(x) + (\pi^2/4a^2)\, f^2(x)\, g^2(\eta)} \; +$$
$$+\, (\pi/4a)\, f(x)\, f'(x)\, g^2(\eta)\, \sin\,(2\pi x/a) - W''(x)\, f(x)\, g(\eta)\, \sin\,(\pi x/a) +$$
$$+\, (1/2)\, f'^2(x)\, g^2(\eta)\, \sin^2\,(\pi x/a)\,, \tag{A 1}$$

$$\boxed{\begin{aligned} \gamma_{yy} = &-\nu\,[U'(x) + eW''(x) + (1/2)\, W'^2(x) + \\ &+ (\pi^2/4a^2)\, f^2(x)\, g^2(\eta)]\, + \\ &+ (1/8b^2)\, f^2(x)\, [-\dot{g}^2(\eta) + g(\eta)\, \ddot{g}(\eta)]\, \cos\,(2\pi x/a) \end{aligned}}\,, \tag{A 2}$$

$$\boxed{2\gamma_{xy} = 0} - \nu y[U''(x) + eW'''(x) + W'(x)\, W''(x)] -$$
$$-\, \nu\, (\pi^2 b/2a^2)\, f(x)\, f'(x) \int\limits_0^\eta g^2(\eta)\, \mathrm{d}\eta + (1/2b)\, f(x) f'(x) \int\limits_0^\eta g(\eta)\, \ddot{g}(\eta)\, \mathrm{d}\eta -$$
$$-\, (1/4b)\, f(x)\, f'(x)\, g(\eta)\, \dot{g}(\eta)\, \cos\,(2\pi x/a)\,. \tag{A 3}$$

In this evaluation we have applied an integration by parts to the integral

$$\int\limits_0^\eta \dot{g}^2(\eta)\, \mathrm{d}\eta = g(\eta)\, \dot{g}(\eta) - \int\limits_0^\eta g(\eta)\, \ddot{g}(\eta)\, \mathrm{d}\eta\,. \tag{A 4}$$

We have framed the significant terms in the membrane strains (A1)—(A3), and we shall neglect all other terms in our evaluation of the stretching energy in the plate. The neglection of the first term in (A3) implies a relative error estimated at $\nu^2\pi^2 B^2/2(1+\nu)\, 12L^2$, i.e. less than 3 percent for a "moderately wide" panel with $B/L \leqq 1$. The relative error involved by the neglection of the other unframed terms in our energy evaluation is of the order b^2/L^2, and we shall ignore similar terms arising from the energy calculation of the framed contributions to the strains. The resulting *stretching energy for a single plate bay* $0 < y < b$ is

$$Ehb/2 \int [\{U'(x) + eW''(x) + (1/2)\, W'^2(x)\}^2 +$$
$$+\, (\pi^2/2a^2)\, f^2(x)\, \{U'(x) + eW''(x) + (1/2)\, W'^2(x)\} \int\limits_0^1 g^2(\eta)\, \mathrm{d}\eta +$$
$$+\, (\pi^4/16a^4)\, f^4(x) \int\limits_0^1 \{g^4(\eta) +$$
$$+\, [a^4/8(1-\nu^2)\, \pi^4 b^4]\, [-\dot{g}^2(\eta) + g(\eta)\, \ddot{g}(\eta)]^2\}\, \mathrm{d}\eta]\, \mathrm{d}x\,. \tag{A 5}$$

The *curvatures* of the plate, associated with the normal deflection (4.1) are given by

$$\boxed{\varkappa_{xx} = W''(x) - (\pi^2/a^2)\, f(x)\, g(\eta) \sin (\pi x/a)} \; +$$
$$+ (2\pi/a)\, f'(x)\, g(\eta) \cos (\pi x/a) + f''(x)\, g(\eta) \sin (\pi x/a), \qquad \text{(A 6)}$$

$$\boxed{\varkappa_{yy} = (1/b^2)\, f(x)\, \ddot{g}(\eta) \sin (\pi x/a)}, \qquad \text{(A 7)}$$

$$\boxed{\varkappa_{xy} = (\pi/ab)\, f(x)\, \dot{g}(\eta) \cos (\pi x/a)} + (1/b)\, f'(x)\, \dot{g}(\eta) \sin (\pi x/a), \quad \text{(A 8)}$$

where the unframed contributions will again be ignored for the same reasons as before. The resulting *bending energy for a single plate bay* $0 < y < b$ is

$$\frac{Ebh^3}{24(1 - \nu^2)} \int \left[W''^2(x) + \frac{\pi^4}{2a^4} f^2(x) \int_0^1 \left\{ g^2(\eta) + \frac{2a^2}{\pi^2 b^2} \dot{g}^2(\eta) + \right. \right.$$
$$\left. \left. + \frac{a^4}{\pi^4 b^4} \ddot{g}^2(\eta) \right\} \, d\eta \right] dx. \qquad \text{(A 9)}$$

The *stretching and bending energy of a stiffener*, due to the overall buckling mode, is

$$EA_s/2 \int [U'(x) + (e - e^*)\, W''(x) + (1/2)\, W'^2(x)]^2 \, dx +$$
$$+ EI_s/2 \int W''^2(x) \, dx. \qquad \text{(A 10)}$$

The *specific twist* of a stiffener, due to the local buckling mode, is given by

$$\boxed{\omega = (\pi/ab)\, f(x)\, \dot{g}(0) \cos (\pi x/a)} + (1/b)\, f'(x)\, \dot{g}(0) \sin (\pi x/a), \quad \text{(A 11)}$$

where the unframed term is again ignored. The *torsion energy in a stiffener* is thus given by

$$\int (\pi^2/4a^2 b^2)\, S_t \dot{g}^2(0)\, f^2(x) \, dx. \qquad \text{(A 12)}$$

The *lateral curvature* of the stiffener is

$$\boxed{\varkappa_{\text{lat}} = - (\pi^2 e^*/a^2 b)\, f(x)\, \dot{g}(0) \sin (\pi x/a)} \; +$$
$$+ (2\pi \, e^*/ab)\, f'(x)\, \dot{g}(0) \cos (\pi x/a) + (e^*/b)\, f''(x)\, \dot{g}(0) \sin (\pi x/a). \quad \text{(A 13)}$$

Let EI_z^* denote the *effective* flexural rigidity of a stiffener for lateral bending associated with torsion. The *lateral bending energy in a stiffener* is then given by

$$\int (\pi^4 e^{*2}/4a^4 b^2)\, EI_z^*\, \dot{g}(0)\, f^2(x) \, dx. \qquad \text{(A 14)}$$

The *energy contribution, due to the presence of a uniform compressive stress σ in the unbuckled fundamental state,* involves the squares of the slopes, including the lateral slope in the stiffener. With the same approximations as before, the result is for a single plate bay and one stiffener

$$- \int \left[(1/2)\, \sigma(bh + A_s)\, W'^2(x) + \right.$$

$$\left. + \left\{ (\sigma bh \pi^2/4a^2) \int_0^1 g^2(\eta)\, d\eta + (\sigma/4)\,(I_s + A_s e^{*2})\,(\pi^2/a^2b^2)\, \dot{g}^2(0) \right\} f^2(x) \right] dx. \quad (A\ 15)$$

The *total energy functional for one plate bay and one stiffener* is given by the sum of (A5), (A9), (A10), (A12), (A14) and (A15). For an imperfect panel with overall imperfections $W_0(x)$, and local imperfections $f_0(x)$ we have to add a further term, obtained from (A15) by the replacement of $W'^2(x)$ by $2W_0'(x)\, W'(x)$ and of $f^2(x)$ by $2f_0(x)\,f(x)$. Stationary values of the energy functional with respect to $U(x)$, $W(x)$ and $f(x)$ characterize the equilibrium configurations, and these are stable, if and only if the stationary value of the energy is a proper relative minimum. Since $U(x)$ occurs only in (A5) and (A10), the minimalization with respect to this function is easily achieved, with the result

$$U'(x) = -\,(1/2)\, W'^2(x) - \frac{bh}{bh + A_s}\, \frac{\pi^2}{4a^2} f^2(x) \int_0^1 g^2(\eta)\, d\eta, \quad (A\ 16)$$

and a reduced energy functional is obtained in the form

$$\int [C_1 W''^2(x) - \sigma C_2 W'^2(x) + (C_3 - \sigma C_4)\, f^2(x) +$$

$$+ C_5 W''(x)\, f^2(x) + C_6 f^4(x)]\, dx, \quad (A\ 17)$$

in which the coefficients are defined by

$$C_1 = (E/2)\,\{bhe^2 + A_s(e - e^*)^2 + I_s + [bh^3/12(1 - \nu^2)]\}, \quad (A\ 18)$$

$$C_2 = (bh + A_s)/2, \quad (A\ 19)$$

$$C_3 = \frac{Ebh^3}{24(1 - \nu^2)}\, \frac{\pi^4}{2a^4} \int_0^1 \left[g^2(\eta) + \frac{2a^2}{\pi^2 b^2} \dot{g}^2(\eta) + \frac{a^4}{\pi^4 b^4} \ddot{g}^2(\eta) \right] d\eta +$$

$$+ \frac{\pi^2}{4a^2b^2} \left[S_t + \frac{\pi^2 e^{*2}}{a^2} EI_z^* \right] \dot{g}^2(0), \quad (A\ 20)$$

$$C_4 = (bh\pi^2/4a^2) \int_0^1 g^2(\eta)\, d\eta + (\pi^2/4a^2b^2)\,(I_s + A_s e^{*2})\, \dot{g}^2(0), \quad (A\ 21)$$

$$C_5 = (Ebh\,\pi^2 e/4a^2) \int_0^1 g^2(\eta)\, d\eta, \quad (A\ 22)$$

$$C_6 = Ebh \, \frac{\pi^4}{32a^4} \left[\int_0^1 \left\{ g^4(\eta) + \frac{a^4}{8(1-\nu^2)\,\pi^4 b^4} \, [-\dot{g}^2(\eta) + g(\eta)\,\ddot{g}(\eta)]^2 \right\} \mathrm{d}\eta - \right.$$

$$\left. - \frac{bh}{bh+A_s} \left\{ \int_0^1 g^2(\eta)\, \mathrm{d}\eta \right\}^2 \right]. \tag{A 23}$$

References

1. Koiter, W. T.: Over de stabiliteit van het elastisch evenwicht. Thesis Delft. Amsterdam: H. J. Paris 1945. English translations published as NASA TT F-10, 833 (1967) and AFFDL Report TR 70-25 (1970).
2. Koiter, W. T.: Elastic Stability and Post-Buckling Behaviour. Proc. Symp. Nonlinear Problems, pp. 257—275. University of Wisconsin Press (1963).
3. Hutchinson, J. W.: Imperfection Sensitivity of Externally Pressurized Spherical Shells. J. Appl. Mech. **34**, 49—55 (1967).
4. Koiter, W. T.: The Nonlinear Buckling Problem of a Complete Spherical Shell under Uniform External Pressure. Proc. Kon. Ned. Ak. Wet. **B 72**, 40—123 (1969).
5. Hutchinson, J. W.; Koiter, W. T.: Postbuckling Theory. Appl. Mech. Revs. **23**, 1353—1366 (1970).
6. Campus, F.; Massonnet, C.: Colloque sur le comportement postcritique des plaques utilisées en construction métallique. Interventions de W. T. Koiter et M. Skaloud, pp. 64—68, 103, 104. Mémoires de la Société Royale des Sciences de Liège, 5-me série, tome VIII, fasc. 5 (1963).
7. van der Neut, A.: The Interaction of Local Buckling and Column Failure of Thin-Walled Compression Members. Proc. 12th Int. Congr. Appl. Mech. Stanford University (1968). Berlin, Heidelberg, New York: Springer 1969, pp. 389—399.
8. Graves Smith, T. R.: The Ultimate Strength of Locally Buckled Columns of Arbitrary Length. In: Thin-Walled Steel Constructions. London: Crosby Lockwood 1967.
9. Meyer, J. J.; van der Neut, A.: The Interaction of Local Buckling and Column Failure of Imperfect Thin-Walled Compression Members. VTH Report 160, Delft (1970).
10. Koiter, W. T.; Kuiken, G. D. C.: The Interaction Between Local Buckling and Overall Buckling on the Behaviour of Built-up Columns. WTHD Report 23, Delft (1971).
11. Thompson, J. M. T.; Lewis, G. M.: On the Optimum Design of Thin-Walled Compression Members. J. Mech. Phys. Solids **20**, 101—109 (1972).
12. Tvergaard, V.: Imperfection-Sensitivity of a Wide Integrally Stiffened Panel under Compression. Int. J. Solids and Structures **9**, 177—192 (1973).
13. van der Neut, A.: The Sensitivity of Thin-Walled Compression Members to Column Axis Imperfection. Int. J. Solids and Structures **9**, 999—1011 (1973).
14. Tvergaard, V.: Influence of Post-Buckling Behaviour on Optimum Design of Stiffened Panels. Int. J. Solids and Structures **9**, 1519—1534 (1973).
15. Thompson, J. M. T.; Hunt, G. W.: A General Theory of Elastic Stability. John Wiley 1973.
16. Besseling, J. F.: Post-Buckling and Non-Linear Analysis by the Finite Element Method as a Supplement to a Linear Analysis. Report of Laboratory of Engineering Mechanics, Delft (1974).
17. Bridget, F. J.; Jerome, C. C.; Vosseler, A. B.: Some New Experiments on Buckling of Thin-Wall Construction. ASME Transactions **56**, 569—578 (1934).

18. Koiter, W. T.: The Effective Width at Stresses far in Excess of the Critical Stress for Various Edge Restraints (in Dutch with an English summary). Report S287, NLR, Amsterdam (1943).
19. Maquoi, R.; Massonnet, C.: Interaction Between Local Plate Buckling and Overall Buckling in Thin-Walled Compression Members. Theories and Experiment.
20. van der Neut, A.: Mode Interaction with Stiffened Panels.
21. Tvergaard, V.; Needleman, A.: Mode Interaction in an Eccentrically Stiffened Elastic-Plastic Panel under Compression.
22. Thompson, J. M. T.; Tulk, J. D.; Walker, A. C.: An Experimental Study of Imperfection-Sensitivity in the Interactive Buckling of Stiffened Plates.

An Experimental Study of Imperfection-Sensitivity in the Interactive Buckling of Stiffened Plates

J. M. T. Thompson, J. D. Tulk, A. C. Walker

University College London, London, England

Summary

The results of carefully-controlled small-scale experiments on the elastic buckling of stiffened plates are presented. Overall and local imperfections have been systematically varied to provide detailed curves of imperfection-sensitivity for geometries spanning the range of interaction between Euler and local plate buckling. The interaction of a third mode associated with the torsional buckling of relatively thin stiffeners is also examined.

The analysis of a Shanley model is used to suggest a possible engineering approach to the design problem. This approach is based on a proposed application of the well-known Perry formula in which the material yield stress is simply replaced by the relevant local buckling stress, together with the predicted asymptotic approach to a reduced modulus load.

1. Introduction

Engineers have long been aware that the failure of certain structural forms may involve a combination of local and overall buckling. Often in the past these have been designed on the basis of empirical formulations which ensured that local buckling was eliminated. Unfortunately, since optimization often seems to call for the simultaneity of failure loads, this approach is unduly wasteful and must be replaced by a more rational procedure if the commercial viability of the structural forms is to be fully exploited. Designers will need a greater understanding of their physical behaviour, and the expected imperfection-sensitivity must be assessed to allow the placing of intelligent fabrication tolerances.

Significant interactive effects in the failure of plated structures were demonstrated analytically by van der Neut [1], and the relevance to optimization was examined by Thompson and Lewis [2]. The mechanics of these phenomena were explored by Koiter and Kuiken [3] and Thompson and Supple [4].

In this paper we present some results from an experimental investigation of the interactive buckling of uniformly-compressed eccentrically-stiffened plates which have been shown to be imperfection-

sensitive over a range of imperfection magnitudes by the detailed analytical study of Tvergaard [5, 6].

2. Experimental Procedure

The stiffened plate configuration with which we have worked is shown
in Fig. 1. It consists of a thin plate with a number of longitudinal
rectangular stiffeners attached to one side. The ends of the panel

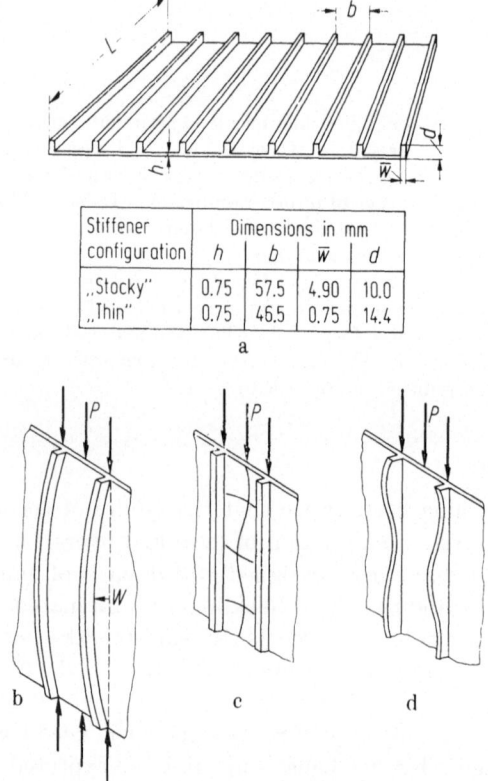

Stiffener	Dimensions in mm			
configuration	h	b	$\bar{w}$	d
„Stocky"	0.75	57.5	4.90	10.0
„Thin"	0.75	46.5	0.75	14.4

a

b   c   d

Fig. 1a—d. The panel and its buckling modes. a) Panel geometry; b) Euler
mode; c) Plate mode; d) Stiffener mode.

assembly at which the loads are applied are simply supported, while the
longitudinal edges are free. Thus the overall buckling mode is essentially
that of a broad Euler column. A second mode involves the local buckling of the skin between adjacent stiffeners, and the interaction between
this and the Euler mode for the case of stocky stiffeners form the major
part of our study.

The interaction between these two modes and a third mode associated with the local torsional buckling of thin stiffeners is briefly examined.

The models were easily and economically made from an epoxy plastic and the low modulus of elasticity allowed them to be tested on a simple hand-operated rig (Fig. 9). This material has the further advantage of remaining elastic up to large strains so that yielding was avoided and the models could be repeatedly tested to their maximum load.

The local critical stress of the panels is determined by the thickness of the plate and the spacing and torsional rigidity of the stiffeners, while the overall critical stress is dependent on the overall flexural stiffness of the panels and on the overall length. Thus by successively reducing the length of a fabricated specimen it was possible to vary systematically the ratio of these two critical stresses without altering any of the other dimensions of the configuration. This ensured that the results of successive tests would be directly comparable, and the lengths employed ranged from about $L = 200$ to about $L = 400$ mm.

The other parameters varied during the tests were the imperfections in the overall and local modes. The former were varied continuously over a wide range by off-setting the applied load from the plane of the neutral axis.

The local imperfections in the specimens could be altered by taking advantage of another property of the epoxy plastic. It is well known that plastics of this type may suffer from a considerable amount of creep at elevated temperatures. Thus, if a panel was loaded to the point where local buckling occurred and then heated, it was possible to induce permanent local deformations of a geometric form precisely corresponding to the local buckling mode. This was done, and small amounts of overall bending produced in the operation were eliminated by making a compensating adjustment to the overall loading eccentricity.

3. A Shanley Model

Analytical models are invaluable in delineating the nonlinear phenomena of elastic stability [7, 8] and it will be useful at this stage to introduce a simple Shanley model as a guide to the later experimental results. This has a single degree of freedom and is shown in Fig. 2, the bi-linear spring representing the plate post-buckling.

With the notation of the figure adequate linearized equations are

$$cW = l(e_2 - e_1),$$
$$P = F_1 + F_2,$$
$$cF_2 = [(c/2) + W] P.$$

If the local buckling load P_{L}, which manifests itself in the response of the right-hand spring, is relatively high, then the column will buckle in a simple Euler fashion at

$$P_{\mathrm{E}} = (c^2/2l)\,E\,.$$

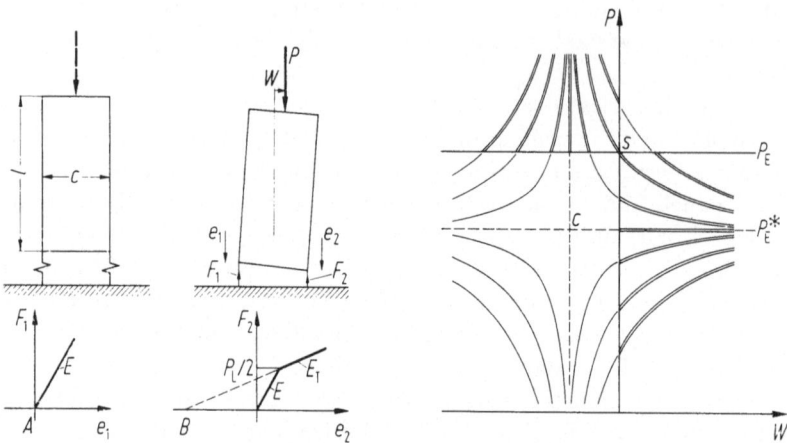

Fig. 2. The Shanley column showing the unloaded and loaded models, and the composite load-deflection picture.

If, however, $P_{\mathrm{L}} < P_{\mathrm{E}}$, continuous deflections of the perfect model will begin at $P = P_{\mathrm{L}}$ and as they grow they will follow the equilibrium paths of a *hypothetical reduced modulus column* which we can imagine to have been loaded from the stress-strain states A and B of the figure. This hypothetical column clearly has an initial deflection W_0 in its unloaded state, and has moreover a second effective imperfection because the central load does not act at its neutral axis. Now W_0 is a function of P_{L} and for $P_{\mathrm{L}} = P_{\mathrm{E}}^{*}$ the two imperfections just cancel out and the hypothetical column bifurcates in a neutral fashion at the reduced modulus load

$$P_{\mathrm{E}}^{*} = \frac{c^2}{2l}\,\frac{2EE_{\mathrm{T}}}{E + E_{\mathrm{T}}}\,.$$

This gives the cross C in the figure, and by varying P_{L} we obtain the whole family of asymptotic curves given by

$$-(c/2)\,[(E - E_{\mathrm{T}})/(E + E_{\mathrm{T}})]\,(P - P_{\mathrm{L}}) = W(P - P_{\mathrm{E}}^{*})\,.$$

Regions of these curves relating to the *real* perfect model are drawn as double lines, and we have obtained a useful composite picture showing the response of the family of real systems generated by varying P_{L}. The bifurcations on the load axis at P_{L} are seen to be everywhere stable except when P_L lies between P_E^* and P_E. Simultaneous buck-ling occurs at S when $P_L = P_E$.

From this model we can infer the behaviour of our plate with thick stiffeners to be essentially as shown in Fig. 3. Here we have drawn four load-deflection diagrams for the cases $P_E^* < P_E < P_L$, $P_E^* < P_E = P_L$, $P_E^* < P_L < P_E$, $P_L < P_E^* < P_E$, with W representing the amplitude of the Euler mode and w the amplitude of the local mode. The behaviour of the perfect plate is shown by the heavy lines and on the essentially neutral Euler path we note the existance of secondary bifurcations of the type predicted by Chilver and Supple.

The presence of an Euler imperfection W_0 gives rise to the light solid lines which bifurcate at the "Perry" load as the local buckling stress is

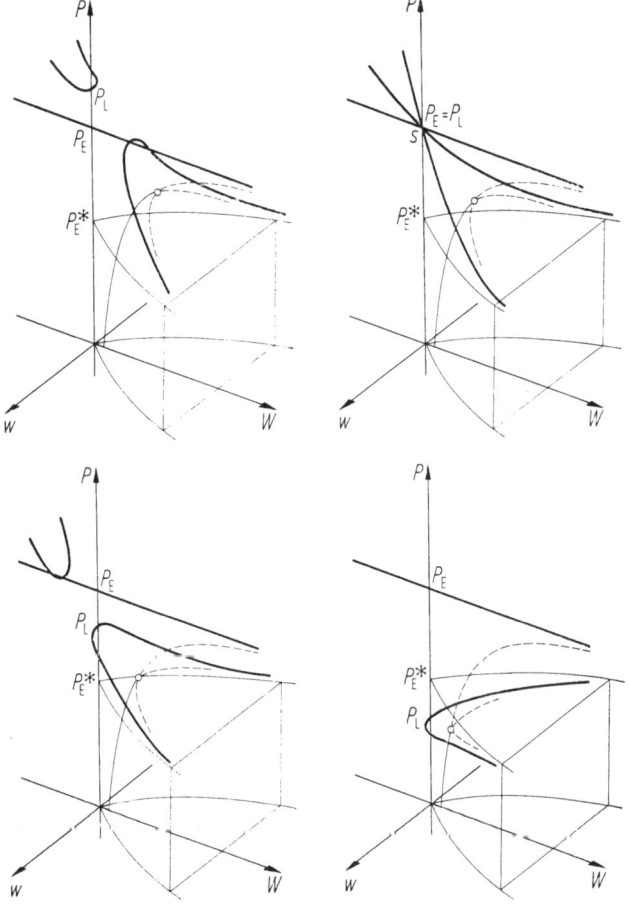

Fig. 3.
Schematic load-deflection behaviour of the panel for four critical load ratios.

reached in the skin, the bifurcations being largely stable below and unstable above P_E^*. Clearly this Perry load and the reduced modulus load
P_E^* are important parameters of the stiffened plate, and we can use
them to sketch the schematic imperfection-sensitivity curves of Fig. 4
relating to an overall Euler imperfection W_0. Here the Perry curves are
roughly parabolic in form and pass through the two relevant critical
loads P_E and P_L: this gives rise to a most severe sensitivity when these
two critical loads coincide as discussed by Thompson and Hunt [8].

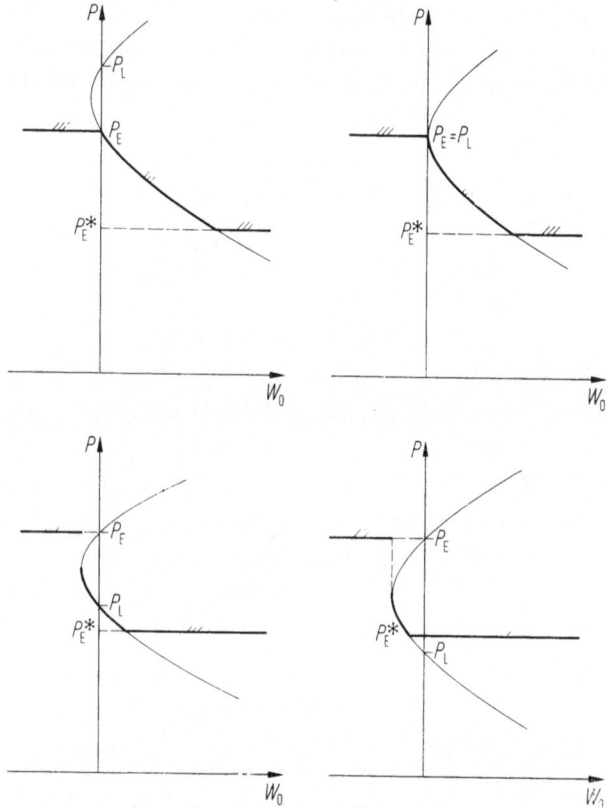

Fig. 4. Schematic imperfection-sensitivity diagrams for the panel for four critical
load ratios.

4. Experimental Results

The results for the panels with stocky stiffeners, which had 8 bays as
drawn in Fig. 1, are shown in Figs. 5 to 7. The results for the panel
with thin stiffeners, which differed slightly by having 9 bays, are shown
in Fig. 8.

To present the results in a useful form we have related the loads to the two critical values P_E and P_L. The former is readily estimated, and for the latter we have used the plate buckling formula

$$\sigma_L = k\,(E\pi^2 h^2/12(1 - \nu^2)\,b^2)$$

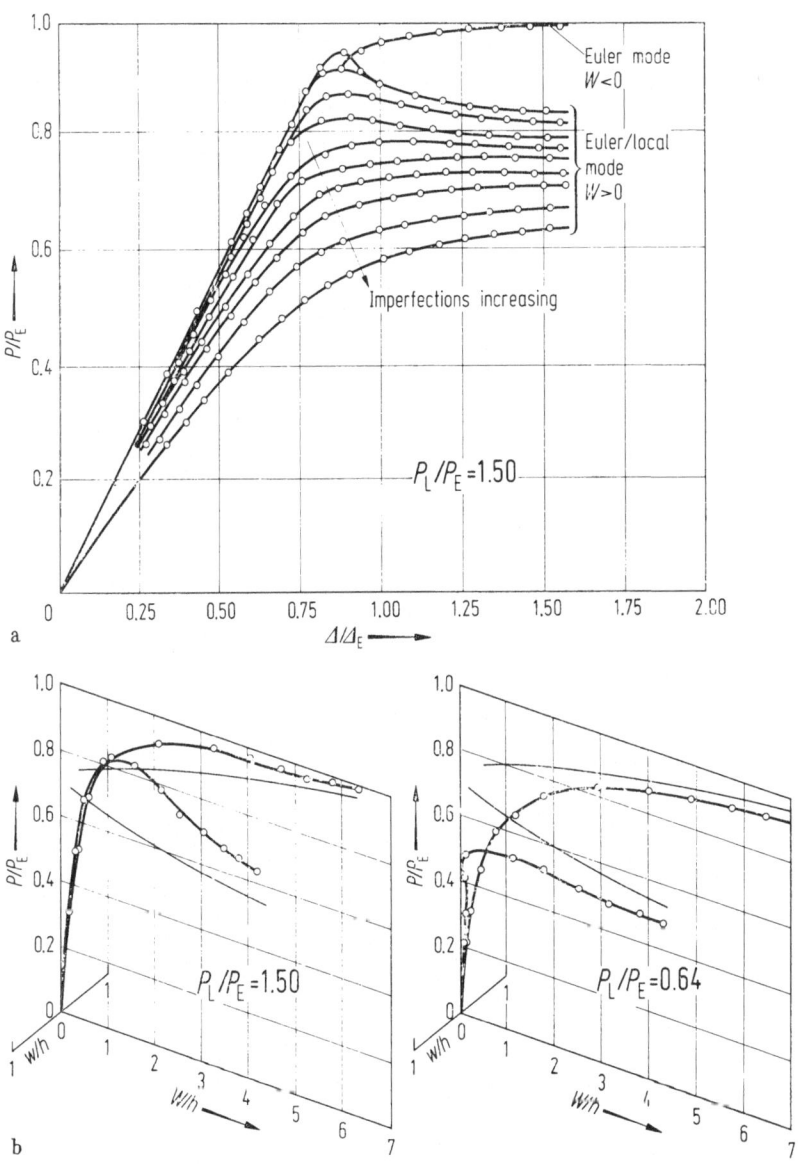

a

b

Fig. 5. Experimental load-deflection curves: panel with stocky stiffeners a) load versus axial deflection, b) load versus local and Euler mode amplitudes.

where k is a constant dependant on the boundary conditions and aspect ratio of the plate. For the specimens with stocky stiffeners k was taken as equal to 6.96 which relates to a long plate clamped at the edges, while for the panel with thin stiffeners it was equated to 4.0 which relates to a long plate with simply-supported edges. The loads are closely confirmed by numerical calculations kindly performed for us by W. H. Wittrick and F. W. Williams of Birmingham University.

Figure 5a shows a family of curves for the load versus its corresponding axial deflection obtained by varying the loading eccentricity which represents our overall Euler imperfection W_0. In this case the other

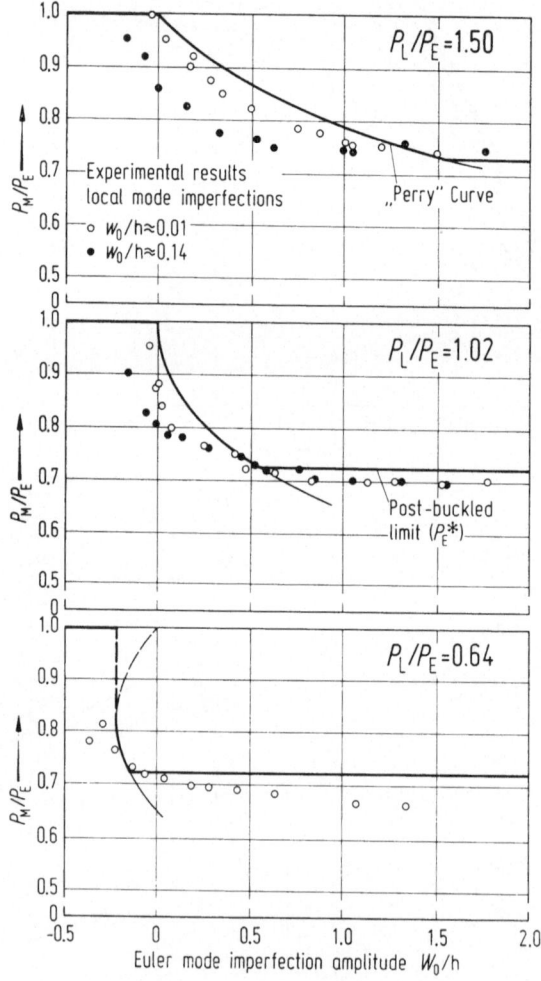

Fig. 6. Experimental imperfection-sensitivity plots for three critical load ratios (panel with stocky stiffeners).

geometric parameters and the local imperfections are held constant and the critical load ratio was $P_L/P_E = 1.50$. The curves show that for small overall imperfections the maximum loads obtained are close to the theoretical Euler critical load of the panel. In the post-buckled range, however, and for larger imperfections the loads tend to the reduced Euler load determined by the residual stiffness of the buckled skin and stiffeners. Also shown on this graph is a curve representing the result obtained when the panel buckles in the opposite direction, that is when the overall curvature tends to put the plate elements into

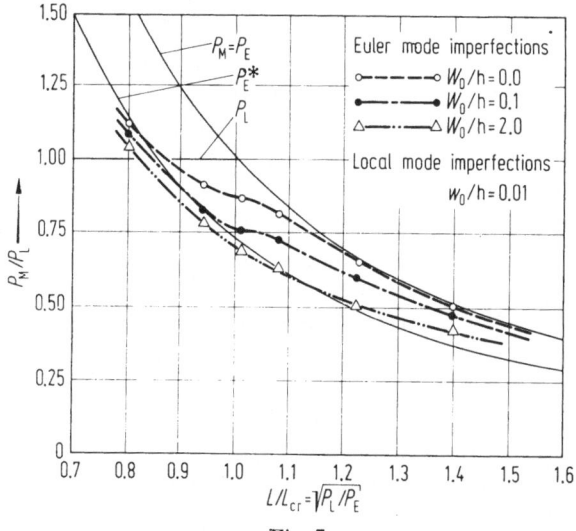

Fig. 7.
Maximum load versus slenderness ratio for the panels with stocky stiffeners.

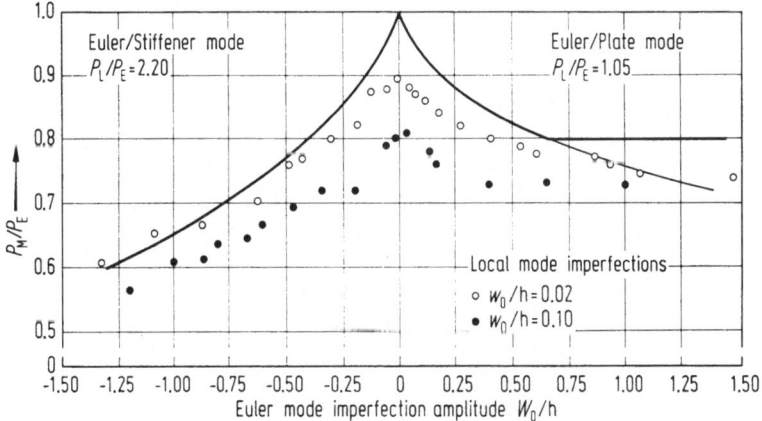

Fig. 8. Experimental imperfection-sensitivity plot for a panel with thin stiffeners.

tension, so that no local buckling occurs. Here, in the absence of local buckling, the panel buckling closely resembles the behaviour of a conventional elastic column.

In Fig. 5b we show experimental load deflection plots for two critical load ratios which are in excellent agreement with our predictions from the Shanley model.

In Fig. 6 we show the experimentally determined imperfection-sensitivity curves for three values of the critical load ratio, P_M representing the observed maximum loads and W_0 the equivalent Euler imperfection amplitudes. The local plate imperfections were as indicated.

Also shown in Fig. 6 are the theoretical Perry curves which give as a function of W_0 the load at which the combination of axial compression and overall bending leads to a local stress at the plate mid-surface equal to the plate buckling stress. These, together with the cut-offs at P_E and P_E^*, give a useful framework on which to view the experimental points. Generally the experimental loads fall below the relevant parts of the Perry curves because these curves make no allowance for the local imperfections present in the tests. For low P_L the points also fall below the P_E^* line because of the continuous fall in post-buckling stiffness as the plate elements enter their advanced post-buckling range.

The experimental data for stocky stiffeners is summarized in the load to slenderness ratio plot of Fig. 7 drawn for various values of W_0 and for the single lower value of w_0. The great value of the reduced modulus load P_E^* is here most apparent, the slight drop below this being possibly due to the just-mentioned post-buckling softening.

Fig. 9. Experimental test rig.

The foregoing data was all obtained with panel specimens which had extremely stocky compact stiffeners, but it is apparent that in many realistic design situations a more efficient use of material would result if thinner and deeper stiffening members were used. A consequence of this approach would be, however, to introduce a third mode of buckling associated with the torsion of the thin outstands. The potential problems of this sort of behaviour were explored in a series of tests on panels with slender outstands, and the results showed that the coupling between this third mode and the Euler mode can be very severe. A typical experimental imperfection-sensitivity plot is shown in Fig. 8 where we have now two theoretical Perry curves but have drawn only one P_E^* cut-off. The "local" mode imperfections were in this case a mixture of plate and stiffener buckling modes, but we give only a measure of the plate amplitudes.

5. Concluding Remarks

At present in civil engineering practice the allowable imperfections specified for stiffened plates are based on a few ad hoc experiments and on analysis that does not fully recognize the interactive buckling. The present research is part of a programme aimed at elucidating the mechanics of the problem as a step towards a more rational determination of the levels of imperfections that may economically be tolerated in practical structures.

More details of the experiments reported here may be found in a companion paper [9].

References

1. van der Neut, A.: The Interaction of Local Buckling and Column Failure of Thin-Walled Compression Members. Proc. XII Internat. Cong. Appl. Mech., Stanford. Berlin, Heidelberg, New York: Springer 1968.
2. Thompson, J. M. T.; Lewis, G. M.: On the Optimum Design of Thin-Walled Compression Members. J. Mech. Phys. Solids 20, 101 (1972).
3. Koiter, W. T.; Kuiken, G. D. C.: The Interaction between Local Buckling and Overall Buckling on the Behaviour of Built-up Columns. Rpt. No. 447, Laboratory of Engineering Mechanics, Delft, May 1971.
4. Thompson, J. M. T.; Supple, W. J.: Erosion of Optimum Designs by Compound Branching Phenomena. J. Mech. Phys. Solids 21, 135 (1973).
5. Tvergaard, V.: Imperfection-Sensitivity of a Wide Integrally Stiffened Panel under Compression. Int. J. Solids Structures 9, 177 (1973).
6. Tvergaard, V.: Influence of Post-Buckling Behaviour on Optimum Design of Stiffened Panels. Int. J. Solids Structures 9, 1519 (1973).
7. Croll, J. G. A.; Walker, A. C.. Elements of Structural Stability. London: Macmillan 1972.
8. Thompson, J. M. T.; Hunt, G. W.: A General Theory of Elastic Stability. London: Wiley 1973.
9. Tulk, J. D.; Walker, A. C.: Model Studies of the Elastic Buckling of a Stiffened Plate, submitted to Experimental Mechanics.

Mode Interaction in an Eccentrically Stiffened Elastic-Plastic Panel under Compression

V. Tvergaard

Technical University of Denmark, Lyngby, Denmark

A. Needleman

Massachusetts Institute of Technology, Cambridge, Massachusetts, U.S.A.

Abstract

For an elastic-plastic, wide panel with eccentric stiffeners plastic buckling due to axial compression is investigated. Special attention is directed to panels designed so that overall buckling as a wide column and local buckling of the plate between the stiffeners occur simultaneously or nearly simultaneously. The bifurcation behaviour for a perfect panel compressed into the plastic range is determined analytically, and the asymptotic initial post-bifurcation analysis is mentioned briefly. Mode interaction and imperfection-sensitivity is investigated numerically by application of an incremental method. Computations for panels that bifurcate in the plastic range show a considerable imperfection-sensitivity, in some cases due to mode interaction and in some cases due to a single mode.

1. Introduction

For a wide, integrally stiffened panel under axial compression the critical modes of bifurcation are either buckling of the panel as a wide Euler column or local buckling of the plate between the stiffeners. Panels designed so that buckling occurs simultaneously in these two modes, attract special attention, since such designs usually maximize the bifurcation load for a given amount of material. The interaction between local buckling and column failure, in the elastic range, has been considered by van der Neut [1], Koiter and Kuiken [2] and Thompson and Lewis [3] for thin-walled columns. For eccentrically stiffened panels buckling in the elastic range Tvergaard [4, 5] has shown, that simultaneous buckling designs are more imperfection-sensitive than other designs due to a strong mode interaction. However, in practice stiffened panels are often designed so that they bifurcate in the plastic range, and here we shall investigate the effect of mode interaction on plastic buckling of panels. Mode interaction in an elastic-plastic structure has been considered by Graves Smith [6] who studied the effect of local mode imperfections on the buckling of box columns.

In the present paper we shall study the behaviour of an eccentrically stiffened wide panel made of an elastic-plastic material with isotropic strain hardening. The bifurcation behaviour at the tangent modulus load is determined analytically, and results of the asymptotic post-bifurcation theory given by Hutchinson [7, 8] are shown. The behaviour of panels with initial imperfections both in the shape of the wide column buckling mode and that of the local buckling mode is investigated numerically. In the present paper the bifurcation and post-bifurcation analyses are only mentioned briefly, but a more detailed description together with more details on the numerical procedure are given in [10].

2. Problem Formulation

The integrally stiffened panel is assumed to be infinitely wide in the x_2-direction with a constant spacing b between the stiffeners and the distance a between the simply supported edges on which the compressive load acts (Fig. 1). The plate thickness is h, and the eccentricity e of the stiffeners is positive in the x_3-direction.

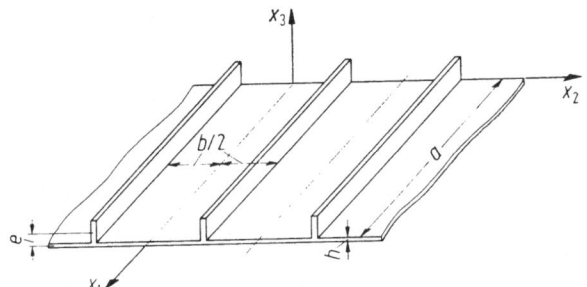

Fig. 1. Part of the integrally stiffened panel.

The displacements of the plate middle surface in the x_1, x_2 and x_3 directions are denoted u_1, u_2 and w, respectively. Then, with the usual assumptions of von Kármán plate theory, the strain rates of the plate middle surface $\dot{\varepsilon}_{\alpha\beta}$ and the bending strain rates $\dot{\varkappa}_{\alpha\beta}$ are taken to be

$$\dot{\varepsilon}_{\alpha\beta} = (\dot{u}_{\alpha,\beta} + \dot{u}_{\beta,\alpha} + w_{,\alpha}\dot{w}_{,\beta} + \dot{w}_{,\alpha}w_{,\beta})/2, \quad \dot{\varkappa}_{\alpha\beta} = \dot{w}_{,\alpha\beta}, \tag{2.1}$$

where a dot denotes differentiation with respect to some monotonically increasing parameter that characterizes the load history. The centre line strain rate and the bending strain rate of a stiffener are taken to be

$$\dot{\varepsilon}_s = \dot{u}_{1,1} - e\dot{w}_{,11} + w_{,1}\dot{w}_{,1} + e^2 w_{,12}\dot{w}_{,12}, \quad \dot{\varkappa}_s = \dot{w}_{,11} \tag{2.2}$$

as in [4]. Here subscript s refers to a stiffener, and Greek indices range from 1 to 2.

Small-strain J_2 flow-theory with isotropic hardening is employed, and the particular uniaxial stress-strain curve chosen is a power hardening law with a well defined yield stress σ_y and continuous tangent modulus

$$\varepsilon = \begin{cases} \sigma/E, & \text{for } \sigma \leqq \sigma_y, \\ \dfrac{\sigma_y}{E}\left(\dfrac{1}{n}\left(\dfrac{\sigma}{\sigma_y}\right)^n - \dfrac{1}{n} + 1\right), & \text{for } \sigma > \sigma_y. \end{cases} \tag{2.3}$$

In the approximately plane state of stress in the plate only the inplane stresses enter into the stress-strain relations, and we can write (with L_{ijkl} denoting the active branch of the three dimensional moduli)

$$\dot{\sigma}_{\alpha\beta} = \hat{L}_{\alpha\beta\gamma\delta}\dot{\eta}_{\gamma\delta}, \quad \hat{L}_{\alpha\beta\gamma\delta} = L_{\alpha\beta\gamma\delta} - (L_{\alpha\beta33}L_{\gamma\delta33}/L_{3333}). \tag{2.4}$$

The plate theory approximation to the strain increments at distance x_3 from the middle surface is

$$\dot{\eta}_{\alpha\beta} = \dot{\varepsilon}_{\alpha\beta} - x_3\dot{\varkappa}_{\alpha\beta}. \tag{2.5}$$

Using this and the usual definition of the membrane stress tensor $N_{\alpha\beta}$ and the moment tensor $M_{\alpha\beta}$, we find the incremental relations

$$\dot{N}_{\alpha\beta} = H^{(1)}_{\alpha\beta\gamma\delta}\dot{\varepsilon}_{\gamma\delta} + H^{(2)}_{\alpha\beta\gamma\delta}\dot{\varkappa}_{\gamma\delta}, \quad \dot{M}_{\alpha\beta} = H^{(2)}_{\alpha\beta\gamma\delta}\dot{\varepsilon}_{\gamma\delta} + H^{(3)}_{\alpha\beta\gamma\delta}\dot{\varkappa}_{\gamma\delta}, \tag{2.6}$$

where

$$H^{(i)}_{\alpha\beta\gamma\delta} = \int_{-h/2}^{h/2} \hat{L}_{\alpha\beta\gamma\delta}(-x_3)^{i-1}\, dx_3. \tag{2.7}$$

For the stiffeners we only account for plasticity due to uniaxial stresses σ_s parallel with the centre line. Then with the instantaneous uniaxial modulus $\hat{L}_s$, the incremental stress-strain relationship and the expression of strain increments are

$$\dot{\sigma}_s = \hat{L}_s\dot{\eta}_s, \quad \dot{\eta}_s = \dot{\varepsilon}_s - (x_3 - e)\,\dot{\varkappa}_s. \tag{2.8}$$

The incremental relations for the axial stiffener force N_s and the bending moment M_s are

$$\dot{N}_s = H^{(1)}_s\dot{\varepsilon}_s + H^{(2)}_s\dot{\varkappa}_s, \quad \dot{M}_s = H^{(2)}_s\dot{\varepsilon}_s + H^{(3)}_s\dot{\varkappa}_s, \tag{2.9}$$

$$H^{(i)}_s = b_s \int_{e-(h_s/2)}^{e+(h_s/2)} \hat{L}_s(e - x_3)^{i-1}\, dx_3. \tag{2.10}$$

Here the height and the width of the stiffener are denoted h_s and b_s, respectively. The increment of twisting moment in a stiffener is taken to be given by the elastic expression

$$\dot{M}_{vs} = G'_s K^i_s \dot{w}_{,12}. \tag{2.11}$$

This approximation is made in order to avoid a full solution of the mixed compression-bending-twisting problem for the stiffeners in the plastic range. This does not mean any approximation in the case of buckling as a wide column, in which the stiffeners stay untwisted, and neither does it affect any bifurcation load based on a flow theory of plasticity with no corners on the yield surface. However, in computations of post-buckling behaviour or behaviour of imperfect panels, in which the stiffeners twist, Eq. (2.11) overestimates the torsional stiffness somewhat.

3. Plastic Bifurcation and Post-Bifurcation Behaviour

The stress state in a perfect panel prior to bifurcation is a pure membrane state with the only nonvanishing stress component being a constant axial stress $\sigma_{11} = \lambda\sigma_{11}^0$ at every point of the plate and the stiffeners, such that the prebuckling unit membrane stress and stiffener force are $N_{11}^0 = \sigma_{11}^0 h$ and $N_s^0 = \sigma_{11}^0 A_s$, respectively. Thus, in general the relationship between the in-plane stress rates and strain rates stops being isotropic as soon as the absolute value of $\lambda\sigma_{11}^0$ exceeds the yield stress σ_y. Then the lowest bifurcation load can be determined as that of an elastic stiffened orthotropic plate with moduli equal to the instantaneous plastic moduli.

In the infinitely wide periodic structure, the buckling mode displacements u_1 and w are symmetric about the centre line $x_2 = 0$ between two stiffeners, an u_2 is antisymmetric about this line. Taking the simple support boundary conditions at the edges $x_1 = 0, a$ to be

$$u_{1,1} = u_2 = w = w_{,11} = 0, \tag{3.1}$$

we find that the exact bifurcation mode takes the form [10]

$$w = \{c_1 \cosh(r_1 x_2) + c_2 \cos(r_2 x_2)\} \sin(k\pi x_1/a), \tag{3.2}$$
$$u_1 = \{c_3 \cosh(r_3 x_2) \cos(r_4 x_2) + c_4 \sinh(r_3 x_2) \sin(r_4 x_2)\} \cos(k\pi x_1/a), \tag{3.3}$$
$$u_2 = \{(c_3 b_1 + c_4 b_2) \sinh(r_3 x_2) \cos(r_4 x_2) +$$
$$+ (-c_3 b_2 + c_4 b_1) \cosh(r_3 x_2) \sin(r_4 x_2)\} \sin(k\pi x_1/a), \tag{3.4}$$

in which k is a positive integer, c_1 to c_4 are constants that may be different in two neighbouring plate sections, and the constants r_1, r_2, r_3, r_4, b_1 and b_2 depend on k, u and the instantaneous moduli. Furthermore, r_1 and r_2 are functions of the load parameter λ_c. An iterative procedure is used to determine the smallest critical bifurcation load λ_c and the corresponding instantaneous moduli, for which the bifurcation equations are satisfied.

Buckling as a wide Euler column, with $k = 1$ and identical modes for all plate sections between two stiffeners, is critical as long as the stiffeners are relatively weak. When the bending stiffness of the stiffeners is sufficiently high, local buckling of the plate between the stiffeners occurs first, usually with a value of k somewhat above a/b. The

164   V. Tvergaard and A. Needleman

local buckling displacement fields in two neighbouring plate sections are usually identical with opposite sign. However, in cases of relatively large torsional stiffness of the stringers, the mode shapes are identical for all plate sections between two stiffeners.

Bifurcation loads for some panels are shown in Fig. 2. Compared with a panel used in [5], these panels have been given a relatively larger plate thickness and stiffener eccentricity, to make them bifurcate at plastic strains. The ratio between the amount of material

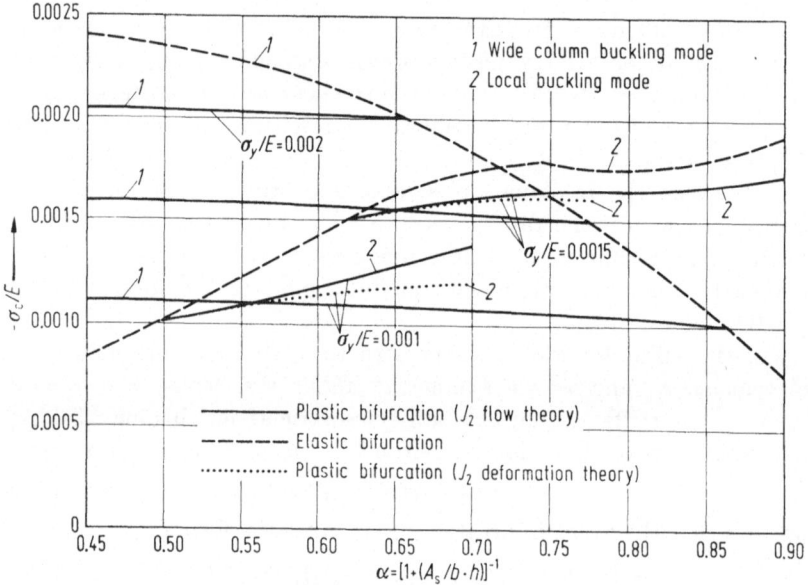

Fig. 2. Bifurcation stress for panels specified by $a/b = 4$, $e/b = 0.1$, $(A_s + hb)/b^2 = 0.0256$ and $\nu = 0.3$; all plastic bifurcation loads are taken for a strain hardening parameter $n = 10$.

in the plate and that in the whole panel is specified by a design parameter $\alpha = (1 + A_s/bh)^{-1}$. The plastic bifurcation loads are given for different ratios of yield stress and Young's modulus, and in all cases the strain hardening parameter $n = 10$ has been chosen. For $\sigma_y/E = 0.001$ the local bifurcation load is only indicated for values of α smaller than about 0.70. For larger values of α the local minimum of λ_c as a function of the wave parameter k vanishes so that a well defined local bifurcation load does not exist. Fig. 2 also compares the bifurcation loads predicted by J_2 flow theory with results of J_2 deformation theory. For the wide column buckling mode these results are indistinguishable from one another, and for the local buckling mode there is only a minor difference in our range of interest. Thus, the imperfection-sensitivity exhibited in the numerical results to be presented below is not due to the discrepancy between the bifurcation predictions of flow and deformation theory, as was the imperfection-sensitivity found by Onat and Drucker [11] for the cruciform column.

An asymptotic theory of the post-bifurcation behaviour of structures in the plastic range has been developed by Hutchinson [7, 8], which extends the bifurcation analysis of Hill [12, 13] into the initial post-bifurcation range. In the vicinity of the bifurcation point the load is expanded in terms of the buckling mode displacement amplitude, as is done in Koiter's theory for elastic post-buckling behaviour [14]. However, in the plastic range elastic unloading plays a crucial role after bifurcation.

An asymptotically exact expression for the load parameter λ is obtained in terms of the amplitude ξ of the buckling mode displacement of the form

$$\lambda = \lambda_c + \lambda_1 \xi + \lambda_2 \xi^{1+\beta} + \cdots \tag{3.5}$$

for $\xi \geqq 0$. For mode displacements in the opposite direction, we change the sign of ξ, and then expression (3.5) still applies with different constants λ_1, λ_2 and β. The buckling mode is normalized so that for $\xi = 1$ the maximum deflection w is equal to the plate thickness.

The constant λ_1 is determined from the Shanley condition, such that when the buckling mode starts to grow, plastic loading occurs everywhere in the current plastic zone, except in at least one point where neutral loading takes place. The constants β and λ_2 are determined from the lowest order terms in an asymptotic expansion of the principle of virtual work. In all cases we find $0 < \beta < 1$, $\lambda_1 > 0$ and $\lambda_2 < 0$, and for three different cases the shape of the elastic unloading zones that propagate into the material is indicated in Fig. 3 together with the corresponding values of β, 1/3, 2/5 and 2/7 [10].

At local buckling, with unloading zones propagating as shown in Fig. 3a, the post-bifurcation behaviour is symmetric. One of the most

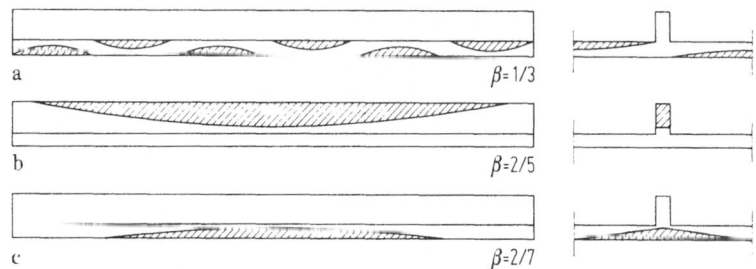

Fig. 3a—c. Shape of elastic unloading zones (plate thickness exaggerated). a) Local buckling with 6 half sine waves in x_1-direction; b) Wide column buckling in positive x_3-direction; c) Wide column buckling in negative x_3-direction.

interesting predictions of the post-bifurcation analysis, which will be related to the numerical analysis discussed below, is that at wide column mode buckling the post-bifurcation behaviour is asymmetric, and quite opposite to the behaviour of the linearly elastic panel, the plastic panel has its lowest carrying capacity when the buckling mode grows in the negative x_3-direction, with unloading zones propagating as shown in Fig. 3c. The detailed post-bifurcation predictions are given in [10].

4. Mode Interaction Behaviour for Imperfect Panels

For some examples of panels that bifurcate in the plastic range, the behaviour is determined by an incremental procedure. At each stage of the loading history, the normal deflection w, and the membrane forces $N_{\alpha\beta}$ and N_s are known. Incremental equilibrium is expressed in terms of the following variational principle: Among all displacement increment fields that satisfy the kinematical boundary conditions, the actual displacement increments satisfy $\delta I = 0$, where

$$I = (1/2) \int_0^a \int_0^b [H^{(1)}_{\alpha\beta\gamma\delta}\dot\varepsilon_{\alpha\beta}\dot\varepsilon_{\gamma\delta} + 2H^{(2)}_{\alpha\beta\gamma\delta}\dot\varepsilon_{\alpha\beta}\dot\varkappa_{\gamma\delta} +$$

$$+ H^{(3)}_{\alpha\beta\gamma\delta}\dot\varkappa_{\alpha\beta}\dot\varkappa_{\gamma\delta} + N_{\alpha\beta}\dot w_{,\alpha}\dot w_{,\beta}]\, dx_1\, dx_2 + (1/2) \int_0^a [H^{(1)}_s\dot\varepsilon_s\dot\varepsilon_s +$$

$$+ 2H^{(2)}_s\dot\varepsilon_s\dot\varkappa_s + H^{(3)}_s\dot\varkappa_s\dot\varkappa_s + G_sK_s\dot w^2_{,12} + N_s(\dot w^2_{,1} + e^2\dot w^2_{,12})]\, dx_1 -$$

$$- \dot\lambda\left[\int_0^b N^0_{11}\dot u_1\, dx_2 + N^0_s(\dot u_1 - e\dot w_{,1})\right]_0^a. \qquad (4.1)$$

Here quantities associated with the stiffener are evaluated at $x_2 = b/2$, $\dot\lambda$ is the prescribed increment of the load parameter, the moduli $H^{(i)}_{\alpha\beta\gamma\delta}$ and $H^{(i)}_s$ are defined by (2.7) and (2.10), respectively, and the incremental strain quantities are given in terms of the displacements by (2.1) and (2.2).

An approximate solution of (4.1) is obtained by a combined Rayleigh Ritz-finite element procedure [15, 10], in which the increment of normal displacement $\dot w$, is expanded in terms of smooth functions as in the standard Rayleigh Ritz method,

$$\dot w = \sum_{j=1}^N \dot\xi_j^* \hat w^j, \qquad (4.2)$$

where the $\hat w^j$ are the assumed functions for the Rayleigh Ritz solution and $\dot\xi_j^*$ are the incremental mode amplitudes. The in-plane displacements $\hat u_\alpha$ are determined by a finite element calculation which is described in more detail in [10].

Solving the variational equation first for the in-plane displacements, we find a displacement field $\hat{u}_\alpha^j$ corresponding to each $\hat{w}^j$, in which we take $\hat{u}_2^j = 0$ at $x_2 = 0, b$ and the edges $x_1 = 0, a$ free. In addition we use the finite element method to find the pure in-plane mode $\hat{u}_\alpha^{N+1}$ due to the external loads N_{11}^0 and N_{s}^0 on the edges $x_1 = 0, a$ with $\hat{u}_2^{N+1} = 0$ at $x_2 = 0, b$, and to find the pure in-plane mode $\hat{u}_\alpha^{N+2}$ due to prescribed uniform displacements $\hat{u}_2^{N+2}$ at $x_2 = 0, b$ and the edges $x_1 = 0, a$ free. Thus, the in-plane displacement functions are given by

$$\dot{u}_\alpha = \sum_{j=1}^{N+2} \xi_j^* \hat{u}_\alpha^j. \tag{4.3}$$

Equations (4.2) and (4.3) now give the trial functions employed in the Rayleigh Ritz procedure for a panel free to slide tangentially along the simply supported edges $x_1 = 0, a$. The use of a different boundary condition $u_2 = 0$ for the bifurcation mode (3.1) has very little effect on the buckling stress in our range of interest.

For the smooth out-of-plane functions $\hat{w}^j$ we choose the wide column buckling mode, the local buckling mode corresponding to the smallest buckling stress, the local buckling mode with two more half sine waves in the x_1-direction than the critical one, and a mode with $k = 1$ that is able to relatively decrease or increase the bulging of the plate between the stiffeners in the wide column buckling mode. Similar modes were used in [5] for a Galerkin solution of an elastic panel. In addition to these four $\hat{w}^j$ modes several others have been tried without finding any that had an appreciable effect.

For each increment of the present computations the moduli $H_{\alpha\beta\gamma\delta}^{(i)}$ are computed at 384 points on the plate middle surface, and the moduli $H_s^{(i)}$ are computed at 48 points on the stiffener centre line. For each of these points the stress history is remembered at 7 points through the plate thickness or through the stiffener height.

For some of the panels for which the bifurcation loads are given in Fig. 2, the behaviour due to various imperfections is shown in Figs. 4 to 6. The panels bifurcate in the plastic range with the wide column mode being critical in Fig. 4, a local mode being critical in Fig. 5 and these two modes being simultaneously critical in Fig. 6. Initial imperfections are considered in the shape of the wide column buckling mode and the local buckling mode, and the ratios between their amplitudes and the plate thickness are denoted $\bar{\xi}_w$ and $\bar{\xi}_l$, respectively. In the following ξ_w and ξ_l denote the additional growth of these two modes, and in the figures we only show the relationship between ξ_w and the load parameter λ.

The plastic post-bifurcation behaviour of the perfect panels is

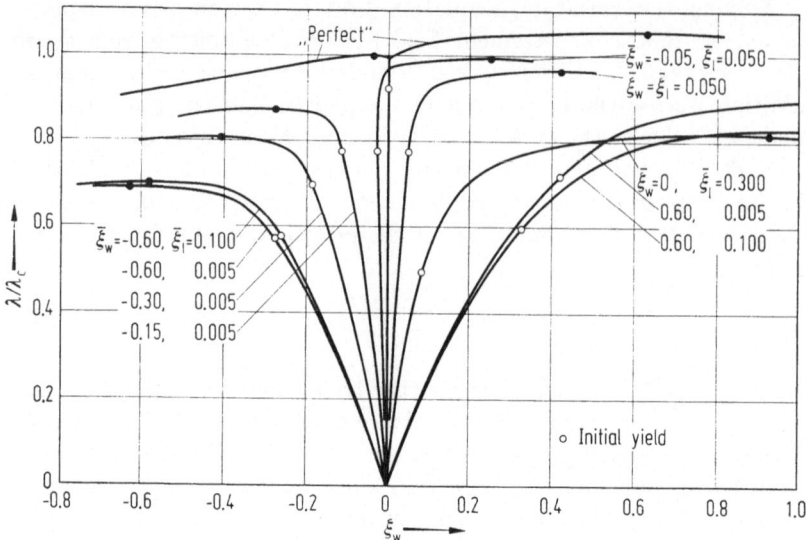

Fig. 4. Load versus wide column mode displacement for panel that bifurcates plastically in wide column buckling mode ($\sigma_y/E = 0.001$, $n = 10$, $\alpha = 0.65$, $a/b = 4$, $e/b = 0.1$, $(A_s + hb)/b^2 = 0.0256$, $\nu = 0.3$).

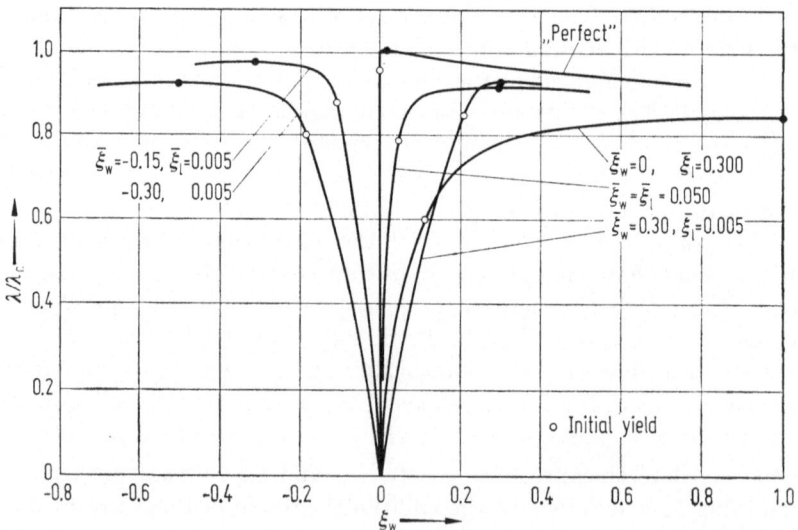

Fig. 5. Load versus wide column mode displacement for panel that bifurcates plastically in local buckling mode ($\sigma_y/E = 0.001$, $n = 10$, $\alpha = 0.525$, $a/b = 4$, $e/b = 0.1$, $(A_s + hb)/b^2 = 0.0256$, $\nu = 0.3$).

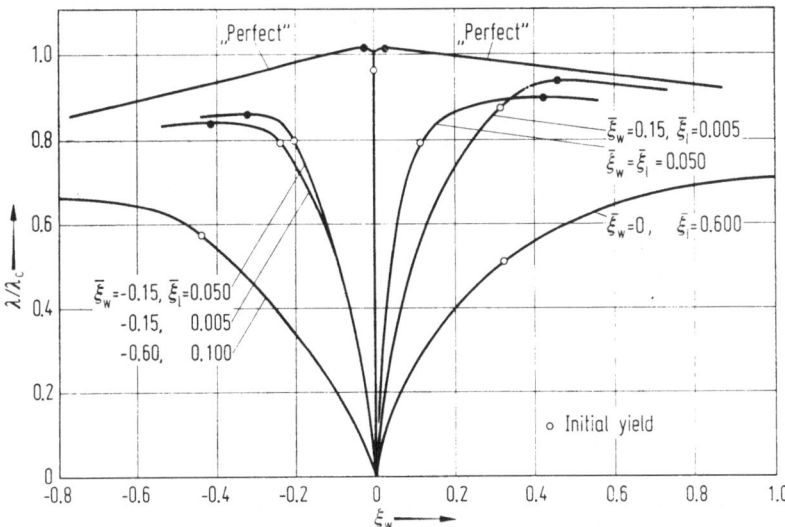

Fig. 6. Load versus wide column mode displacement for panel with simultaneous plastic bifurcation in wide column and local buckling modes ($\sigma_y/E = 0.0015$, $n = 10$, $\alpha = 0.653$, $a/b = 4$, $e/b = 0.1$, $(A_s + hb)/b^2 = 0.0256$, $v = 0.3$).

computed numerically as the behaviour of panels with very small initial imperfections. These numerical results and results of the asymptotic theory agree in predicting that the maximum load is only slightly above the bifurcation load, although maximum loads obtained from the truncated series (3.5) are generally a little smaller than the numerical results. Also the asymmetric post-bifurcation behaviour in the case of wide column buckling agrees with the asymptotic results, and the shapes of elastic unloading regions predicted by the asymptotic analysis agree well with the numerical results. For the panel in Fig. 5 the post-bifurcation behaviour is symmetric with respect to the bifurcation mode amplitude ξ_l as predicted by the asymptotic analysis, and both for negative and positive ξ_l the column mode amplitude ξ_w grows in the positive direction. In the simultaneous buckling case shown in Fig. 6 the post-bifurcation growth of ξ_w in positive direction occurs simultaneously with a considerable growth of ξ_l. The post-bifurcation growth of ξ_w in negative direction results in a very small initial growth of ξ_l that starts to decrease before the maximum is reached.

Due to the form of the asymmetric post-bifurcation behaviour at wide column buckling the structure is expected to have stronger sensitivity to negative $\bar{\xi}_w$ than to positive $\bar{\xi}_w$ in the plastic range (when $\bar{\xi}_l = 0$), although the corresponding linearly elastic panel is not at all sensitive to negative $\bar{\xi}_w$. This behaviour is clearly seen by comparing the curves for $\bar{\xi}_w = -.6$ and those for $\bar{\xi}_w = .6$ in Fig. 4. However, local mode imperfections (positive or negative $\bar{\xi}_l$) promote an interaction between the local mode and the column mode, which increases the sensitivity to positive $\bar{\xi}_w$ but opposes the sensitivity to negative $\bar{\xi}_w$. This last effect is seen from te curve $\bar{\xi}_w = -.05$, $\bar{\xi}_l = .05$ in Fig. 4, where ξ_w initially grows in the direction of $\bar{\xi}_w$ but finally is forced by the local imperfections to change sign just before the maximum load is reached. For sufficiently large negative wide column imperfections such as $\bar{\xi}_w = -.6$ Fig. 4 shows that the magnitude of $\bar{\xi}_l$ has very little effect on the load carrying capacity.

The panel in Fig. 5, for which a local buckling mode is critical, has a carrying capacity above the bifurcation load in the elastic range. When bifurcation takes place in the plastic range, the panel is seen to be imperfection-sensitive, although less sensitive than the panel in Fig. 4. Also, in Fig. 5 the panel is seen to be relatively less sensitive to wide column mode imperfections than to local mode imperfections. The curve for $\bar{\xi}_w = 0$, $\bar{\xi}_l = 0.3$ shows that local mode imperfections promote failure by causing a rapid growth of ξ_w.

In the simultaneous buckling case of Fig. 6 the yield stress had been chosen somewhat higher than in the previous figures and thus somewhat closer to the elastic bifurcation stress. Therefore, the mode displacements prior to initial yielding are relatively large, and this is part of the reason why the panel in Fig. 6 is a little more imperfection-sensitive than the panel in Fig. 4. As in Fig. 4 the considerable imperfection-sensitivity found in Fig. 6 is in some cases due to interaction between the local mode and a positive column mode and in some cases due solely to a negative growth of the column mode.

References

1. van der Neut, A.: Proc. 12th Int. Congr. Appl. Mech., Stanford University 1968, Berlin, Heidelberg, New York: Springer 1969.
2. Koiter, W. T.; Kuiken, G. D. C.: Delft lab. Report-WTHD 23, May (1971).
3. Thompson, J. M. T.; Lewis, G. M.: J. Mech. Phys. Solids **20**, 101 (1972).
4. Tvergaard, V.: Int. J. Solids Structures **9**, 177 (1973).
5. Tvergaard, V.: Int. J. Solids Structures **9**, 1519 (1973).
6. Graves Smith, T. R.: Int. J. Mech. Sci. **13**, 911 (1971).
7. Hutchinson, J. W.: J. Mech. Phys. Solids **21**, 163 (1973).
8. Hutchinson, J. W.: Advances in Appl. Mech., **14**, 67 (1974).
9. Hutchinson, J. W.: J. Mech. Phys. Solids **21**, 191 (1973).
10. Tvergaard, V.; Needleman, A.: Int. J. Solids Structures **11**, 647 (1975).
11. Onat, E. T.; Drucker, D. C.: J. Aero. Sci. **20**, 181 (1953).
12. Hill, R.: J. Mech. Phys. Solids **6**, 236 (1958).

13. Hill, R.: Problems of Continuum Mechanics (S.I.A.M. Philadelphia), p. 155 (1961).
14. Koiter, W. T.: Thesis, Delft, H. J. Paris, Amsterdam 1945. English translation issued as NASA TT F-10, 833, 1967.
15. Ohtsubo, H.: Advances in Computational Methods in Structural Mechanics and Design (ed. J. T. Oden et al). University of Alabama Press 1972, p. 439.

Buckling of Stochastically Imperfect Structures [1]

J. C. Amazigo

Rensselaer Polytechnic Institute, Troy, New York, U.S.A.

Abstract

Some analyses of the static and dynamic buckling of imperfection-sensitive structures with small random imperfections are reviewed and extended. Statistical
properties such as the mean of buckling load and the probability of failure at
specified loads are presented for structures whose response and buckling loads can
be characterized by one imperfection parameter. For structures for which ergodicity assumptions are reasonable an analysis is presented for determing the
dynamic buckling loads for impulsive loading. These buckling loads are compared
with results for static and step loading.

1. Introduction

Two methods currently used for the analysis of the buckling of stochastically imperfect structures depend on the characterization of these imperfections. In one case the imperfections—geometric or otherwise—are
assumed to depend on a relatively few number of parameters. It is
further assumed that the response of the structure and the buckling
load can be determined as functions of the finite number of random
parameters. It then becomes routine to calculate statistical properties
of the response and buckling load based on some assumption of the
joint probability of these random parameters. This has been the approach of Bolotin [1, 2], Thompson [3], and Roorda [4]. We use this
approach here to determine the failure probability and the mean of
the buckling load of cubic and quadratic structures [5] for zero mean
normally distributed imperfections. An application of this theory to
externally pressurized cylindrical shells is presented, and these results
are compared with experimental results.

The second method is applicable to structures for which it is
reasonable to evoke the ergodicity assumption. For such structures the
analysis is based on the properties of stochastic differential equations.
This method has been used by Fraser, Budiansky and Amazigo [5 to

---

[1] This work was supported in part by the National Science Foundation under
Grant GP-33679X.

13]. In these studies various approximate techniques have been used to investigate the dependence of the buckling load on the autocorrelation or spectral density of the imperfections. The methods include equivalent linearization, truncated hierarchy, regular and singular perturbation. Here we give only a brief exposition of the application of Poincaré perturbation method to the dynamic buckling under impulsive loading of an imperfect axially compressed column resting on nonlinear softening foundation.

2. Statistical Problems Based on Single Mode Results

Koiter's theory [15] of static buckling of structures that takes into account initial imperfections in the shape of the classical buckling mode is now well-known. More readily accessible redevelopment and extension of this theory to dynamic buckling is contained in [5, 14]. We shall omit details of these analyses and present a brief summary of results. For structures for which single mode Galerkin-type solution is adequate there exists the simple relation

$$(1 - \lambda/\lambda_c)\,\xi + a\xi^2 + b\xi^3 + \cdots = \bar{\xi}\lambda/\lambda_c \tag{1}$$

between the load parameter λ, a measure ξ of the deformation such as the amplitude of the component of the deformation in the shape of the classical buckling mode, and $\bar{\xi}$, the imperfection amplitude. λ_c is the classical buckling load. The coefficients a and b depend on the structure under consideration. We follow Budiansky [5] in defining a structure as quadratic if $a \neq 0$ and cubic if $a = 0$, $b \neq 0$. The static buckling load λ_s is found for sufficiently small imperfections by maximizing λ with respect to ξ neglecting $0(\xi^3)$ terms for quadratic structures and neglecting $0(\xi^4)$ terms for cubic structures. The results of the maximization are

$$(1 - \lambda_s/\lambda_c)^2 = Q\bar{\xi}\lambda_s/\lambda_c, \tag{2}$$

and

$$(1 - \lambda_s/\lambda_c)^{3/2} = C\,|\bar{\xi}|\,\lambda_s/\lambda_c, \tag{3}$$

where $Q = -4a$, $C = 3(-3b)^{1/2}/2$ for $b < 0$. For $b > 0$ the structure is imperfection insensitive.

Budiansky and Hutchinson [5, 14] have extended these results to dynamic buckling. For buckling under step loading Eq. (3) is replaced by

$$(1 - \lambda_D/\lambda_c)^{3/2} = 2^{1/2}C\,|\bar{\xi}|\,\lambda_D/\lambda_c. \tag{4}$$

It is worth noting that such simple results may also be obtained without the restrictive assumption that the imperfection be in the shape of the classical buckling mode (see [16, 17]).

We present here some details of the calculations for cubic structures whose buckling loads are governed by (3) and (4). Since the subsequent analysis can very readily be extended to quadratic structures only the results will be presented. Furthermore since the statistical analysis based on (4) will obviously be a repetition of the analysis based on (3) we carry out the Bolotin formulation only for static buckling. Similar calculations have been done in [1, 2, 4, 6]. We limit ourselves to three statistical properties of the buckling load λ_s for imperfections $\bar{\xi}$ which have zero mean Gaussian distribution. These properties are the mean $\langle \lambda_s \rangle$, the probability density $p_{\lambda_s}(\lambda_s)$, and the probability of failure P_F defined by

$$P_F(\lambda') = 1 - \text{probability that } \lambda_s/\lambda_c \text{ lies between } \lambda' \text{ and } 1.$$

It is clear that these properties are well defined and can readily be calculated for any assigned probability density for $\bar{\xi}$. A choice for such distribution is best determined from experiments. Lacking such information we choose a zero mean Gaussian distribution

$$p(\bar{\xi}) = \sigma^{-1} (2\pi)^{-1/2} \exp(-\bar{\xi}^2/2\sigma) \tag{5}$$

motivated by the central limit theorem. σ is the standard deviation of the imperfection. Now for (3)

$$P_F(\lambda') = 1 - 2 \int_0^\alpha p(\bar{\xi}) \, d\bar{\xi} = \text{erfc} \{\alpha/\sqrt{2\sigma}\}, \tag{6}$$

where

$$\alpha = (1 - \lambda')^{3/2}/C\lambda'. \tag{7}$$

The probability density function of λ_s is obtained from (6) by differentiation. Thus with $\bar{\lambda} = \lambda_s/\lambda_c$

$$p_{\lambda_s}(\lambda_s) = -\left. \frac{dP_F}{d\lambda'} \right|_{\lambda'=\bar{\lambda}} = \frac{(1 - \bar{\lambda})^{1/2} (2 + \bar{\lambda})}{C\sigma\bar{\lambda}^2 \sqrt{2\pi}} \exp\left\{ -\frac{(1 - \bar{\lambda})^3}{2C^2\sigma\bar{\lambda}^2} \right\}. \tag{8}$$

Consequently the mean $\langle \lambda_s/\lambda_c \rangle$ of the normalized buckling load is given by

$$\langle \bar{\lambda} \rangle = \int_0^{\lambda_c} \lambda_s P_{\lambda_s}(\lambda_s) \, d\lambda_s. \tag{9}$$

Figure 1 gives the dependence of the probability of failure P_F and the mean buckling load on the product of the structures physical parameter C and the standard deviation. We conjecture that P_F is a more meaningful design criterion than $\langle \lambda_s \rangle$. Thus for an ensemble of similar structures for which, say $C\sigma = 0.1$ it is more useful to know that the probability of failure at a loading of $\lambda = 0.75\lambda_c$ is 10% than that the average buckling load is $0.85\lambda_c$.

Figure 2 represents similar results for the quadratic structure except that $P_F = 1 -$ probability that $\lambda_s/\lambda_c > \lambda'$, since the structure does not buckle for $Q\bar{\xi} < 0$. In determining $\langle\bar{\lambda}\rangle$ only the structures for which $Q\bar{\xi} > 0$ are considered.

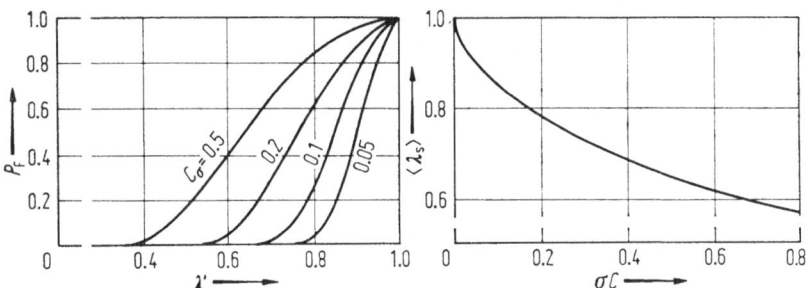

Fig. 1. Probability of failure and mean buckling load—cubic structure.

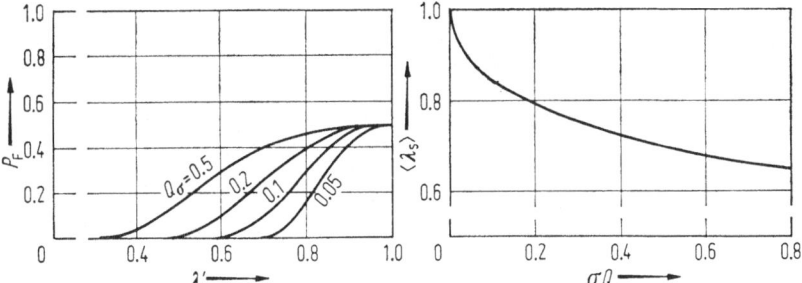

Fig. 2. Probability of failure and mean buckling load—quadratic structure.

We note that the above analysis can be readily extended to structures for which λ_s depends on a finite number of imperfection parameters $\bar{\xi}_1, \bar{\xi}_2, \ldots, \bar{\xi}_n$; thus

$$\lambda_s = \lambda_s(\bar{\xi}_1, \bar{\xi}_2, \ldots, \bar{\xi}_n). \tag{10}$$

It is however a nontrivial problem to obtain relation on (10) and perform the above analysis for $n > 2$ say. It is this difficulty that limits the effectiveness of this method.

We conclude this section by discussing an application of this method of analysis to the buckling under external pressure of imperfect circular cylindrical shells. The classical nondimensional hydrostatic pressure $\lambda_c RL^2/\pi^2 D$ was calculated by Batdorf [18] as a function of a curvature parameter $Z = L^2(1 - \nu^2)^{1/2}/Rh$. Here λ_c is buckling pressure, R, h, and L are shell radius, thickness and length respectively, $D = Eh^3/[12(1 - \nu^2)]$ is shell bending stiffness, and ν is Poisson's ratio. The static buckling load and dynamic buckling load for step loading are given [19, 17] by (3) and (4) respectively where $\bar{\xi}h$ is the coefficient of the classical buckling mode in the Fourier series expansion of the imperfection. The dependence of b on Z is plotted in [19]. From these results we compute $\langle \lambda_{\mathrm{s}} R L^2 / \pi^2 D \rangle$ from (9) for three values of the standard deviation σ. These results are plotted for $10 \leq Z \leq 1\,000$ n Fig. 3 and compared with buckling tests from several sources as collected by Dow [20]. Note that the curve for $\sigma = 0$ corresponds to the classical buckling load. These theoretical and experimental results are in excellent qualitative agreement. However we emphasize that the choice of $p(\bar{\xi})$ and values of σ are not based on experimental results.

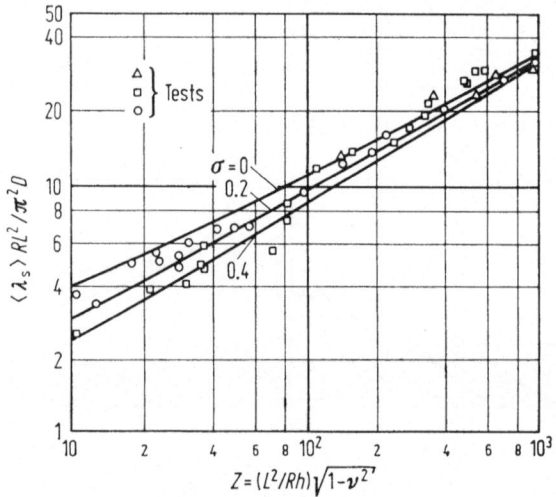

Fig. 3. Externally pressurized cylinder.

3. Statistical Problems Based on Ergodicity Hypothesis

Let us characterize the imperfections on similar structures by an ensemble of functions $\bar{v}(x, x_i)$ where at least one dimension x is infinite in extent. x_i are other independent variables. If the ensemble average $\langle \bar{v} \rangle$ is independent of x and the covariance function $\langle \bar{v}(x, x_i)\, \bar{v}(y, y_j) \rangle$ depends only on $|x - y|$, x_i and y_j and not explicitly on x or y, it is reasonable to evoke the ergodicity hypothesis. That is that ensemble avereags $\langle \cdots \rangle$ and averages taken with respect to $x, \{\cdots\}$, for any member of the ensemble are equal, Thus

$$\langle \bar{v}(x, x_i) \rangle = \lim_{X \to \infty} (1/2X) \int_{-X}^{X} \bar{v}(x, x_i)\, \mathrm{d}x \equiv \{\bar{v}(x, x_i)\},$$

$$\langle \bar{v}(x, x_i)\, \bar{v}(x + \zeta, y_j) \rangle = \lim_{X \to \infty} (1/2X) \int_{-X}^{X} \bar{v}(x, x_i)\, \bar{v}(x + \zeta, y_j)\, \mathrm{d}x \equiv$$

$$\equiv R(\zeta, x_i, y_j), \tag{11}$$

where R is called the autocorrelation. Based on the assumption of ergodicity *deterministic* buckling loads have been obtained for numerous structures using approximate methods such as equivalent linearization [5, 6, 8, 9], truncated hierarchy [7, 9], and perturbation [10, 12, 13]. We will not review these analyses here, but will discuss the application of an extension of Poincaré perturbation method to the determination of the dynamic buckling load of a model structure.

We consider an infinitely long imperfect column resting on nonlinear softening foundation subjected to axial impulse. Buckling of this model structure under step loading is discussed in [13]. Neglecting axial inertia and nonlinear geometric effects the governing initial boundary value problem for the nondimensional lateral deflection $w(x, \tau)$ is

$$w_{\tau\tau} + w_{xxxx} + 2\Lambda(\tau)\, w_{xx} + w - w^3 = -2\Lambda(\tau)\, \varepsilon \overline{w}_{xx}(x) - \infty < x < \infty,$$
$$\tau > 0^-,$$
$$w(x, 0^-) = w_\tau(x, 0^-) = 0, \quad -\infty < x < \infty, \tag{12}$$
$$w, w_x \text{ bounded},$$

where $(\ )_x = \partial(\ )/\partial x$, $(\ )_\tau = \partial(\ )/\partial\tau$. For impulsive loading $\Lambda(\tau) = I\, \delta(\tau)$ where $\delta(\tau)$ is the delta function. The nondimensional axial coordinate x, stress free initial displacement $\overline{w}$, additional lateral displacement w, axial impulse parameter I, and time τ are related to the corresponding physical quantities by

$$x = (k_1/E\overline{I})^{1/4}\, X, \quad \varepsilon\overline{w}(x) = (k_3/k_1)^{1/2}\, \overline{W}(x),$$
$$w = (k_3/k_1)^{1/2}\, W, \quad I = P/2(E\overline{I}k_1)^{1/2}, \quad \tau = (k_1/m)^{1/2}\, T.$$

$E\overline{I}$ is the bending stiffness of the column, and ε is a small positive parameter made definite by the condition

$$\langle w(x)\, w(x) \rangle = 1.$$

The lateral deflection is restrained by a continuous elastic foundation that produces a restoring forces per unit length of $k_1 W - k_3 W^3$, with $k_1, k_3 > 0$. m is the mass per unit length of the column.

The imperfection $\overline{w}(x)$ are assumed to be ergodic zero-mean Gaussian random functions. Thus $\langle \overline{w}(x) \rangle = 0$ and the autocorrelation $R_{\overline{w}}(\zeta)$ is

$$R_{\overline{w}}(\zeta) = \langle \overline{w}(x)\, \overline{w}(x + \zeta) \rangle = \lim_{X \to \infty} (1/2X) \int_{-X}^{X} \overline{w}(x)\, \overline{w}(x + \zeta)\, dx.$$

The corresponding spectral density of $\overline{w}$ is

$$S(\omega) = (1/2\pi) \int R_{\overline{w}}(\zeta)\, e^{-i\omega\zeta}\, d\zeta.$$

Integrals with unspecified limits are evaluated from $-\infty$ to ∞. We now seek a relationship between I and the limit as $\tau \to \infty$ of the mean square $\Delta^2(\tau)$ of the deflection. The buckling load is obtained by maximizing I with respect to this limit.

By integrating (12) from $\tau = 0^-$ to $\tau = 0^+$ we have

$$w_{\tau\tau} + w_{xxxx} + w - w^3 = 0, \quad -\infty < x < \infty, \quad \tau > 0^+,$$

$$w(x, 0^+) = 0, \quad w_\tau(x, 0^+) = -2I\varepsilon\overline{w}_{xx}(x), \quad -\infty < x < \infty, \tag{13}$$

w, w_x bounded.

Following the scheme developed in [13] and omiting details we introduce a new time scale t by

$$\tau = t(1 + \varkappa_2\varepsilon^2 + \varkappa_4\varepsilon^4 + \cdots) \tag{14}$$

and expand u in a power series in ε

$$w(x, \tau) \equiv u(x, t; \varepsilon) = \sum_{j=1}^{\infty} \varepsilon^j u^{(j)}(x, t). \tag{15}$$

Substituting (15) into (13) and equating powers of ε gives the sequence of equations:

$$Nu^{(1)} \equiv u_{tt}^{(1)} + u_{xxxx}^{(1)} + u^{(1)} = 0, \tag{16}$$

$$Nu^{(2)} = 0, \tag{17}$$

$$Nu^{(3)} = 2\varkappa_2 u_{tt}^{(1)} + (u^{(1)})^3, \text{ etc.}, \tag{18}$$

and

$$u^{(1)}(x, 0^+) = 0, \quad u_t^{(1)}(x, 0^+) = -2I\overline{w}_{xx}(x), \tag{19}$$

$$u^{(2)}(x, 0^+) = u_t^{(2)}(x, 0^+) = 0, \tag{20}$$

$$u^{(3)}(g, 0^+) = 0, \quad u_t^{(3)}(x, 0^+) = -2Iu_{xx}^{(2)}(x, 0^+) + \varkappa_2 u_t^{(1)}(x, 0^+). \tag{21}$$

Let

$$\Delta^2 = \lim_{\tau \to \infty} \{w(x, \tau)\, w(x, \tau)\}, \tag{22}$$

$$R_{ij}(\zeta, t, t') = \{u^{(i)}(x + \zeta, t)\, u^{(j)}(x, t')\} \tag{23}$$

and

$$\Delta_{ij} = \lim_{t \to \infty} R_{ij}(0, t, t). \tag{24}$$

Substitution of (15) into (22) and noting that $u^{(2)} \equiv 0$ gives

$$\Delta^2 = \varepsilon^2 \Delta_{11}(I; S) + 2\varepsilon^2 \Delta_{13}(I; S) + \cdots. \tag{25}$$

Thus, as shown in [10], the buckling load, I_D satisfies the equation

$$8(\Delta_{13}(I_D; S)/\Delta_{11}(I_D; S))\, \varepsilon^2 = 1. \tag{26}$$

The solution of (16) and (19) satisfying boundedness condition is

$$u^{(1)}(x, t) = -2I \int G(x - y, t)\, \overline{w}_{yy}(y)\, \mathrm{d}y, \tag{27}$$

where

$$G(x - y, t) = \begin{cases} (1/2\pi) \int (\xi^2 + 1)^{-1/2} \sin\left[(\xi^4 + 1)^{1/2} t\right] e^{-i\xi(x-y)} \, d\xi, \ t > 0, \\ 0 \qquad\qquad\qquad\qquad\qquad\qquad\qquad\qquad\qquad\quad t < 0. \end{cases}$$

$$(28)$$

The solution for $u^{(3)}$ is

$$u^{(3)}(x, t) = \int\limits_0^t \int G(x - y, t - t') \left[2\varkappa_2 u_{t't'}^{(1)}(y, t') + (u^{(1)}(y, t'))^3\right] dt' \, dy +$$

$$+ \varkappa_2 \int G(x - y, t) \, u_t^{(1)}(y, 0^+) \, dy. \tag{29}$$

From (28) and (29) and the definitions (23) and (24) and using the properties of Gaussian random functions and applying the method of stationary phase [21] we obtain after lengthy calculation

$$\Delta_{11} = 2I^2 \int \frac{\omega^4 S(\omega)}{\omega^4 + 1} \, d\omega, \tag{30}$$

$$\Delta_{13} = 6I^4 \int \frac{\omega^4 S(\omega)}{(\omega^4 + 1)^2} \, d\omega \cdot \int \frac{\xi^4 S(\xi)}{\xi^4 + 1} \, d\xi. \tag{31}$$

In deriving these results we assumed that $S(\omega)$ is bounded everywhere and that $\int \omega^4 S(\omega) \, d\omega$ exists. Under these restrictions the results are independent of $\varkappa_2$. For more general conditions $\varkappa_2$ is chosen to ensure that R_{13} is bounded as $t \to \infty$.

Substituting these results into (26) gives

$$2\sqrt{6} \, I_{\mathrm{D}} \left[\int \frac{\omega^4 S(\omega)}{(\omega^4 + 1)^2} \, d\omega\right]^{1/2} \varepsilon = 1. \tag{32}$$

For the purpose of comparison we exhibit the static buckling load λ_{s} and buckling load under step loading λ_{D} obtainable from the analyses in [7] and [13] respectively; namely

$$4\sqrt{6} \, \lambda_{\mathrm{s}} \left[\int \frac{\omega^4 S(\omega)}{(\omega^4 - 2\lambda_{\mathrm{s}}\omega^2 + 1)^3} \, d\omega\right]^{1/2} \varepsilon = 1 \tag{33}$$

and

$$4\sqrt{3} \, \lambda_{\mathrm{D}} \left[\int B_4(\omega) \, d\omega\right]^{-1/2} \left[3 \int B_4(\omega) \, d\omega \int R_6(\xi) \, d\xi + \right.$$

$$\left. + \iint \frac{B_4(\omega) \, B_4(\xi)}{4a^2(\omega) - a^2(\xi)} \, d\xi \, d\omega\right]^{1/2} \varepsilon = 1, \tag{34}$$

where $B_{2j}(\omega) = \omega^4 S(\omega)/(\omega^4 - 2\lambda_{\mathrm{D}}\omega^2 + 1)^j$, $a^2(\omega) = \omega^4 - 2\lambda_{\mathrm{D}}\omega^2 + 1$. Equation (33) has also been obtained by use of Khasminskii's theorem [23]. As shown in [7, 13] there exist simple asymptotic expressions for λ_{s} and λ_{D} for $\lambda_{\mathrm{s}}, \lambda_{\mathrm{D}} \to 1^-$; namely

$$(1 - \lambda_{\mathrm{s}})^{5/4} = 3(2)^{-1/4} \left[\pi S(1)\right]^{1/2} \lambda_{\mathrm{s}}\varepsilon \tag{35a}$$

and

$$(1 - \lambda_D)^{5/4} = 3.14[\pi S(1)]^{1/2} \lambda_D \varepsilon. \tag{35b}$$

However no such expression exists for I_D since I_D depends on a weighted integral of the spectral density $S(\omega)$.

Figure 4 gives the dependence of λ_s on ε for several values of c in the spectral density

$$S(\omega) = e^{-c|\omega|}. \tag{36}$$

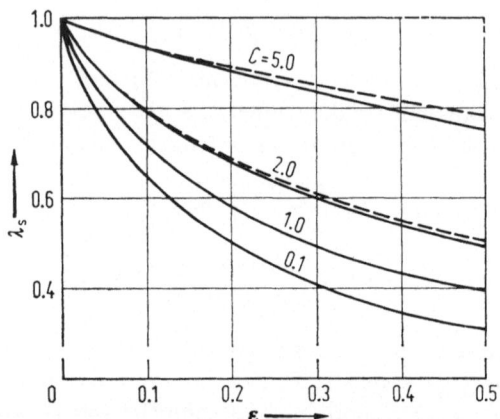

Fig. 4. Infinite column—static loading.

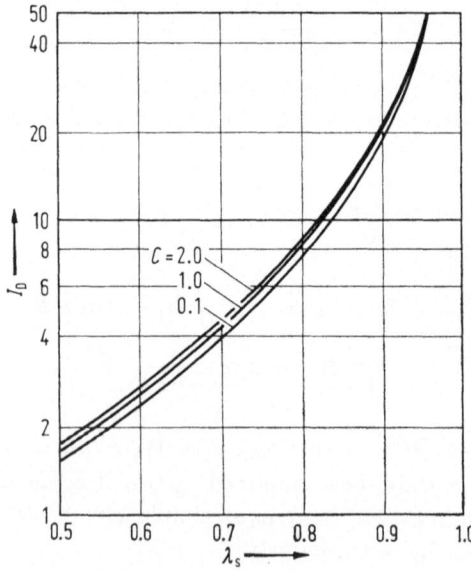

Fig. 5. Infinite column—impulsive loading.

The choice of this spectral density is motivated by experimental results reported in [22] for axisymmetric imperfections in cylindrical shells. Figure 5 and 6 give the relations between the pairs I_D, λ_s and λ_D, λ_s based on (32) to (34). The integrals in these equations were evaluated numerically. The asymptotic results based on (35) are shown as broken curves.

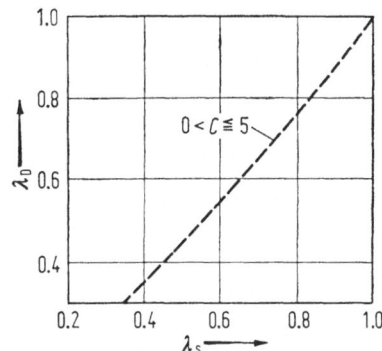

Fig. 6.
Infinite column — step
loading.

The method discussed here which is based on the ergodicity hypothesis leads to the conclusion that the structure will buckle statically or dynamically at the corresponding load given by (32) to (34) with probability 1. This result may appear paradoxical. However, to dispel the apparent contradiction we note that no matter how the origin of an infinitely long column is defined the buckling load for such columns with imperfections $\overline{w}(x) = \sin(x + \phi)$ is independent of ϕ and hence independent of any probabilistic distribution we may assign to ϕ.

4. Concluding Remarks

Two modes of approach to the analysis of the buckling under static or time dependent loads of imperfection — sensitive structures with random imperfections have been discussed. These methods are applicable to a large class of problems. The probability of failure criterion obtained for one class of structures and the simple asymptotic formulas obtained for deterministic buckling load for the other class may prove useful in establishing design criteria. It should prove worthwhile to determine experimentally the statistical distribution of imperfections so that the calculations done here will be carried out for these distributions thus yielding quantitatively useful results.

References

1. Bolotin, V. V.: Statistical Methods in the Nonlinear Theory of Elastic Shells. NASA TT F-85, 1962.
2. Bolotin, V. V.: Statistical Methods in Structural Mechanics, Chapter 4. Holden-Day, 1965.

182   Buckling of Stochastically Imperfect Structures

3. Thompson, J. M. T.: Toward a General Statistical Theory of Imperfection-Sensitivity in Elastic Post-Buckling. J. Mech. Phys. Solids 15, 413—417 (1967).
4. Roorda, J.: Some Statistical Aspects of the Buckling of Imperfection-Sensitive Structures. J. Mech. Phys. Solids 17, 111—123 (1969).
5. Budiansky, B.: Dynamic Buckling of Elastic Structures: Criteria and Estimates, in Dynamic Stability of Structures, edited by G. Herrmann. New York: Pergamon 1966, pp. 83—106.
6. Fraser, W. B.: Buckling of Structures with Random Imperfections. Ph. D. thesis, Harvard University, 1965.
7. Amazigo, J. C.: Buckling under Axial Compression of Long Cylindrical Shells with Random Axisymmetric Imperfections. Quart. Appl. Math. 26, 4, 537—566 (1969).
8. Fraser, W. B.; Budiansky, B.: The Buckling of a Column with Random Initial Deflections. J. Appl. Math. 36, 233—240 (1969).
9. Amazigo, J. C.; Budiansky, B.; Carrier, G. F.: Asymptotic Analyses of the Buckling of Imperfect Columns on Nonlinear Elastic Foundations. Int. J. Solids Struct. 6, 883—900 (1971).
10. Amazigo, J. C.: Buckling of Stochastically Imperfect Columns on Nonlinear Elastic Foundations. Quart. Appl. Math. 29, 403—409 (1971).
11. Amazigo, J. C.; Budiansky, B.: Asymptotic Formulas for the Buckling Stresses of Axially Compressed Cylinders with Localized or Random Axisymmetric Imperfections. J. Appl. Mech. 39, 179—184 (1972).
12. Amazigo, J. C.: Asymptotic Analysis of the Buckling of Externally Pressurized Cylinders with Random Imperfections. Quart. Appl. Math. 31, 429—442 (1974).
13. Amazigo, J. C.: Dynamic Buckling of Structures with Random Imperfections. Stochastic Problems in Mechanics. Ed. H. Leipholz, Univ. of Waterloo Press 1974, pp. 243—254.
14. Budiansky, B.; Hutchinson, J. W.: Dynamic Buckling of Imperfection-Sensitive Structures. Proc. XI Internat. Cong. Appl. Mech., Munich, 1964.
15. Koiter, W. T.: On the Stability of Elastic Equilibrium (in Dutch). Thesis, Delft, Amsterdam, 1945; English translation: NASA TT F-10, 1967, pp. 883.
16. Amazigo, J. C.; Frank, D.: Dynamic Buckling of an Imperfect Column on Nonlinear Foundation. Quart. Appl. Math. 31, 1—9 (1973).
17. Lockhart, D.; Amazigo, J. C.: Dynamic Buckling of Externally Pressurized Imperfect Cylindrical Shells. J. Appl. Mech. 42, 316—320 (1975).
18. Batdorf, S. B.: A Simplified Method of Elastic-Stability Analysis for Thin Cylindrical Shells. NACA Report 874, 1947.
19. Budiansky, B.; Amazigo, J. C.: Initial Post-Buckling Behavior of Cylindrical Shells under External Pressure. J. Math. and Physics 47, 3, 233—235 (1968).
20. Dow, D. A.: Buckling and Post-Buckling Tests of Ring-Stiffened Cylinders Loaded by Uniform "External Pressure". NASA Langley Research Center TN D-3111, 1965.
21. Erdelyi, A.: Asymptotic Expansions. Dover 1956, p. 51.
22. Arbocz, J.: The Effect of Initial Imperfections on Shell Stability in Thin-Shell Structure. Ed. Y. C. Fung and E. E. Sethler, Prenctice Hall 1974.
23. Videc, B. P.; Sanders, J. L.: Application of Khas'minskii's Limit Theorem to the Buckling Problem of a Column with Random Initial Deflection to appear in Quart. Appl. Math.

Reliability of Slender Columns: Comparison of Different Approximations

G. Augusti
Università di Firenze, Florence, Italy

A. Baratta
Università di Napoli, Naples, Italy

Abstract

Following the general lines set forth in [1] for the introduction of probabilistic concepts into the practical design of imperfection-sensitive structures, numerical results are presented pertaining to the probability of failure, under random axial load, of an elasto-plastic column studied earlier under given load [2]. Several possible approximate procedures to reduce the needed amount of calculations are proposed and discussed.

List of Symbols

$F(\sigma) = \text{Prob} \{\sigma_c \leqq \sigma\}$	Cumulative probability distribution of critical stress
F_1, F^*, etc.	Approximations of $F(\sigma)$
$f(\sigma) = \mathrm{d}F/\mathrm{d}\sigma$	Probability density of critical stress
$g(\sigma)$	Probability density of applied stress
r	Radius of gyration of column cross-section
s_a, s_c, s_y, s_γ, s_λ	Standard deviation of σ_a, σ_c, σ_y, γ and λ respectively
x_i, $\bar{x}_i$	Generic random variable and its average
y_{om}	Max. initial out-of-straightness of column
$\gamma = y_{om}/r$	Geometrical imperfection parameter
γ_1	Absolute value of γ
$\lambda = l/r$	Slenderness ratio of column (l free buckling length)
$\sigma_a = N/A$	Applied (mean) stress
σ_c	Critical (collapse) stress
$\sigma_E = \pi^2 E/\lambda^2$	Elastic buckling stress
σ_{nom}	Critical stress of ideally perfect column
σ_y	Yield stress
$\mathfrak{P}_c$	Probability of failure (collapse) of column

Introduction

Probabilistic considerations are gaining an increasing place in the rational approach to design in structural, as well as in other branches of engineering: thus, the key design factor, rather than some ill-defined factor of safety, becomes the reliability, or probability of success of the design in the uncertain conditions that can be encountered during its lifetime.

For most structures, the main source of random uncertainty lies in the applied loads, but sometimes the randomness of the parameters affecting the strength is not negligible. The latter is most often the case for imperfection-sensitive structures, not only because of the inherent character of random variable of any imperfection, but also because of the consequent well known dispersion of the actual critical stress (or load-carrying capacity) of the structure. And it is soon to be noted that all structures that are prone to buckling do exhibit—in a larger or lesser degree—imperfection sensitivity and scatter of critical loads, when tested in realistic conditions: this has been well demonstrated by the recent tests of as-received rolled steel colmns carried out under the sponsorship of the European Steel Construction Committee [3].

Unfortunately, as discussed in detail by one of the writers in a recent paper [1], the probabilistic studies of buckling structures that can give useful indications for actual design, are still in their infancy, and no generally accepted procedure has been recognized. In this respect, it must soon be noted that the present state of knowledge would not justify, in actual applications, the great amount of numerical computations that are at the basis of many research papers in this field. In fact, only very small probabilities of failure can be accepted in structural engineering, and this implies a great importance of the tails of the probability distributions of the independent variables, tails that most often can only be inferred by implication and/or extrapolation rather than obtained from statistical measurements.

Moreover, in this range of probabilities, very small variations in the design load (another ill-defined quantity of the deterministic philosophy) may cause the expected probability of failure to change by some order of magnitude (see e.g. [2]).

For all these reasons, it is evident why the development and exploitation of approximate procedures is of the utmost importance at the present stage of development of probabilistic structural analysis and design. It is also clear that, often, at this stage, an approximation, if simple enough, must be considered satisfactory when yielding probabilities of failure of the same order of magnitude as those numerically computed on the basis of assumed (theoretical or extrapolated) probability-distribution functions.

The purpose of this paper is therefore to present some further qualitative and quantitative investigations on the above cited problems, along the lines set forth in [1] and in particular to propose and discuss a number of possible approximations. Throughout the paper, use is made of the "imperfect column" model already presented in [2] and here reproduced in Fig. 1: namely a compressed column made of elastic-perfectly plastic material and initially bent in the shape of a

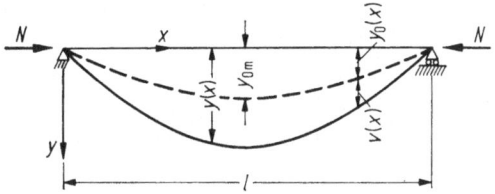

Fig. 1. Column model.

sinusoidal half-wave of maximum amplitude y_{om} (geometrical imperfection). It was shown in [2] that with good approximation the (geometrical) imperfection parameter γ, the mean compressive stress at collapse (critical stress) σ_c, the yield stress σ_y and the elastic buckling stress $\sigma_E = \pi^2 E/\lambda^2$ are related by

$$\gamma_1 = 3(\sigma_y/\sigma_c - 1)(1 - \sqrt[3]{\sigma_c/\sigma_E}). \tag{1}$$

In particular, the relationship between γ_1 and σ_c has the typical concave-upward shape common to all the analogous relationships of imperfection—sensitive structures: σ_c is maximum when $\gamma = 0$ (ideally perfect structure), then decreases with increasing γ_1, at first rapidly, then more and more slowly. Since this variation of the load-carrying capacity with the geometrical imperfection is the most important and typical characteristic of imperfection-sensitive structures, it seems fair to deduce that the results obtained for this particular model apply, at least qualitatively, to other structural elements.

In [2] γ, σ_y, and λ were considered to be normally-distributed random variables, and the cumulative probability distribution function of σ_c, $F(\sigma)$, was calculated numerically under several, apparently realistic, assumptions as to the values of averages and variances of the distributions of γ, σ_y and λ: in particular, the average $\bar{\gamma}$ was assumed to be zero, i.e. equal to the desired value, as seemed logical, although different from the assumption of most other authors ([1], footnote in Sec. 2).

1. Second-Moment Approximation of $F(\sigma)$

As discussed in detail in [1] Sec. 1, many authors who have dealt with similar problems have limited their concern to the determination of the average $\bar{\sigma}_c$ and standard deviation s_c of the critical stress σ_c, obtained from approximate (first-order) formulae that are basically power series expansions of the functional relationship

$$\sigma_c = \sigma_c(x_1, x_2, .., x_n) \tag{2}$$

in the average point of the independent variables $(\bar{x}_1, \bar{x}_2, \ldots, \bar{x}_n)$.

The good approximation of the average, and the acceptable approximation of the standard deviation of the critical stress obtained in this way, for the column model used herein, have been confirmed by new detailed calculations summarized in Fig. 2a, b[1].

However, calculations limited to the average and standard deviation of the sought variable σ_c are significant only if coupled with the assump-

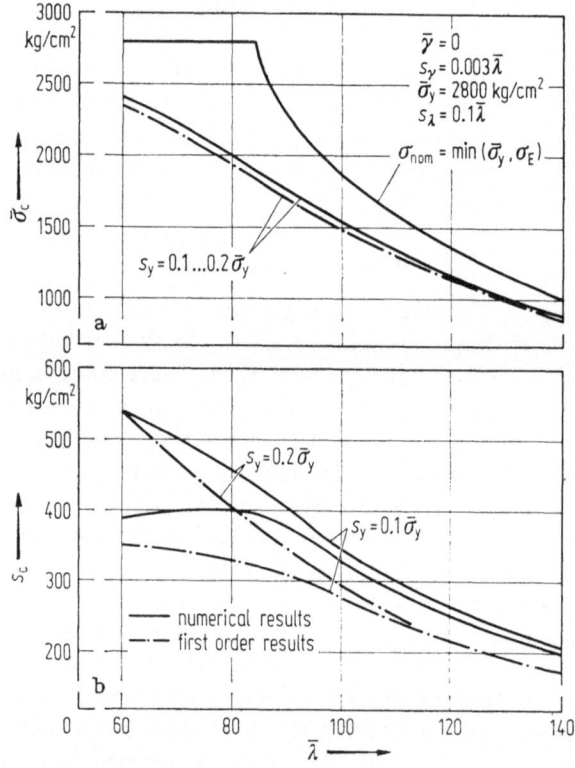

Fig. 2 a and b. Comparison of numerically exact and first order values of average critical stress $\bar{\sigma}_c$ and its standard deviation s_c; the values of $\bar{\sigma}_c$ obtained with $s_y = 0.1\bar{\sigma}_y$ and $s_y = 0.2\bar{\sigma}_y$ are graphically indistinguishable; Fig. 2a shows also the graph of σ_{nom}, the critical stress of the perfect column ($\gamma = \bar{\gamma}$, $\bar{\lambda} = \lambda$, $\sigma_y = \bar{\sigma}_y$, $s_\gamma = s_\lambda = s_y = 0$).

---

[1] It must be remembered that, having assumed $\bar{\gamma} = 0$, the average point of the independent variables is a singular point of the functional relationship $\sigma_c = \sigma_c(\gamma, \lambda, \sigma_y)$: therefore, as suggested in [1] Sec. 2, the average of the absolute value of γ ($\bar{\gamma}_1 = 0.8s_\gamma$) has been introduced in the expansions.

tion that $F(\sigma)$ can be approximated by the normal (Gaussian) distribution with the same average $\bar{\sigma}_c$ and standard deviation s_c. This type of assumption[1] yields acceptable results in many structural problems, in which the functional relationships are approximately linear in the ranges of interest, so that the assumption of normal distribution of the independent variables implies approximately normal distribution of the dependent variables: indeed, the second-moment treatment has been directly or indirectly accepted by most building codes that have intended to put probabilistic considerations at the basis of ther prescriptions.

Unfortunately, in the case of imperfection-sensitive structures, as already remarked, the relationship between critical stress and imperfection is definitely non-linear; consequently, reliable results cannot be expected from the second-moment approximation, even if the imperfection and other independent variables are normally distributed. This had already been pointed out in [1] in which the writer underlined the skewness of the density functions $f(\sigma) = \mathrm{d}F/\mathrm{d}\sigma$ obtained, experimentally and numerically, in previous researches. This is now definitively confirmed (at least for the model column studied herein) by the results reported in Table 1, where numerically calculated values of $F(\sigma)$ are compared with values of the normal distributions $F_1(\sigma)$ and $F_2(\sigma)$, whose average and standard deviation are respectively the same as $F(\sigma)$ and the values obtained from first-order formulae. In fact, $F_1(\sigma)$ and $F_2(\sigma)$ are in most cases close enough to each other (according to the acceptable approximation obtained from first-order formulae for $\bar{\sigma}_c$ and s_c), but become different by orders of magnitude from $F(\sigma)$, as soon as their values fall below a few percent, i.e. fall in the range most significant in structural engineering.

In conclusion, the writers would not recommend the second-moment approximation for the load capacity of imperfection-sensitive structures, and in Sec. 3 below propose and discuss a number of alternative approximations.

2. Numerical Computation of the Probability of Failure $\mathfrak{P}_c$

As well known, each value of the cumulative probability distribution of the critical stress, $F(\sigma)$, coincides with the (conditional) probability of failure of a column subjected to a well determined axial stress $\sigma = \sigma_a = N/A$. Reference [2] (like Sec. 1 above) was limited to the determination of $F(\sigma)$ and did not consider the random variability of the applied stress σ_a.

---

[1] Usually denoted second moment approximation because the variance s_c is equal to the second moments of the area under the density curve $f(\sigma) = \mathrm{d}F/\mathrm{d}\sigma$.

Table 1. Comparison of distribution functions of critical stress: $F(\sigma)$ numerically s_c, $F_2(\sigma)$ normal with first order $\bar{\sigma}_c$ and s_c. Data as in Fig. 2 ($s_y = 0.1\bar{\sigma}_y$). Note

$\bar{\lambda}$		60			80		
Exact	$\bar{\sigma}_c$	2400 kgf/cm²			2020 kgf/cm²		
	s_c	390 kgf/cm²			400 kgf/cm²		
First	$\bar{\sigma}_c$	2350 kgf/cm²			1930 kgf/cm²		
order	s_c	350 kgf/cm²			330 kgf/cm²		
$\sigma = \sigma_c$		$F(\sigma)$	$F_1(\sigma)$	$F_2(\sigma)$	$F(\sigma)$	$F_1(\sigma)$	$F_2(\sigma)$
350							
450							
550							
650							
750							
850		1.1E-10	3.6E-5	8.9E-6	4.7E-8	1.8E-3	5.4E-4
1050		4.6E-8	2.7E-4	1.0E-4	4.7E-5	7.8E-3	3.9E-3
1250		7.5E-6	1.6E-3	8.0E-4	3.2E-3	2.7E-2	2.0E-2
1450		3.9E-1	7.5E-3	5.1E-3	3.8E-2	7.8E-2	7.3E-2
1650		6.5E-3	2.7E-2	2.3E-2	1.6E-1	1.8E-1	2.0E-1
1850		4.5E-2	7.9E-2	7.6E-2	3.2E-1	3.4E-1	4.0E-1
2050		1.6E-1	1.9E-1	2.0E-1			
2250		3.6E-1	3.5E-1	3.9E-1			

If the latter variability is taken into account, the probability of failure of the column is given by the well known convolution integral

$$\mathfrak{P}_c = \int_0^\infty g(\sigma)\, F(\sigma)\, d\sigma, \tag{3}$$

where $g(\sigma)$ is the probability density function of the applied stress σ_a, which is assumed to be always positive (i.e. compressive).

Some curves of $\mathfrak{P}_c$ vs. $\bar{\sigma}_a/\sigma_{nom}$ have been numerically calculated and are plotted in Figs. 3 and 4 drawn in semilogarithmic scale and limited to the range of small failure probabilities. In these computations, in analogy with the already assumed normal distributions of λ, σ_y and γ, σ_a has been assumed to be normally distributed: its average and standard deviation are indicated by $\bar{\sigma}_a$ and s_a respectively, while σ_{nom} is by definition the critical stress of the perfect column, i.e. (Fig. 2a)

$$\sigma_{nom} = \min(\bar{\sigma}_y; \sigma_E). \tag{4}$$

It is soon to be noted that, for values of the parameters that lead to acceptable values of $\mathfrak{P}_c$, the latter depends almost exclusively on the right-hand side of $g(\sigma)$ (i.e. the integration in Eq. (3) can be started from $\bar{\sigma}_a$ rather than 0 without any significant error) and is sensibly exact, $F_1(\sigma)$ normal with numerically exact average $\bar{\sigma}_c$ and standard deviation that 8.9E-6 stands for 8.9×10^{-6}, etc.

100			120			140		
1550 kgf/cm²			1165 kgf/cm²			900 kgf/cm²		
330 kgf/cm²			525 kgf/cm²			205 kgf/cm²		
1490 kgf/cm²			1124 kgf/cm²			860 kgf/cm²		
280 kgf/cm²			214 kgf/cm²			174 kgf/cm²		
$F(\sigma)$	$F_1(\sigma)$	$F_2(\sigma)$	$F(\sigma)$	$F_1(\sigma)$	$F_2(\sigma)$	$F(\sigma)$	$F_1(\sigma)$	$F_2(\sigma)$
						1.0E-9	3.8E-3	1.7E-3
			7.4E-10	2.4E-3	8.5E-4	2.8E-5	1.5E-2	9.4E-3
			6.0E-6	7.8E-3	3.7E-3	5.0E-3	4.5E-2	3.8E-2
4.6E-8	3.5E-3	1.6E-3	4.9E-4	2.0E-2	1.4E-2	5.5E-2	1.1E-1	1.2E-1
1.4E-5	8.2E-3	4.7E-3	1.0E-3	5.2E-2	4.1E-2	2.3E-1	2.4E-1	2.6E-1
3.4E-4	1.8E-2	1.2E-2	5.7E-2	1.1E-1	1.0E-1	4.7E-1	4.1E-1	4.8E-1
2.4E-2	6.7E-2	6.0E-2	3.6E-1	3.3E-1	3.7E-1			
1.7E-1	1.8E-1	2.0E-1						
4.3E-1	3.8E-1	4.4E-1						

affected by its upper tail: both these aspects decrease with increase in the strength dispersion.

Inspection of the figures shows that, over the range investigated (that corresponds to reasonable values of the parameters) the relative importance of the several parameters involved and of their variations varies very much: for instance note how the sensitivity of $\mathfrak{P}_c$ to both the average and the dispersion of the applied load decreases when the dispersion of the yield stress is increased. Similar effects would derive from increases in the dispersion of the other independent variables affecting the column strength.

It is therefore very difficult, if not impossible, to reduce the number of independent variables to be taken into account without causing excessive errors over a more or less large portion of the significant range. Overall approximations seem therefore more promising that simplifications of the functional relationships (2): approximations of this type are discussed in Sec. 3 below.

In order to investigate the dependance of $\mathfrak{P}_c$ on the type of load probability distribution, Fig. 5 compares two curves obtained under the hypothesis of normal (or Gaussian) distribution $g(\sigma)$, and the analogous curves calculated assuming that σ_a is distributed according to the rectangular (or uniform) density function of average $\bar{\sigma}_a$ and stand-

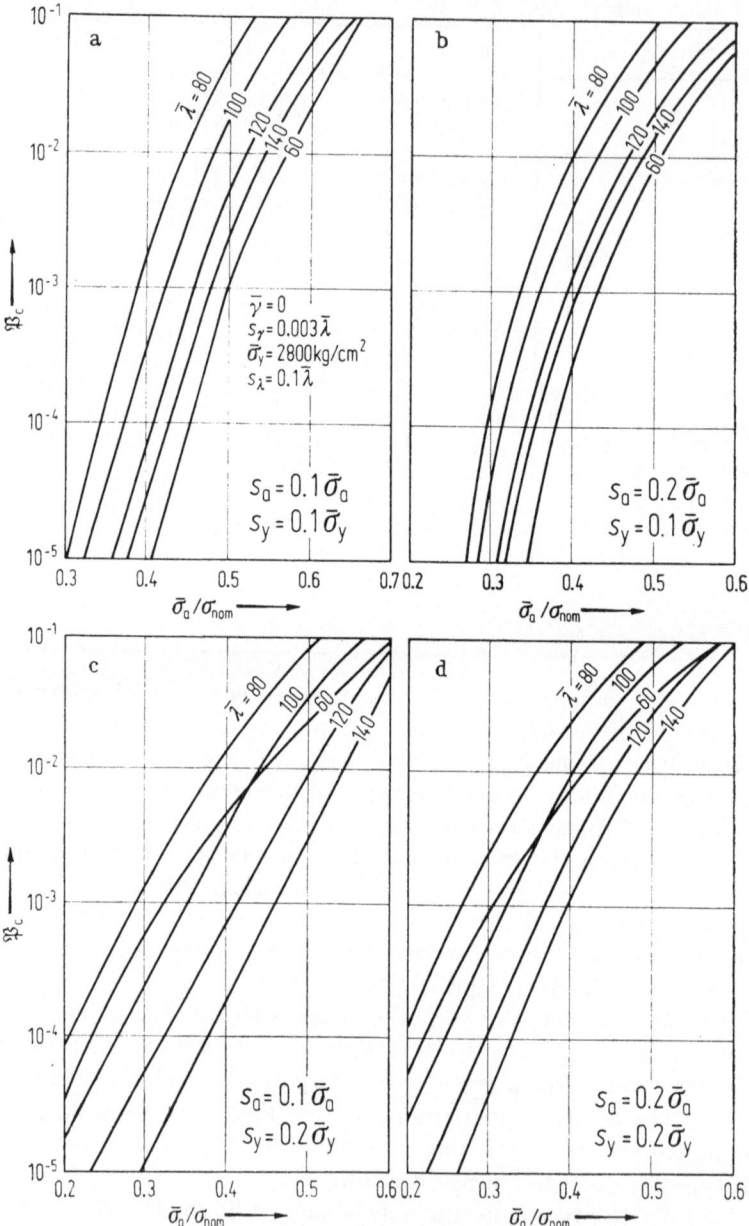

Fig. 3 a—d. Probability of failure $\mathfrak{P}_c$ vs. $\bar{\sigma}_a/\sigma_{nom}$ for normally distributed applied stress σ_a (constant-$\bar{\lambda}$ curves).

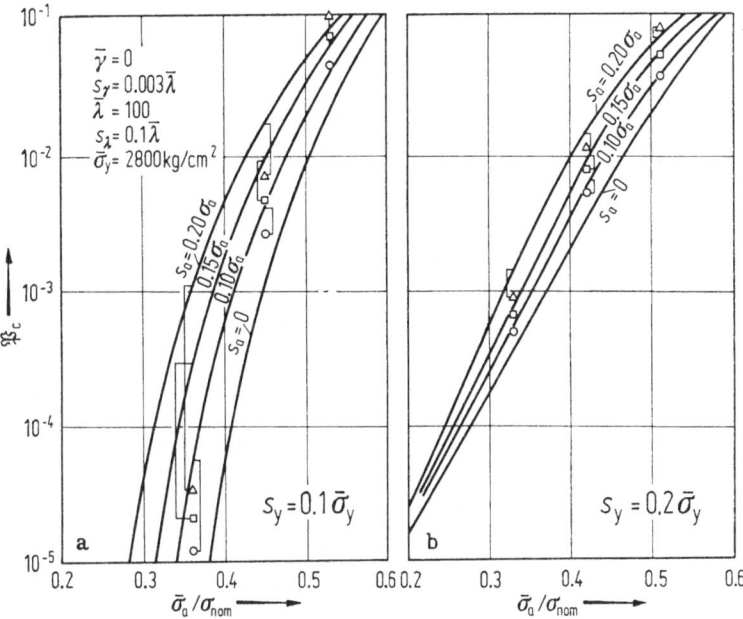

Fig. 4 a and b. Probability of failure $\mathfrak{P}_c$ vs. $\bar{\sigma}_a/\sigma_{nom}$ for normally distributed σ_a (constant-s_a curves); and points calculated from Eq. (12), $N = 2$.

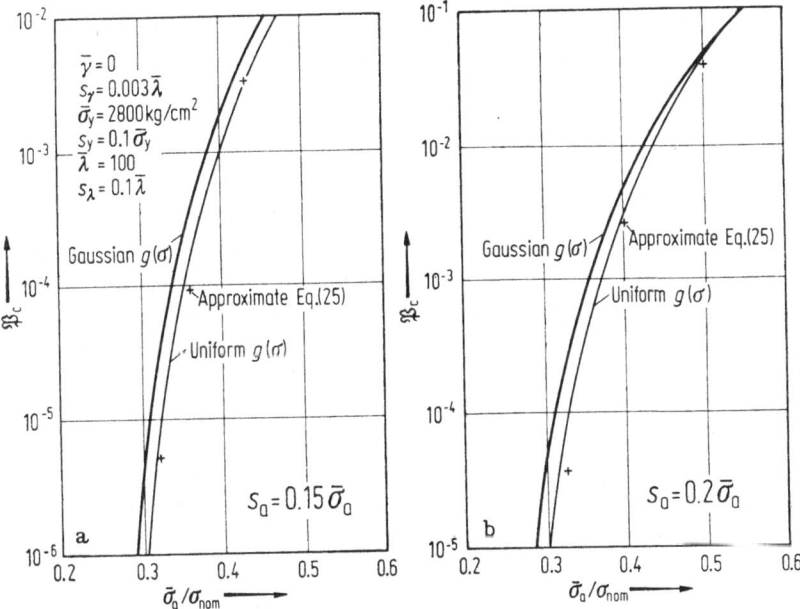

Fig. 5 a and b. Probability of failure $\mathfrak{P}_c$ vs. $\bar{\sigma}_a/\sigma_{nom}$ for normal (Gaussian) and rectangular (uniform) distribution of applied stress σ_a; and points calculated from Eq. (25).

ard deviation s_a

$$g(\sigma) = \begin{cases} 0 & \forall \sigma \in (0, \bar{\sigma}_a - \sqrt{3}\, s_a), \\ 1/(2\sqrt{3}\, s_a) & \forall \sigma \in (\bar{\sigma}_a - \sqrt{3}\, s_a,\ \bar{\sigma}_a + \sqrt{3}\, s_a), \\ 0 & \forall \sigma \in (\bar{\sigma}_a + \sqrt{3}\, s_a, \infty). \end{cases} \qquad (5)$$

Figure 5 shows corresponding curves comparatively close to each other, which would indicate that the actual shape of $g(\sigma)$ (at least as long as it stays symmetrical) is not the most important factor influencing the probability of failure.

3. Approximate Evaluations of $\mathfrak{P}_c$

3.1. "Semiprobabilistic" Approach

A possible way to circumvent the calculation of the convolution integral, Eq. (3), might be an approximation of the type

$$\mathfrak{P}_c = F(\sigma_{ak}), \qquad (6)$$

where σ_{ak} is a specified characteristic value of the applied stress, i.e. a stress such that

$$\text{Prob}\,\{\sigma_a \leqq \sigma_{ak}\} = \int_0^{\sigma_{ak}} g(\sigma)\,d\sigma = k. \qquad (7)$$

In Fig. 6, two $\mathfrak{P}_c$-curves, calculated assuming $g(\sigma)$ normal, are compared with the $F(\sigma)$ curves in which

$$\sigma = \sigma_{ak} = \bar{\sigma}_a + s_a \quad \text{and} \quad \sigma = \sigma_{ak} = \bar{\sigma} + \sqrt{3}\, s_a, \qquad (8)$$

i.e., with the curves (6) corresponding to $k = 84.13\%$ and $k = 95.8\%$ respectively.

Although in the specific examples the curve of $k = 84\%$ is not excessively far from the exact curve, it does not seem that any approximation of the type (6), may yield a priori reliable results; in fact, because of the different inclinations of the curves, an acceptable approximation could be at most reached in a limited range of probabilities.

3.2. Exponential $F(\sigma)$ and further Approximations

In an approximation to the integral, Eq. (3), use can be made of the well known fact [2, 4] that, in the small probability range, any distribution function is well approximated by an exponential function, which in the present case may be written

$$F(\sigma) \approx F^*(\sigma) = F(\bar{\sigma}_a) \exp\,[\alpha(\sigma - \bar{\sigma}_a)], \qquad (9)$$

where

$$\alpha = [dF/d\sigma]_{\sigma=\bar{\sigma}_a} = f(\bar{\sigma}_a). \qquad (10)$$

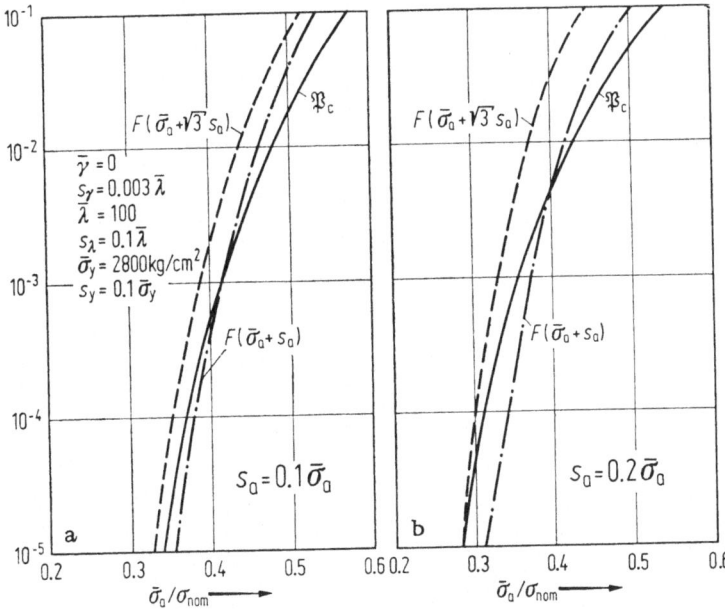

Fig. 6 a and b. Comparison of probability of failure ,from Eq. (3) with conditional probabilities of failure $F(\sigma_{ak})$; normal distribution $g(\sigma)$; $k = 84.1\%$ und 95.8% respectively.

In this way, the exact distribution $F(\sigma)$ and its exponential approximation $F^*(\sigma)$, are made to coincide, with their derivatives, in the neighbourhood of the average $\bar{\sigma}_a$ of the applied stress. Given the small values of $\mathfrak{P}_c$ that are acceptable in structural problems, the exponential approximation $F^*(\sigma)$ is a good approximation in the vicinity of $\sigma = \bar{\sigma}_a$. In the semilogarithmic scale used in the figures of this paper, Eqs. (9) and (10) yield the straight line tangent to $F(\sigma)$ at $\sigma = \bar{\sigma}_a$: an example is shown in Fig. 7.

Expanding $F^*(\sigma)$ into a power series

$$F^*(\sigma) = F(\bar{\sigma}_a)\,[1 + \alpha(\sigma - \bar{\sigma}_a) +$$
$$+ \alpha^2(\sigma - \bar{\sigma}_a)^2/2 + \alpha^3\,(\sigma - \bar{\sigma}_a)^3/6 + \cdots)\quad (11)$$

then substituting into Eq. (3) and performing the integration [3]

$$\mathfrak{P}_c \approx F(\bar{\sigma}_a)\left[1 + \sum_{n=2}^{N}\frac{\alpha^n m_a^{(n)}}{n!}\right],\qquad (N \geqq 2)\qquad (12)$$

where $m_a^{(n)}$ is the n-th moment of the area under the curve $g(\sigma)$ with respect to the axis $\sigma = \bar{\sigma}_a$: when $g(\sigma)$ is symmetrical, all terms with odd n vanish.

Putting $N = 2$ in Eq. (12) (as suggested in [4]) is equivalent to limiting the expansion (11) to the quadratic expression

$$F^*(\sigma) \approx F(\bar{\sigma}_a)\,[1 + \alpha(\sigma - \bar{\sigma}_a) + \alpha^2(\sigma - \bar{\sigma}_a)^2/2] = F^{**}(\sigma) \qquad (11')$$

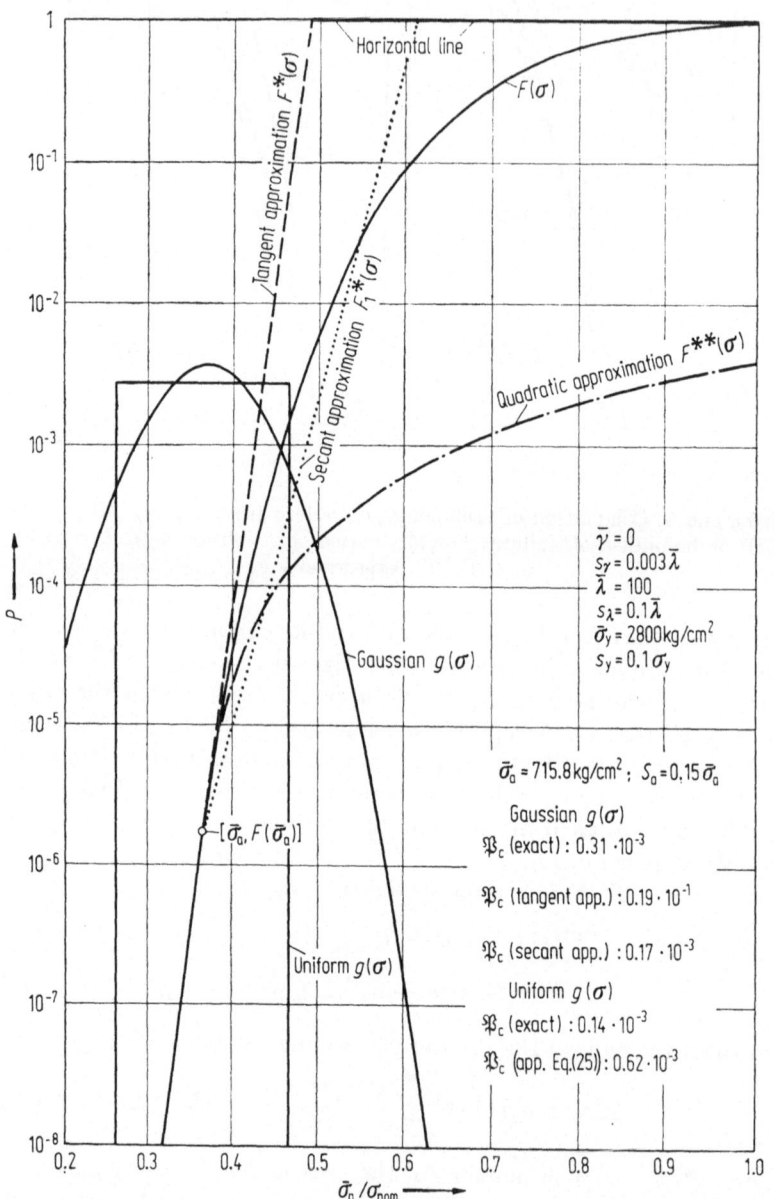

Fig. 7. Example approximations of $F(\sigma)$ and comparison of results.

of which an example is also plotted in Fig. 7. A few values of $\mathfrak{P}_c$ calculated in this way are plotted in Fig. 4; the approximation thus obtained appears to be acceptable in most cases, but can become very unsafe when the dispersion of σ_a prevails, as in the left-hand portion of Fig. 4a.

Inspection of Fig. 7 confirms that, on the whole, acceptable results can be expected from both the exponential approximation $F^*(\sigma)$, Eq. (9), and from the parabolic one $F^{**}(\sigma)$, Eq. (11'), when $F(\bar{\sigma}_a)$ is very small ($F(\bar{\sigma}_a) \leqq 10^{-3}$, say) and the dispersion of the applied load, measured by s_a, is much smaller than the steepness of $F(\sigma)$, measured by $1/\alpha$. In fact, if (but only if) both these conditions hold, nonnegligible contributions to $\mathfrak{P}_c$, Eq. (3), come only from ranges where Eqs. (9) and (11') are good approximations to $F(\sigma)$: moreover, since

$$F^{**}(\sigma) < F(\sigma) < F^*(\sigma) \tag{11''}$$

the two approximations, used in conjunction, would give bounds to the true value of $\mathfrak{P}_c$.

3.3. Exponential $F(\sigma)$ with Upper Cut-Off

If either of the conditions set at the end of 3.2 is not fullfilled, some other approximation to $F(\sigma)$ must be tried.

A first possibility is to continue to accept Eqs. (9) and (10), but with an upper cut-off at the level $F^*(\sigma) = 1$, i.e.

$$F(\sigma) \approx F_1^*(\sigma) = \begin{cases} F(\bar{\sigma}_a) \exp\left[\alpha(\sigma - \bar{\sigma}_a)\right] & \forall \sigma \in (0, \sigma_1) \\ 1 & \forall \sigma \in (\sigma_1, \infty), \end{cases} \tag{13}$$

where α is defined by Eq. (10) and σ_1 by

$$F(\bar{\sigma}_a) \exp\left[\alpha(\sigma_1 - \bar{\sigma}_a)\right] = 1. \tag{14}$$

An alternative is to accept Eqs. (13) and (14), but to calculate α from the condition

$$F(\sigma_2) = F^*(\sigma_2) \tag{15}$$

for some value $\sigma_2 > \bar{\sigma}_a$: i.e.[1]

$$\alpha = \left[\ln F(\sigma_2) - \ln F(\bar{\sigma}_a)\right]/(\sigma_2 - \bar{\sigma}_a). \tag{16}$$

Diagrams of $\mathfrak{P}_c$ calculated with the two approximations of $F(\sigma)$ just described (indicated as "tangent" and "secant" respectively) are plotted in Fig. 8: for the latter, the choice

$$\sigma_2 = \bar{\sigma}_a + 3s_a \tag{15'}$$

---

[1] Note that Eq. (16), with σ_2 close to $\bar{\sigma}_a$, can be used as a numerical approximation to Eq. (10).

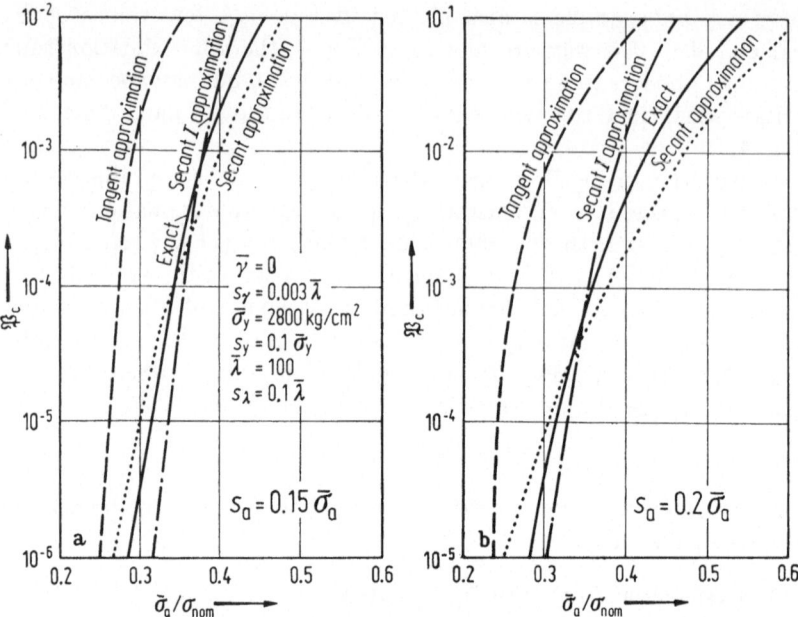

Fig. 8 a and b. Probability of failure $\mathfrak{P}_c$ calculated with normal $g(\sigma)$ and different approximate expressions of $F(\sigma)$.

has been made. A further alternative ("secant I approximation") is to derive α from the condition (Fig. 7)

$$\sigma_1 : F(\sigma_1) = 10^{-1}. \tag{17}$$

In this way, in the large $F(\sigma)$ range, the approximation does not differ from the true $F(\sigma)$ by more than an order of magnitude. However, larger differences can occur between $\bar{\sigma}_a$ and σ_2.

Two curves corresponding to this secant I approximation are also plotted in Fig. 8.

3.4. Exponential $F(\sigma)$ and Rectangular $g(\sigma)$

Most conceptual and numerical difficulties, in the previously discussed approximations, came from the necessity of taking account of the upper tail of the applied load probability density function $g(\sigma)$ (while the lower tail is always negligible for structurally significant values of $\mathfrak{P}_c$). Everything becomes much more simple if the load density can be approximated by a rectangular (or uniform) distribution, Eq. (5).

In this case the order of magnitude of $\mathfrak{P}_c$ is obtained most easily in the following way. Let σ_3 be defined by

$$\sigma_3 = \bar{\sigma}_a + \sqrt{3}\, s_a, \tag{18}$$

and σ_4 by the condition

$$F(\sigma_4) = 0.1 \, F(\sigma_3). \tag{19}$$

Then

$$\mathfrak{P}_c = \int_0^{\sigma_3} g(\sigma) \, F(\sigma) \, d\sigma \approx \int_{\sigma_4}^{\sigma_3} g(\sigma) \, F(\sigma) \, d\sigma, \tag{20}$$

where the difference between two integrals is of the order 10%. But, according to Eq. (5),

$$\int_{\sigma_4}^{\sigma_3} g(\sigma) \, d\sigma = (\sigma_3 - \sigma_4)/(2 \sqrt{3} \, s_a) = 0.289 \, (\sigma_3 - \sigma_4)/s_a, \tag{21}$$

so that upper and lower bounds to Eq. (20) are immediate

$$0.0289 F(\sigma_3) \, (\sigma_3 - \sigma_4)/s_a < \mathfrak{P}_c < 0.289 F(\sigma_3) \, (\sigma_3 - \sigma_4)/s_a . \tag{20'}$$

A better evaluation of $\mathfrak{P}_c$, Eq. (20), is obtained if σ_3 and σ_4 fall in the range where $F(\sigma)$ can be approximated by an exponential function $F^*(\sigma)$ which is most conveniently chosen to satisfy the conditions

$$F^*(\sigma_3) = F(\sigma_3),$$
$$F^*(\sigma_4) = F(\sigma_4), \tag{22}$$

i.e.

$$F^*(\sigma) = F(\sigma_3) \exp \left[\alpha(\sigma - \sigma_3) \right] \tag{23}$$

with α given by

$$F(\sigma_4) = F(\sigma_3) \exp \left[\alpha(\sigma_4 - \sigma_3) \right] \tag{23'}$$

i.e., because of (19),

$$\alpha = \ln 10/(\sigma_3 - \sigma_4). \tag{24}$$

Introducing Eqs. (5) and (24) into the right-hand-side integral in Eq. (20) and performing the integration

$$\mathfrak{P}_c \approx 0.113 F(\sigma_3) \, (\sigma_3 - \sigma_4)/s_a. \tag{25}$$

Some points calculated from Eq. (25) are plotted in Fig. 5. Given the comparatively little amount of computations required, the approximation obtained from Eq. (25) seems quite reasonable[1].

Summary and Conclusions

The probability of failure $\mathfrak{P}_c$ of the elastic-plastic column in Fig. 1 has been investigated, when both the strength parameters and the applied load are random variables. The uncertain statistics of these variables, and the sensitivity of $\mathfrak{P}_c$ to the design load, suggests that approximate

---

[1] Note that the converse formula, obtained taking normal $g(\sigma)$ and rectangular $f(\sigma) = dF/d\sigma$ yields much too small values of $\mathfrak{P}_c$, probably because of the skewness of the actual $f(\sigma)$.

treatments may-be much more appropriate than lengthy numerical calculations.

However, the classical second-moment treatment, based on the assumption that all the relevant quantities are normally distributed, does not appear to be defendable for imperfection-sensitive structures.

Therefore, a number of alternative approximations have been tried and discussed. All of them require the determination of two values of the cumulative probability distribution $F(\sigma)$ of column strength σ_c (or of one value of $F(\sigma)$ and of its derivative $f(\sigma)$) in the low range of probabilities, i.e. not a much greater effort than the determination of the average $\bar{\sigma}_c$ and standard deviation s_c required by the second-moment treatment.

Although none of the approximations suggested can be concluded to be generally acceptable (if anything, because tested only for a specific, albeit typical, structural element), some of them are certainly worth further investigations.

It is the writers' opinion that these kinds of probabilistic investigations are essential for any real progress in the design of imperfection-sensitive structures.

Acknowledgements

The research reported in this paper has been supported by grants of the Italian National Research Council (C.N.R.).

References

1. Augusti, G.: Probabilistic Treatments of Column Buckling Problems. In: Stochastic Problems in Mechanics (Proceedings of a Symposium held on September 24—26, 1973); University of Waterloo Press 1974, pp. 255—274.
2. Augusti, G.; Baratta, A.: Probabilistic Theory of Slender Column Strength. Costruzioni Metalliche 23, No. 1, 1971 (in Italian) and Construction Metallique 8, No. 2, 1971 (in French).
3. Sfintesco, D.: Fondement expérimental des courbes européennes de flambement. Costruzioni Metalliche 22, No. 6, 1970 (in French).
4. Augusti, G.; Baratta, A.: Limit Analysis of Structures with Stochastic Strength Variations. J. Struct. Mech. 1, No. 1, 43—62 (1972).

Some Statistical Aspects of Coupled Buckling Structures

K. C. Johns

Université de Sherbrooke, Sherbrooke, Quebec, Canada

1. Introduction

A statistical approach is taken to the buckling strength of asymmetric two and one degree of freedom structures. The two degree of freedom structures are those studied elsewhere in the idealized [1] and imperfect [2, 3] cases, which display interaction between buckling modes. A goal of the study is to examine buckling load (output) statistics against supposed imperfection (input) statistics.

A statistical attack on the problem seemed justified because a) the theory for the interacting structures indicated widely varying behavior depending on small changes in size of two imperfection parameters, b) the imperfections are small enough to be difficult to measure and will likely vary statistically in any production process of "identical" structures, and c) physical structures which do experience interaction of modes in post-buckling, such as shells, are known to display wide scatter of experimental strength results.

An attempt is also made to compare strength statistics for a general system not experiencing mode interaction, to see, for example, if scatter, or mean load, or probability of failure is worse in the coupled mode structures than in the single mode ones.

2. Probability Density Functions

Studies on the elastic post-buckling behavior of two degree of freedom asymmetric structural systems have shown [2, 3] that imperfection sensitivity can be described by a surface as shown in Fig. 1. This is a surface which can have a polar coordinates description

$$\lambda = 1 - \Lambda^* = [\varepsilon_r / K(\theta')]^{1/2}, \tag{1}$$

where Λ^* is buckling load, λ is change in its value from the idealized case, ε_r is a generalized incremental imperfection parameter and $K(\theta')$ is the curvature of the surface for a given ratio of orthogonal imperfections ε_1, ε_2 or for a given angle θ' (see Fig. 1). This is simply a half-power law sensitivity to imperfections in all ratios and signs, where the coefficient $K(\theta')$ may vary widely and discontinuously. A structure buckling into one mode alone could be thought of as having sensitivity to imperfections defined by

$$\lambda = (\varepsilon_1/K_0)^{1/2}, \tag{2}$$

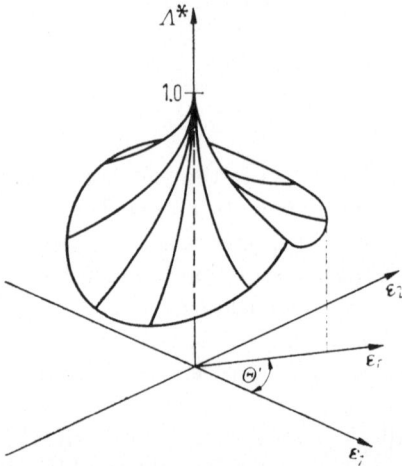

Fig. 1. Imperfection sensitivity surface of coupled system.

which is a cylindrical shape in the space of Fig. 1, and which can be seen in the work of Roorda [4]. The structure is sensitive to imperfection ε_1 and only one sign of it, at that. Figure 2a shows that, in fact, single degree-of-freedom structures will also be sensitive to imperfections $\varepsilon_1 < 0$, which cause collapse due to inelastic deformations. The sensitivity to imperfections is as shown in Fig. 2b. Since the point of this paper is to point out the evils and dangers inherent in systems having interaction effects, we will take the least tendentious stance possible and use the curves of Fig. 2c for a single degree-of-freedom system sensitive to two signs of imperfection. These systems are no more sensitive than this, and some value of Δ^* will probably exist in reality.

Partly for ease of computation, and following other authors, we assume a joint probability distribution for two imperfection parameters ε_1', ε_2' where there is no correlation, such that

$$p(\varepsilon_1', \varepsilon_2') = \frac{1}{2\pi\sigma_1'\sigma_2'} \exp\left\{-\frac{1}{2}\left[\frac{(\varepsilon_1' - m_1')^2}{\sigma_1'^2} + \frac{(\varepsilon_2' - m_2')^2}{\sigma_2'^2}\right]\right\}. \tag{3}$$

We will further assume that either ε_1' or ε_2' may be 'scaled' to give $\sigma_1' = \sigma_2' = \sigma$ and this then becomes

$$p(\varepsilon_1, \varepsilon_2) = \frac{1}{2\pi\sigma^2} \exp\left\{ -\frac{1}{2\sigma^2}((\varepsilon_1 - m_1)^2 + (\varepsilon_2 - m_2)^2) \right\}. \qquad (4)$$

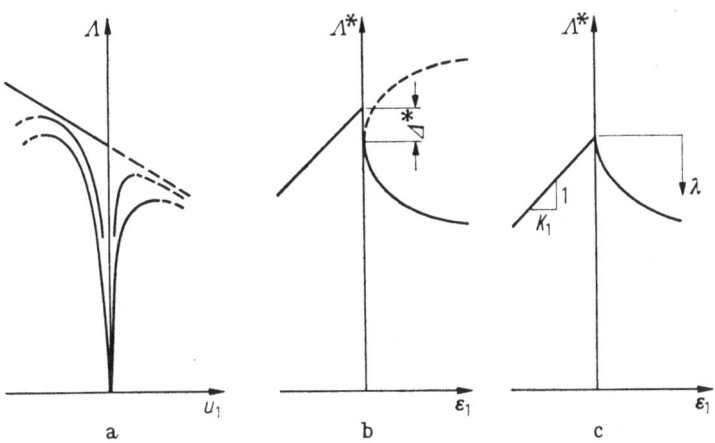

Fig. 2 a—c. Imperfection sensitivity of simple system.

Changing to polar coordinates by the substitution $\varepsilon_1 = \varepsilon_r \cos\theta'$, $\varepsilon_2 = \varepsilon_r \sin\theta'$,

$$p(\varepsilon_r, \theta') = \frac{1}{2\pi\sigma^2} \exp\frac{-1}{2\sigma^2}\{\varepsilon_r^2 - 2\varepsilon_r(m_1 \cos\theta' + m_2 \sin\theta') + (m_1^2 + m_2^2)\}, \qquad (5)$$

which may be written

$$p(\varepsilon_r, \theta) = \frac{1}{2\pi\sigma^2} \exp -\frac{1}{2\sigma^2}\{\varepsilon_r^2 - 2\varepsilon_r\bar{\varepsilon}_r \cos\theta + \bar{\varepsilon}_r^2\}, \qquad (6)$$

where θ, θ', and $\bar{\varepsilon}_r$ are identified in Fig. 3 which shows "contour" lines of equal $p(\varepsilon_1, \varepsilon_2)$ or $p(\varepsilon_r, \theta)$. It seems helpful to write at first that the fraction of structures having change in critical load in a certain incremental range, because of imperfections at two arbitrary angles θ_1, and θ_2:

$$p(\lambda)\,d\lambda\Big|_{\substack{\theta=\theta_1 \\ \theta=\theta_2}} = p(\varepsilon_{r1}, \theta_1)\,d\varepsilon_r\,\varepsilon_{r1}\,d\theta + p(\varepsilon_{r2}, \theta_2)\,d\varepsilon_r\,\varepsilon_{r2}\,d\theta, \qquad (7)$$

so that

$$p(\lambda)\Big|_{\substack{\theta=\theta_1 \\ \theta=\theta_2}} = p(\varepsilon_{r1}, \theta_1)\,\varepsilon_{r1}\,d\theta\,(d\varepsilon_r/d\lambda)\big|_{\theta=\theta_1} + p(\varepsilon_{r2}, \theta_2)\,\varepsilon_{r2}\,d\theta\,(d\varepsilon_r/d\lambda)\big|_{\theta=\theta_2}, \qquad (8)$$

and that the probability density of λ due to all values of θ is simply the integral

$$p(\lambda) = \int\limits_{0}^{2\pi} p(\varepsilon_\mathrm{r}, \theta)\, \varepsilon_\mathrm{r}\, (\mathrm{d}\varepsilon_\mathrm{r}/\mathrm{d}\lambda)\, \mathrm{d}\theta, \qquad (9)$$

where a substitution from (6) for $p(\varepsilon_\mathrm{r}, \theta)$ and from (1) for ε_r and $\mathrm{d}\varepsilon_\mathrm{r}/\mathrm{d}\lambda$ gives

$$p(\lambda) = \frac{\lambda^3}{\pi\sigma^2} \int\limits_{0}^{2\pi} [K(\theta)]^2 \exp\frac{-1}{2\sigma^2} \times$$

$$\times \{[K(\theta)]^2\, \lambda^4 - 2[K(\theta)]\, \lambda^2\bar{\varepsilon}_\mathrm{r} \cos\theta + \bar{\varepsilon}_\mathrm{r}^2)\}\, \mathrm{d}\theta. \qquad (10)$$

This might be appropriate for automatic computation where $K(\theta)$ is obtained by an analytical approach to a specific problem. A cleaner expression results if $K(\theta)$ can be approximated by a single value K for all θ. This gives

$$p(\lambda) = \frac{K^2\lambda^3}{\pi\sigma^2} \exp\left\{\frac{-(K^2\lambda^4 + \bar{\varepsilon}_\mathrm{r}^2)}{2\sigma^2}\right\} \int\limits_{0}^{2\pi} \exp\left\{\frac{K\lambda^2\bar{\varepsilon}_\mathrm{r}}{\sigma^2} \cos\theta\right\} \mathrm{d}\theta. \qquad (11)$$

This can be compared with the integral definition of a modified Bessel function of the first kind of order zero, $I_0(z)$. A series expansion can be used such as is found in [5], and this can be written

$$p(\lambda) = \frac{2K^2\lambda^3}{\sigma^2} \exp\left\{\frac{-(K^2\lambda^4 + \bar{\varepsilon}_\mathrm{r}^2)}{2\sigma^2}\right\} \left\{1 + \left(\frac{K^2\bar{\varepsilon}_\mathrm{r}^2}{\sigma^4 2^2(1!)^2}\right)\lambda^4 + \right.$$

$$\left. + \left(\frac{K^4\bar{\varepsilon}_\mathrm{r}^4}{\sigma^8 2^4(2!)^2}\right)\lambda^8 + \left(\frac{K^6\bar{\varepsilon}_\mathrm{r}^6}{\sigma^{12} 2^6(3!)^2}\right)\lambda^{12} + \cdots\right\}, \qquad (12)$$

which is an initially cubic and higher order function of λ. The first moment, or mean, is

$$\bar{\lambda} = \int\limits_{0}^{\infty} \lambda p(\lambda)\, \mathrm{d}\lambda, \qquad (13)$$

into which may be substituted expression (12). This may be integrated in terms of gamma functions, giving

$$\bar{\lambda} = 2^{1/4} \sqrt{\frac{\sigma}{K}} \exp\left(\frac{-\bar{\varepsilon}_\mathrm{r}^2}{2\sigma^2}\right) \left\{\Gamma(1^1/_4) + \frac{\bar{\varepsilon}_\mathrm{r}^2}{2(1!)^2\,\sigma^2}\, \Gamma(2^1/_4) + \right.$$

$$\left. + \frac{\bar{\varepsilon}_\mathrm{r}^4}{2^2(2!)^2\,\sigma^4}\, \Gamma(3^1/_4) + \frac{\bar{\varepsilon}_\mathrm{r}^6}{2^3(3!)^2\,\sigma^6}\, \Gamma(4^1/_4) + \cdots\right\}, \qquad (14)$$

which converges rapidly for the values of the parameters used in calculation here. The variance may be calculated from the second moment about zero. This is

$$\mathscr{I}_0 = \int\limits_{0}^{\infty} \lambda^2 p(\lambda)\, \mathrm{d}\lambda. \qquad (15)$$

Again substituting from (12) for $p(\lambda)$, this is

$$\mathscr{I}_0 = 2^{1/2} \frac{\sigma}{K} \exp\left(\frac{-\bar{\varepsilon}_r^2}{2\sigma^2}\right) \left\{ \Gamma(1^1/_2) + \frac{\bar{\varepsilon}_r^2}{2(1!)^2 \sigma^2} \Gamma(2^1/_2) + \right.$$

$$\left. + \frac{\bar{\varepsilon}_r^4}{2^2(2!)^2 \sigma^4} \Gamma(3^1/_2) + \frac{\bar{\varepsilon}_r^6}{2^3(3!)^2 \sigma^6} \Gamma(4^1/_2) + \cdots \right\}. \qquad (16)$$

The variance, η_2, is then simply

$$\eta_2 = \mathscr{I}_0 - \bar{\lambda}^2. \qquad (17)$$

To compare this with a single degree-of-freedom system which has a probability density function for the only imperfection which "counts", say ε_1, of

$$p(\varepsilon_1) = \frac{1}{\sqrt{2\pi}\,\sigma} \exp\left\{\frac{-(\varepsilon_1 - m_1)^2}{2\sigma^2}\right\}, \qquad (18)$$

a sensitivity to imperfections must be assumed. The one chosen corresponds to Fig. 2c and is

$$\lambda = [\varepsilon_1/K_0]^{1/2}, \quad \varepsilon_1 \geqq 0,$$

$$\lambda = -\varepsilon_1/K_1, \quad \varepsilon_1 \leqq 0. \qquad (19)$$

The probability density of λ is then calculated by perceiving that the fraction of structures buckling at a certain load value is made up of the sum of the fraction of structures having imperfections at two different positive and negative values of ε_1. With a technique similar to that used above, it is possible to write

$$p(\lambda) = \frac{1}{\sqrt{2\pi}\,\sigma} \left[2K_0\lambda \exp\left\{\frac{-(K_0\lambda^2 - m_1)^2}{2\sigma^2}\right\} + \right.$$

$$\left. + K_1 \exp\left\{\frac{-(K_1\lambda + m_1)^2}{2\sigma^2}\right\} \right]. \qquad (20)$$

Restricting consideration to cases which buckle elastically only, i.e. for $\varepsilon_1 \geqq 0$, the first term in square parentheses only will apply, and it is necessary to divide by N, which is the fraction of structures buckling elastically,

$$N = \int_0^\infty p(\varepsilon_1)\, d\varepsilon_1. \qquad (21)$$

This gives

$$p(\lambda) = \frac{2K_0\lambda}{N\sqrt{2\pi}\,\sigma} \exp\left\{\frac{-(K_0\lambda^2 - m_1)^2}{2\sigma^2}\right\}. \qquad (22)$$

This is essentially Roorda's [4] line of attack, and it avoids the arbitrary approach to inelastic failure taken here. An explicit expression for $\bar{\lambda}$ in either of these single degree-of-freedom cases is not available, but for the case of m_1 equal to zero it is possible to establish that

$$\bar{\lambda}\,\big|_{m_1=0} = 2^{-3/4}\sqrt{\frac{\sigma}{\pi K_0}}\,\Gamma(3/4) + \frac{\sigma}{\sqrt{2\pi K_1}}, \tag{23}$$

for the case of the system treated in expression (20). The second moment about zero is

$$\mathcal{I}_0\,\big|_{m_1=0} = \frac{\sigma}{\sqrt{2\pi}\,K_0} + \frac{\sigma^2}{\sqrt{\pi K_1^2}}\,\Gamma(1^1/_2). \tag{24}$$

It is necessary to compute results for some typical cases in order to compare anything about the two types of systems, but one thing apparent from a look at (20), or (22), and (12) is that the single degree-of-freedom systems give a probability density function for λ, change in buckling load, which is initially linear in λ, whereas the two degree-of-freedom systems have initially cubic curves.

3. Computational Results

It was deemed useful to compare probability density functions and their first and second moments, for change in buckling load for the two types of system, for different imperfection distribution parameters. Calculations were based on the arbitrary selection of $\bar{\varepsilon}_r$ varying from 0 through to 0.200, and σ values from 0 through to 0.250. For comparison purposes, m_1 was taken in the same range as $\bar{\varepsilon}_r$, even though Fig. 3 shows that m_1 is in fact the ε_1 component of $\bar{\varepsilon}_r$. Unity was assigned to the values of the sensitivity coefficients K, K_0, and K_1 was taken as 0.5. A system originally buckling into one mode which is optimized to give interaction-type buckling will not necessarily have the same curvature of the sensitivity surface, K, as it had curvature of the

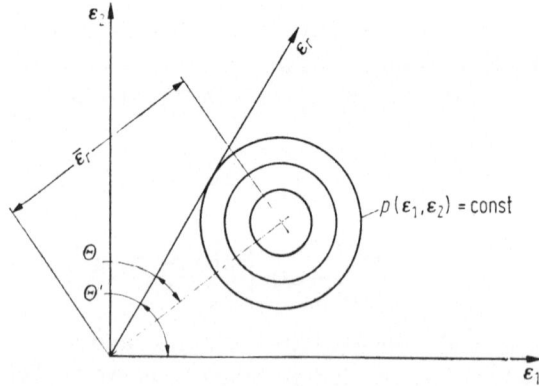

Fig. 3. Joint probability density of imperfections.

sensitivity curve K_0. Indeed, the theory [3] shows that K varies with θ, so any general comparison of this type involves a necessary over-simplification.

Figure 4 shows the probability density function for the two degree-of-freedom system, i.e. Eq. (12), as a solid curve, the density function of Eq. (20) as dot-dashes and Eq. (22) as long dashes. The same

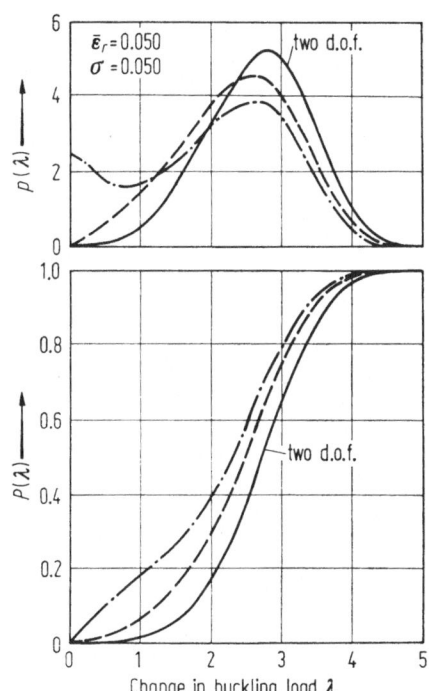

Fig. 4. Probability density
and cumulative probability
of load change.

convention is used in all subsequent figures. The abscissa is λ, or change in buckling load from the ideal, so that by Eq. (1), buckling load Λ^* may read as $(1 - \lambda)$. The extreme left abscissa is the idealized buckling load. The curve of dot-dashes has a finite ordinate at zero, which may be seen by inspection from Eq. (20). This is associated with the finite slope of the sensitivity curve shown in Fig. 2c. The coupled system density curve, and cumulative probability curve shown below, are shifted to the right with respect to the others. Abscissas corresponding to cumulative probability of the order of 0.02 show roughly double the change in buckling load, λ. Figure 5 shows the same curves for larger input statistics $\bar{\varepsilon}_r$ and σ. In this case there is only a 2% chance of finding buckling strength within 18.6% of the ideal value in the coupled, or two degree-of-freedom system, compared to 9.9% of the ideal in the

single degree-of-freedom system associated with Eq. (22), and 2.5%
of the ideal in the system experiencing inelastic effects.

Figures 6 and 7 show output statistics $\bar{\lambda}$, or average change in buckling strength, and η_2, its variance, against input statistics. These
demonstrate that $\bar{\lambda}$ is higher, i.e. buckling load is, on average, lower for

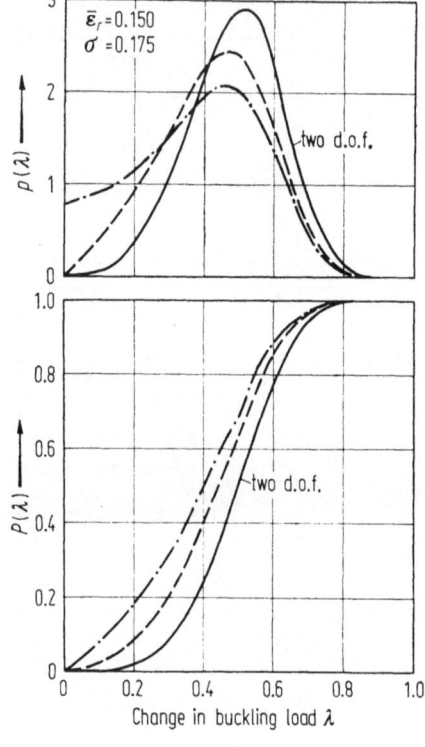

Fig. 5. Probability density
and cumulative probability
of load change.

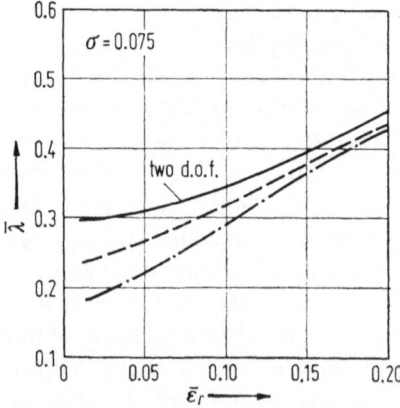

Fig. 6. Mean load change
against mean imperfection.

the coupled system, but that variance is not higher. The hypothetical, general system adopted does not, therefore, prove that scatter as measured by variance is greater for coupled systems. Perhaps this can be established by examination of a specific structure. The imperfection sensitivity surface will then undulate more, circumferentially, and this will create further scatter. This seems a logical next step for this research.

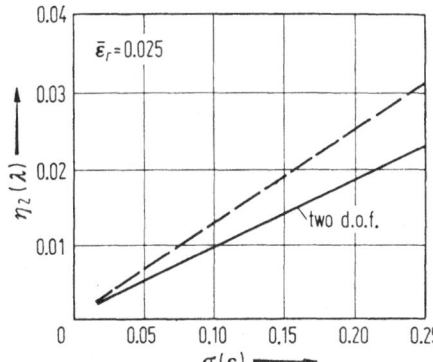

Fig. 7. Variance of load change against standard deviation of imperfections.

References

1. Chilver, A. H.: J. Mech. Phys. Solids **15**, 15 (1966).
2. Ho, D.: Int. J. Non-Linear Mech. **7**, 311 (1971).
3. Johns, K. C.: Int. J. Non-Linear Mech. **9** (1974) (in press).
4. Roorda, J.: J. Mech. Phys. Solids **17**, 111 (1969).
5. Gradshteyn, I. S.; Ryzhik, I. M.: Tables of Integrals, Series and Products, Academic Press 1965.

Some Remarks on Liapunov Stability of Elastic Dynamical Systems

H. H. E. Leipholz

University of Waterloo, Waterloo, Ontario, Canada

1. Introduction

It is well known [1] that for a conservative elastic system, the Hamiltonian can be used as a Liapunov functional. Aim of this paper is to show that for certain kinds of nonconservative systems either this fact remains true or at least a kind of generalized Hamiltonian can be found, which serves the same purpose.

In order to remind the reader of some basic features of Liapunov stability, firstly, a brief review of this theory will be given in the context of conservative systems. Secondly, some specific types of nonconservative systems will be considered for which a Liapunov functional can be specified.

2. Conservative Systems

Let

$$L = \int_V \mathscr{L} \, dV \tag{1}$$

be the Lagrangian of the system occupying the volume V and having the Lagrangian density $\mathscr{L}$. Let q be the coordinate vector and $\delta L/\delta q$ be the functional derivative of L. Then,

$$\frac{d}{dt} \frac{\delta L}{\delta \dot{q}} - \frac{\delta L}{\delta q} = 0 \tag{2}$$

are the Lagrangian equations of the system. Introducing as new variables the impulses p, the Hamiltonian can be defined as

$$H = \int_V p\dot{q} \, dV - L(q, p). \tag{3}$$

Applying (2) not only to q but also to the new coordinate vector p, and using in addition (3), the set of equations

$$\frac{d}{dt} \frac{\delta L}{\delta \dot{q}} - \frac{\delta L}{\delta q} = \dot{p} + \frac{\delta H}{\delta q} = 0, \quad \frac{d}{dt} \frac{\delta L}{\delta \dot{p}} - \frac{\delta L}{\delta p} = -\dot{q} + \frac{\delta H}{\delta p} = 0 \tag{4}$$

is obtained, which can be rewritten to yield the canonical equations

$$\frac{\delta H}{\delta q} = -\dot{p}, \quad \frac{\delta H}{\delta p} = \dot{q}. \tag{5}$$

The time derivative of H is

$$\dot{H} = \int_V \left[\frac{\delta H}{\delta q} \dot{q} + \frac{\delta H}{\delta p} \dot{p} \right] dV. \tag{6}$$

Using (5) in (6) yields

$$\dot{H} = 0. \tag{7}$$

Hence, H can be used to ensure stability of the system as long as it is positive definite. This results from the following

Theorem: Let γ_0 and γ_1 be two appropriate positive constants. Let ϱ_0 and ϱ_1 be two norms in the abstract space in which the posed problem is being considered and let ϱ_1 depend continuously on ϱ_0. Let $H \geq \gamma_1 \varrho_1$ (i.e., H is positive definite) and $H \leq \gamma_0 \varrho_0$ (i.e., H admits an infinitesimal upper bound with respect to ϱ_0) be valid by definition. Then, the system is stable with respect to ϱ_0 und ϱ_1 if

$$\dot{H} \leq 0. \tag{8}$$

Proof: Let $\varrho_0 < \delta$ for $t = 0$. Then, due to $H \geq \gamma_1 \varrho_1$, $H \leq \gamma_0 \varrho_0$, and (8),

$$\gamma_1 \varrho_1 \leq H \leq \gamma_0 \varrho_0 \tag{9}$$

for any $t > 0$. Stability requires $\varrho_1 < \varepsilon$ for any $t > 0$. This can be ensured according to (8) and (9) by chosing $\varrho_0 < \delta = \gamma_1 \varepsilon / \gamma_0$ for $t = 0$, Hence, stability prevails if H is positive definite and if (7) holds good, a condition which complies with (8).

For linear systems, the integrand of H is a quadratic form. For nonlinear systems,

$$H = H_0 + H^*, \tag{10}$$

where the integrand of H_0 is again a quadratic form and represents the Hamiltonian which corresponds to the linearized part of the actually nonlinear system. Hence, the integrand of H^* contains the terms in the integrand of H which are of higher than second order. Let η_i, $i = 1, 2, \ldots, n$, be the variables of the abstract space in which the posed problem is being considered (the η_i represent the coordinates q and their spatial derivatives as well as the $\dot{q}$). Then, stability can be predicted in the first approximation, i.e., H_0 determines already stability, if the integrand $\mathscr{H}^*$ of H^* satisfies condition

$$\mathscr{H}^* \leq M(|\eta_1|^{3+\alpha} + |\eta_2|^{3+\beta} + \cdots + |\eta_n|^{3+\omega}), \tag{11}$$

where $M, \alpha, \beta, \ldots, \omega$ are appropriate positive constants.

3. Nonconservative Systems

In this case, the vector Q of polygenic forces, i.e., of forces which are not derivable from a potential, enters into the canonical equations which now read

$$\dot{p} = -\frac{\delta H}{\delta q} + Q, \quad \dot{q} = \frac{\delta H}{\delta p}. \tag{12}$$

Hence, (7) changes into

$$\dot{H} = \int\limits_V \left[\frac{\delta H}{\delta q} \dot{q} + \frac{\delta H}{\delta p} \dot{p} \right] dV = \int\limits_V \dot{q} Q \, dV \equiv 0 \tag{13}$$

according to (6) and (12).

3.1. Let the nonconservative system be purely dissipative, i.e., let there exist a dissipative Rayleigh's functional

$$R = (1/2) \int\limits_V \mathscr{R} \, dV, \quad \mathscr{R} = \beta \dot{q}^2, \beta > 0. \tag{14}$$

Then,

$$Q = -(1/2) \, (\partial \mathscr{R}/\partial \dot{q}) = -\beta \dot{q}, \tag{15}$$

and

$$\dot{H} = -\int\limits_V \beta \dot{q}^2 \, dV < 0, \tag{16}$$

which follows from (13), (14) and (15). Hence, for systems of this type, condition (8) of the stability theorem holds true. Obviously, the Hamiltonian H can be used as a Liapunov functional as long as H is positive definite.

3.2. Let the nonconservative system involve follower forces. For systems of this type,

$$Q = \alpha q_{x_i}, \quad q_{x,i} = \partial q/\partial_{x,i}, \quad i = 1, 2, 3, \quad \alpha = \text{const}. \tag{17}$$

Condition (13) yields

$$\dot{H} = \int\limits_V \alpha \dot{q} q_{x_i} \, dV. \tag{18}$$

In order to predict whether (8) is satisfied or whether (18) is at least periodic and bounded, (which also would ensure stability if H were positive definite), one had to know the solution of q, which is out of question, or one had to make additional assumptions with respect to the nature of q like in [1, 2, 3], or [4].

However, besides this, a special class of nonconservative systems with follower forces can be defined, for which in place of the Hamiltonian another functional, the so called "meta-energy", is conservative. Systems of this class shall be called "conservative of the second kind". Obviously, the "meta-energy", a sort of generalized Hamiltonian, may be used as a Liapunov functional.

3.3. Conservative systems of the second kind.

(a) *Definitions, Concepts and Propositions*

Given function space $\mathcal{S}$. Let $\phi(x) \in \mathcal{S}$, $\psi(x) \in \mathcal{S}$ be elements in $\mathcal{S}$ with domain V of x.

Consider operators P and Q in domain V.

Definition 1.: Functional

$$\mathcal{F}(\phi, \psi) = \int_V P(\phi)\, Q(\psi)\, \mathrm{d}V \equiv \langle P(\phi), Q(\psi) \rangle \tag{19}$$

is said to be the *vigor of P in V with respect to Q.*

Definition 2: Functional

$$\mathcal{F}_\mathrm{v}(\phi, \psi) = \langle P(\phi), \mathrm{d}Q(\psi) \rangle \tag{20}$$

is said to be the *virtual vigor of P in V with respect to Q.* Functional

$$\mathcal{F}_\mathrm{v}^*(\phi, \psi) = \langle \mathrm{d}P(\phi), Q(\psi) \rangle \tag{21}$$

is said to be the *complementary virtual vigor of P in V with respect to Q.*

In (20) and (21), $\mathrm{d}P$ and $\mathrm{d}Q$ are Fréchet-differentials of operators P and Q.

Definition 3: Operator P is said to be symmetric with respect to Q, if

$$\langle \mathrm{d}P(\phi), Q(\psi) \rangle \equiv \langle P(\psi)\, \mathrm{d}Q(\phi) \rangle, \tag{22}$$

and ϕ, ψ are in the domain of definition of P, Q. According to this definition, symmetry is equivalent with the state of virtual vigor and complementary virtual vigor being equal.

Corollary 3a: If

$$Q \equiv I, \quad I \text{ unit operator}, \tag{23}$$

then operator P satisfying (22) under assumption (23) is said to be *symmetric with respect to itself.*

Corollary 3b: For a linear operator, for example for P being linear,

$$\mathrm{d}P(\phi) = P(\phi + k) - P(\phi) \equiv P(k), \ \phi \text{ and } k \in \mathcal{S}. \tag{24}$$

Hence, for P being symmetric with respect to Q, and P and Q being linear, relationship (22) yields

$$\langle P(k), Q(\psi) \rangle \equiv \langle P(\psi)\, Q(k) \rangle. \tag{25}$$

Corollary 3c: For a linear operator P which is symmetric to itself (i.e., $Q \equiv I$, I unit operator), relationship (22) yields

$$\langle P(k), \psi \rangle \equiv \langle P(\psi), k \rangle. \tag{26}$$

Definition 4: If a pair of linear operators P, Q does satisfy condition (25), then operator P is said to be *selfadjoint with respect to operator Q.*

Definition 5: If a linear operator P satisfies condition (26), then it is said to be *selfadjoint* (with respect to itself).

Definition 6: P is said to be the gradient of functional $\mathscr{F}(\phi, \phi)$ with respect to Q if

$$\frac{\mathrm{d}\mathscr{F}}{\mathrm{d}t} = 2 \int\limits_V P(\phi) \frac{\mathrm{d}Q(\phi)}{\mathrm{d}t} \, \mathrm{d}V, \quad \phi = \phi(x, t). \tag{27}$$

Proposition 1: For P to be the gradient of $\mathscr{F}(\phi, \phi)$ with respect to Q it is sufficient and necessary that P be symmetric with respect to Q.
Proof: (α) Assume P to be symmetric with respect to Q. Then, (22) holds. Also,

$$\frac{\mathrm{d}\mathscr{F}}{\mathrm{d}t} = \int\limits_V \left[\frac{\mathrm{d}P}{\mathrm{d}t} Q + P \frac{\mathrm{d}Q}{\mathrm{d}t} \right] \mathrm{d}V. \tag{28}$$

Using (19) and (22) in (28),

$$\frac{\mathrm{d}\mathscr{F}}{\mathrm{d}t} = 2 \int\limits_V P \frac{\mathrm{d}Q}{\mathrm{d}t} \, \mathrm{d}V.$$

According to (27), this proves that symmetry of P with respect to Q is sufficient for P to be the gradient of $\mathscr{F}$.
(β) Assume P to be the gradient of $\mathscr{F}$. Then, (27) holds. At the same time (28) holds. For (27) and (28) to hold simultaneously, it follows necessarily from equating (27) and (28) that

$$\int\limits_V P \frac{\mathrm{d}Q}{\mathrm{d}t} \, \mathrm{d}V = \int\limits_V \frac{\mathrm{d}P}{\mathrm{d}t} Q \, \mathrm{d}V = \left\langle P \frac{\mathrm{d}Q}{\mathrm{d}t} \right\rangle \equiv \left\langle \frac{\mathrm{d}P}{\mathrm{d}t} Q \right\rangle$$

must hold. Hence, (22) is necessary for (27) to hold true, i.e., symmetry of P with respect to Q is necessary for P to be the gradient of $\mathscr{F}$.

Consider a dynamic system with volume V having the trajectory $w(x, t)$ in the x, t space.

Definition 7: Functional

$$\mathscr{T}(\dot{w}, w) = \int\limits_V \mu \dot{w} \frac{\mathrm{d}T(w)}{\mathrm{d}t} \, \mathrm{d}V \tag{29}$$

is said to be the kinetic vigor in V with respect to operator T, when $\dot{w} = \mathrm{d}w/\mathrm{d}t$, and μ is the mass density of the system.

Definition 8: Functional

$$\mathscr{P}(w, w) = \int\limits_V F(w) \, T(w) \, \mathrm{d}V \tag{30}$$

is said to be the potential vigor in V with respect to operator T, when $F(w)$ is the differential operator representing all internal and external forces acting on the system.

Corrolary 8: If $T \equiv I$, I unit operator, $\mathscr{T}$ is twice the kinetic energy and $\mathscr{P}$ is twice the potential energy of the system.

Proposition 2: Let

$$\mu \ddot{w} + F(w) = 0 \tag{31}$$

be the differential equation of the dynamic system yielding as its solution the trajectory w of the system. Moreover, let operator $\mathrm{d}T(w)/\mathrm{d}t$ be the gradient of $\mathscr{T}$ with respect to unit operator I, and operator $F(w)$ be the gradient of $\mathscr{P}$ with respect to operator T. Then, the functional

$$\mathscr{E}(\dot{w}, w) = \mathscr{T}(\dot{w}, w) + \mathscr{P}(w, w), \tag{32}$$

called "meta-energy", is *conservative*, i.e., $\mathrm{d}\mathscr{E}/\mathrm{d}t = 0$, $\mathscr{E} = $ const.

Corrolary 2a: If $\mathscr{E}$ is conservative, the corresponding dynamic system for which $\mathscr{E}$ holds is said to be *conservative of the second kind*. If $T \equiv I$, I unit operator, $\mathscr{E}$ becomes the mechanical energy E of the system. If E is conservative, the corresponding dynamic system is said to be *conservative of the first kind*, i.e., conservative in the classical sense.

Proof of Proposition 2: According to (32) and definition 6, and by assumption,

$$\frac{\mathrm{d}\mathscr{E}}{\mathrm{d}t} = \frac{\mathrm{d}\mathscr{T}}{\mathrm{d}t} + \frac{\mathrm{d}\mathscr{P}}{\mathrm{d}t} = 2 \int_V \mu \frac{\mathrm{d}T(w)}{\mathrm{d}t} \frac{\mathrm{d}I(\dot{w})}{\mathrm{d}t} \mathrm{d}V + 2 \int_V F(w) \frac{\mathrm{d}T(w)}{\mathrm{d}t} \mathrm{d}V.$$

However,

$$\frac{\mathrm{d}I(\dot{w})}{\mathrm{d}t} \equiv \ddot{w}.$$

Hence,

$$\frac{\mathrm{d}\mathscr{E}}{\mathrm{d}t} = 2 \int_V [\mu \ddot{w} + F(w)] \frac{\mathrm{d}T(w)}{\mathrm{d}t} \mathrm{d}V. \tag{33}$$

Since (31) holds, relationship (33) yields

$$\frac{\mathrm{d}\mathscr{E}}{\mathrm{d}t} = 0, \quad \text{i.e.,} \quad \mathscr{E} = \text{const},$$

i.e., $\mathscr{E}$ is conservative.

Corrolary 2b: Under the conditions in proposition 2, functional $\mathscr{E}$ can serve as a Liapunov-functional if it is definite.

Proof: Let $\mathscr{E}(\dot{\phi}, \phi) = \mathscr{T}(\dot{\phi}, \phi) + \mathscr{P}(\phi, \phi)$ be a functional in the space $\mathscr{S}$ in which ϕ is an arbitrary element. Then, due to the conditions in proposition 2,

$$\frac{\mathrm{d}\mathscr{E}}{\mathrm{d}t} = 2 \int_V [\mu \ddot{\phi} + F(\phi)] \frac{\mathrm{d}T(\phi)}{\mathrm{d}t} \mathrm{d}V.$$

Moreover, let $\mathcal{E}$ be definite by assumption. For $\phi \to w$, $d\mathcal{E}/dt = 0$, according to (31) and (33). Hence, $\mathcal{E}$ definite is sufficient for stability of the dynamic system according to the stability theorem in section 2.

Corrolary 2c: Proposition 2 holds true if (31) holds true, and if in addition $dT(w)/dt$ is symmetric with respect to $I(\dot{w})$, and $F(w)$ is symmetric with respect to $T(w)$.

Proof: According to proposition 1, $dT(w)/dt$ is then the gradient of $\mathcal{T}$ with respect to $I(\dot{w})$ and $F(w)$ is then the gradient of $\mathcal{P}$ with respect to $T(w)$. Also, (31) holds true by assumption. Hence, all the conditions of proposition 2 are satisfied.

(b) *Extension of Hamilton's Principle*

Consider the "meta-Lagrangian"

$$\mathcal{L}^* = \int\limits_V \left[\mu \dot{w} \frac{dT(w)}{dt} - F(x, w)\, T(w) \right] dV = \mathcal{T} - \mathcal{P}. \qquad (34)$$

Proposition 3: If $dT(w)/dt$ is symmetric with respect to $I(\dot{w})$, $F(x, w)$ is symmetric with respect to $T(w)$, and if

$$[\delta(T(w))]_{t_0}^{t_1} = 0, \qquad (35)$$

then the following variational principle

$$\delta \int\limits_{t_0}^{t_1} \mathcal{L}^* \, dt = 0, \qquad (36)$$

holds true.

Proof:

$$\delta \int\limits_{t_0}^{t_1} \mathcal{L}^* \, dt = \int\limits_{t_0}^{t_1} \int\limits_V \left[\mu \dot{w}\, \delta\left(\frac{dT(w)}{dt} \right) + \mu \frac{dT(w)}{dt} \delta \dot{w} - \delta(F(x, w))\, T(w) - \right.$$
$$\left. - F(x, w)\, \delta(T(w)) \right] dV \, dt. \qquad (37)$$

Also, using (35),

$$\int\limits_{t_0}^{t_1} \int\limits_V \mu \dot{w}\, \delta\left(\frac{dT(w)}{dt} \right) dV \, dt = \int\limits_{t_0}^{t_1} \int\limits_V \mu \dot{w} \frac{d}{dt}(\delta T(w)) \, dV \, dt =$$
$$= \int\limits_V \left[\int\limits_{t_0}^{t_1} \mu \dot{w} \frac{d}{dt}(\delta T(w)) \, dt \right] dV = \int\limits_V \left\{ [\mu \dot{w}\, \delta T(w)]_{t_0}^{t_1} - \int\limits_{t_0}^{t_1} \mu \ddot{w}\, \delta T(w)\, dt \right\} dV$$
$$= \int\limits_{t_0}^{t_1} \left[\int\limits_V [-\mu \ddot{w}\, \delta T(w)] \, dV \right] dt. \qquad (38)$$

Moreover, because of symmetry,

$$\int\limits_V \mu \dot{w}\, \delta\left(\frac{dT(w)}{dt} \right) dV = \int\limits_V \mu\, \delta \dot{w} \frac{dT(w)}{dt} \, dV, \qquad (39)$$

and

$$\int_V \delta F(x, w) \, T(w) \, \mathrm{d}V = \int_V F(x, w) \, \delta T(w) \, \mathrm{d}V. \tag{40}$$

Hence, using (38), (39) and (40) in (37) yields

$$\delta \int_{t_0}^{t_1} \mathscr{L}^* \, \mathrm{d}t = -2 \int_{t_0}^{t_1} \int_V \{[\mu \ddot{w} + F(x, w)] \, \delta T(w)\} \, \mathrm{d}V \, \mathrm{d}t. \tag{41}$$

Let (31) be the differential equation of the dynamic system, then (41) yields (36).

Let the variation of $\delta T(w)$ be arbitrary, then the principle (36) together with (41) yields the differential equation (31) of the dynamic system.

(c) *Extension of Rayleigh's Principle*

Consider a dynamic system having the differential equation

$$\mu \ddot{w} + \lambda F(x, w) = 0, \tag{42}$$

where λ is an eigenvalue. Moreover, consider the functional

$$\int_{t_0}^{t_1} \mathscr{L}^* \, \mathrm{d}t = \int_{t_0}^{t_1} \int_V \left[\mu \dot{w} \frac{\mathrm{d}T(w)}{\mathrm{d}t} - \lambda F(x, w) \, T(w) \right] \mathrm{d}V \, \mathrm{d}t \tag{43}$$

under the assumption that $\mathrm{d}T/\mathrm{d}t$ is symmetric with respect to $I(\dot{w})$, and F is symmetric with respect to T. Finally, consider the "meta-Rayleighian"

$$R_g = \left[\int_{t_0}^{t_1} \int_V \mu \dot{\phi} \frac{\mathrm{d}T(\phi)}{\mathrm{d}t} \, \mathrm{d}V \, \mathrm{d}t \right] \Big/ \left[\int_{t_0}^{t_1} \int_V F(x, \phi) \, T(\phi) \, \mathrm{d}V \, \mathrm{d}t \right], \tag{44}$$

where ϕ is an arbitrary element in $\mathscr{S}$ which lies in the domain of definition of operators F and T.

Proposition 4: The eigenvalue λ in (42) is the extremum of the meta-Rayleighian R in the subspace of $\mathscr{S}$, which comprises all ϕ under the assumption (35).

Proof: The variation of (44) is

$$\delta R_g = 2 \left[\int_{t_0}^{t_1} \int_V \mu \dot{\phi} \frac{\mathrm{d}}{\mathrm{d}t} (\delta T) \, \mathrm{d}V \, \mathrm{d}t \right] \cdot \left[\int_{t_0}^{t_1} \int_V FT \, \mathrm{d}V \, \mathrm{d}t \right] - \tag{45}$$

$$- 2 \left[\int_{t_0}^{t_1} \int_V \mu \dot{\phi} \frac{\mathrm{d}T}{\mathrm{d}t} \, \mathrm{d}V \, \mathrm{d}t \right] \cdot \left[\int_{t_0}^{t_1} \int_V F \, \delta T \, \mathrm{d}V \, \mathrm{d}t \right] \Big/ \left[\int_{t_0}^{t_1} \int_V FT \, \mathrm{d}V \, \mathrm{d}t \right]^2.$$

In deriving (45) from (44), the symmetry of $\mathrm{d}T/\mathrm{d}t$ and F has been taken into account.

Since (35) holds by assumption, also (38) holds true. Using (38) and (44) in (45) yields the relationship

$$\delta R_g \cdot \int_{t_0}^{t_1} \int_V FT \, \mathrm{d}V \, \mathrm{d}t = -2 \left\{ \int_{t_1}^{t_0} \int_V (\mu \ddot{\phi} + R_g F) \, \delta T \, \mathrm{d}V \, \mathrm{d}t \right\}. \tag{46}$$

216   H. H. E. Leipholz

For $\phi \equiv w$, $R_g \equiv \lambda$. Hence, according to (42),

$$\delta R_g = 0 \quad \text{for} \quad \phi \equiv w, \qquad (47)$$

which proves that R_g has an extremum for $\phi \equiv w$. However, as already stated, $R_g \equiv \lambda$ for $\phi \equiv w$. Therefore, λ is the extremum of R_g.

(d) *Examples*

The theory of "conservative systems of the second kind" were void if there were not relevant examples to which it could be applied. Such examples do exist, and may be found in [5]. One example may be mentioned here: Pflüger's rod. For this system,

$$F(x, w) = -\lambda \ddot{w} + \alpha w'^v + g(l - x) w'', \quad T(w) = w'', \qquad (48)$$

and in order to check the symmetry properties of F and T as required in Corrolary 2c, the integration has to be carried out over the length of the rod from $x = 0$ to $x = l$.

One will find them satisfied. Hence, with

$$F(x, w) = -\lambda \ddot{w} + \alpha w'^v + g(l - x) w'' = 0, \qquad (49)$$

the differential equation of Pflüger's rod which corresponds to (31), and according to corrolary 2c, proposition 2 and corrolary 2b are valid, i.e., Pflüger's rod is a conservative system of the second kind and possesses a Liapunov function (32). Also, because of the symmetry of operators F and T, a "meta-Rayleighian" (44) does exist, permitting the calculation of eigenvalues λ and g by means of an extremum principle.

Acknowledgement

This research was carried out under the support of the National Research Council of Canada under Grant No. A7297.

References

1. Knops, R. J.; Wilkes, E. W.: Theory of Elastic Stability. Encyclopedia of Physics Vol. VIa/3, Mechanics of Solids III. Chief Editor S. Flügge, Editor C. Truesdell. Berlin, Heidelberg, New York: Springer 1973, pp. 125—302.
2. Leipholz, H.: Application of Liapunov's Direct Method to the Stability Problem of Rods Subjected to Follower Forces. Instability of Continuous Systems. Berlin, Heidelberg, New York: Springer 1971, pp. 1—10.
3. Leipholz, H.: Über die Anwendung von Liapunov's direkter Methode auf Stabilitätsprobleme kontinuierlicher, nichtkonservativer Systeme. Ing. Arch. **39**, 257—268 (1970).
4. Leipholz, H.; Huseyin, K.: On the Stability of One-Dimensional Continuous Systems with Polygenic Forces. Meccanica **VI**, 253—257 (1971).
5. Leipholz, H.: On Conservative Elastic Systems of the First and Second Kind. Ing. Arch. **43**, 255—271 (1974).

An Investigation of the Stability and Vibration of a Hyperbolic Shell of one Sheet under Axisymmetric Loading

M. Kozarov

Applied Mechanics and Aerospace Sciences, Sofia, Bulgaria

M. Kishkilov

Technical Institute for Civil Engineering, Sofia, Bulgaria

Abstract

This article deals with the stability and oscillation of an isotropic hyperbolic shell of one sheet. The loading is assumed axisymmetric and uniformly distributed along the edges of the shell. It is asumed that the load is of the type $P(t) = P_0 + P_t \cos \theta t$. The shell is simply supported. The basic equations are derived from Vlasov's general theory of shells, by applying the Ostrogradsky-Hamilton variational principal. The basic system of differential equations is solved by using the Bubnov-Galerkin variational method.

In the most general case a system of differential equations of the Mathieu type is obtained. As a special case, the vibration and static stability of the structure are considered. At the conclusion a numerical example is solved.

1. Introduction

The present article deals with some aspects of the stability and vibration of a simply supported, thin shell in the form of a hyperboloid of one sheet. It has been assumed that the material is elastic and isotropic. The loading is axisymmetric and uniformly distributed along the edges of the shell. Similar problems have been considered by a number of authors, but in most of the cases, the shell is either cylindrical or conical. Both of these cases are noted for the simplicity of their geometry. This article offers for the first time an investigation of the static and dynamic stability and vibrations of a hyperboloid of one sheet. This particular type of shell is of considerable practical interest.

The basic equations are derived according to Vlasov's general theory of thin elastic shells by means of the Ostrogradsky-Hamilton variational principle. The problem is assumed to be linear. The basic system of differential equations is solved by the Bubnov-Galerkin method. A system of differential equations of the Mathieu type is obtained. A formula for the critical stress under static loading is derived. The critical stresses have been computed for different values of the ratio of the principle radii of curvature and a comparison has been made to corresponding results for a cylindrical shell. The frequency equation for the free vibrations of the shell has also been derived.

2. Derivation of the Basic Differential Equations

A thin, elastic, isotropic, hyperbolic shell of revolution is considered (Fig. 1). According to [3], the expression for the strain components of a thin shell in terms of the assumed notations are:

$$
\begin{aligned}
\varepsilon_1 &= \frac{1}{A_1}\frac{\partial u}{\partial \phi} + \frac{1}{A_1 A_2}\frac{\partial A_1}{\partial \beta}v + \frac{w}{\varrho_1}, \\
\varepsilon_2 &= \frac{1}{A_2}\frac{\partial v}{\partial \beta} + \frac{1}{A_1 A_2}\frac{\partial A_2}{\partial \phi}u + \frac{w}{\varrho_2}, \\
\gamma &= \frac{A_1}{A_2}\frac{\partial}{\partial \beta}\left(\frac{u}{A_1}\right) + \frac{A_2}{A_1}\frac{\partial}{\partial \phi}\left(\frac{v}{A_2}\right), \\
\varkappa_1 &= -\frac{1}{A_1}\frac{\partial}{\partial \phi}\left(\frac{1}{A_1}\frac{\partial w}{\partial \phi}\right) - \frac{1}{A_1 A_2^2}\frac{\partial A_1}{\partial \beta}\frac{\partial w}{\partial \beta}, \\
\varkappa_2 &= -\frac{1}{A_2}\frac{\partial}{\partial \beta}\left(\frac{1}{A_2}\frac{\partial w}{\partial \beta}\right) - \frac{1}{A_1^2 A_2}\frac{\partial A_2}{\partial \phi}\frac{\partial w}{\partial \phi}, \\
\tau &= -\frac{1}{A_1 A_2}\left(\frac{\partial^2 w}{\partial \phi\, \partial \beta} - \frac{1}{A_1}\frac{\partial A_1}{\partial \beta}\frac{\partial w}{\partial \phi} - \frac{1}{A_2}\frac{\partial A_2}{\partial \phi}\frac{\partial w}{\partial \beta}\right),
\end{aligned}
\tag{1}
$$

where ε_1, ε_2, w are the linear and angular components of strain at the middle surface of the shell, $\varkappa_1$, $\varkappa_2$ and τ characterize the change of curvature and its torsion. A_1 and A_2 are Lame's coefficients. For a hyperbolic shell they are of the following type:

$$
A_1 = \varrho_1, \quad A_2 = \varrho_2 \cos \phi.
\tag{2}
$$

The principal radii of curvature are respectively

$$
\varrho_1 = \frac{b^2}{a}\frac{1}{(1 - \varepsilon^2 \sin^2 \phi)^{3/2}}, \quad \varrho_2 = \frac{a}{(1 - \varepsilon^2 \sin^2 \phi)^{3/2}}
\tag{3}
$$

and they satisfy the Gauss-Codazzi equation.

$$
\frac{\partial}{\partial \phi}(\varrho_2 \cos \phi) = -\varrho_1 \sin \phi.
\tag{4}
$$

By substituting Eq. (2) and (4) into (1) we obtain

$$
\begin{aligned}
\varepsilon_1 &= \frac{1}{\varrho_1}\frac{\partial u}{\partial \phi} - \frac{w}{\varrho_1}, \\
\varepsilon_2 &= \frac{1}{\varrho_2 \cos \phi}\frac{\partial v}{\partial \beta} - \frac{\tan \phi}{\varrho_2}u + \frac{w}{\varrho_2}, \\
\gamma &= \frac{1}{\varrho_2 \cos \phi}\frac{\partial u}{\partial \beta} + \frac{1}{\varrho_1}\frac{\partial v}{\partial \phi},
\end{aligned}
\tag{5}
$$

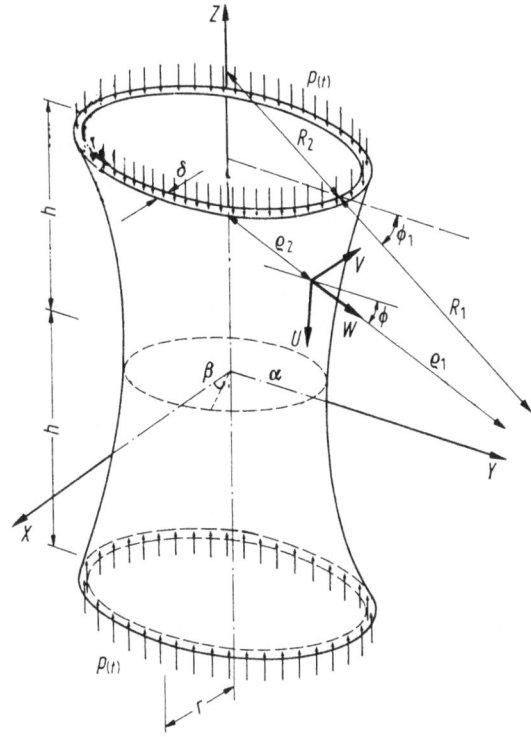

Fig. 1

$$\varkappa_1 = -\frac{1}{\varrho_1^2}\frac{\partial^2 w}{\partial\phi^2},$$

$$\varkappa_2 = -\frac{1}{\varrho_2^2\cos^2\phi}\frac{\partial^2 w}{\partial\beta^2} + \frac{\tan\phi}{\varrho_1\varrho_2\cos\phi}\frac{\partial w}{\partial\phi},$$

$$\tau = -\frac{1}{\varrho_1\varrho_2\cos\phi}\frac{\partial^2 w}{\partial\phi\,\partial\beta} - \frac{\tan\phi}{\varrho_2^2\cos\phi}\frac{\partial w}{\partial\beta}.$$

Both Eqs. (5) and (4) take into account the different signs of curvature. According to the membrane theory of hyperbolic shells [3], the stress in an undisturbed state is defined by the following expressions:

$$N_\phi = \frac{1}{ab}\cos\phi\sqrt{a^2\sin^2\phi + b^2}\,\frac{\partial\varLambda}{\partial\beta},$$

$$N_\beta = \frac{u\cos^3\phi}{b\sqrt{a^2\sin^2\phi + b^2}}\frac{\partial\varLambda}{\partial\beta}, \tag{6}$$

$$N_{\phi\beta} = -\frac{1}{b}\cos^2\phi\,\frac{\partial\varLambda}{\partial\phi},$$

where Λ is the stress function which must satisfy the differential equation

$$\frac{\partial^2 \Lambda}{\partial \phi^2} - \frac{\partial^2 \Lambda}{\partial \beta^2} = 0 \qquad (6')$$

with the following boundary conditions at $\phi = \pm \phi_1$,

$$N_\phi = -p\, \delta,$$
$$N_{\phi\beta} = 0. \qquad (6'')$$

For the rather limited investigation of a hyperboloid with slight curvature which frequently occurs in engineering practice, the expressions (6) assume a simplified form (7)

$$N_\phi = -\frac{p\, \delta}{\cos \phi_1} \cos \phi, \quad N_\beta = -\frac{a^2 p\, \delta}{b^2 \cos \phi_1} \cos^3 \phi, \quad N_{\phi\beta} = 0. \qquad (7)$$

By a hyperbola with a "slight" curvature we shall define one with eccentricity $\varepsilon \geq 5$.

According to the theory of thin elastic shells [4], the strain energy of the shell is defined by the expression

$$\Pi = \frac{E\, \delta}{2(1 - \mu^2)} \iint_F [(\varepsilon_1 + \varepsilon_2)^2 - 2(1 - \mu)\,(\varepsilon_1 \varepsilon_2 - w^2/4)]\, A_1 A_2\, \mathrm{d}\phi\, \mathrm{d}\beta +$$

$$+ \frac{E\, \delta^3}{24(1 - \mu^2)} \iint_F [(\varkappa_1 + \varkappa_2)^2 - 2(1 - \mu)\,(\varkappa_1 \varkappa_2 - \tau^2)]\, A_1 A_2\, \mathrm{d}\phi\, \mathrm{d}\beta, \qquad (8)$$

where integration in both cases is over the entire middle surface of the shell.

The potential of the external axial load, applied along the edges of the shell is defined by the work done by the fictitious transverse load. The stress in the undisturbed state (7) of the shell is used [1]. The fictitious transverse load is defined by the formula [2]

$$\bar{q} = -(N_\phi \varkappa_1 + N_\beta \varkappa_2 + 2N_{\phi\beta}\tau). \qquad (9)$$

The potential of the external axial load will then be defined by:

$$\Pi = (1/2) \iint_F \bar{q} w A_1 A_2\, \mathrm{d}\phi\, \mathrm{d}\beta. \qquad (10)$$

The kinetic energy of the shell is

$$T = \frac{m^*}{2} \iint_F \left[\left(\frac{\partial u}{\partial t} \right)^2 + \left(\frac{\partial v}{\partial t} \right)^2 + \left(\frac{\partial w}{\partial t} \right)^2 \right] A_1 A_2\, \mathrm{d}\phi\, \mathrm{d}\beta. \qquad (11)$$

By applying the Ostrogradsky-Hamilton variational principle we obtain

$$\delta \int_{t_1}^{t_2} L\, \mathrm{d}t = \delta \int_{t_1}^{t_2} (\Pi + \Pi_P - T)\, \mathrm{d}t = \delta \int_{t_1}^{t_2} \int_{-\phi_1}^{+\phi_2} \int_{-\pi}^{+\pi} \Omega\, \mathrm{d}\phi\, \mathrm{d}\beta\, \mathrm{d}t = 0. \qquad (12)$$

The unknown functions u, v and w are obtained according to [5] as integrals of the Euler-Langrange differential equations for (12)

$$\frac{\partial\Omega}{\partial u} - \frac{\partial}{\partial\phi}\left(\frac{\partial\Omega}{\partial u_\phi}\right) - \frac{\partial}{\partial\beta}\left(\frac{\partial\Omega}{\partial u_\beta}\right) - \frac{\partial}{\partial t}\left(\frac{\partial\Omega}{\partial u_t}\right) +$$

$$+ \frac{\partial^2}{\partial\phi^2}\left(\frac{\partial^2\Omega}{\partial u_{\phi\phi}}\right) + \frac{\partial^2}{\partial\beta^2}\left(\frac{\partial^2\Omega}{\partial u_{\beta\beta}}\right) + \frac{\partial^2}{\partial\phi\,\partial\beta}\left(\frac{\partial^2\Omega}{\partial u_{\phi\beta}}\right) = 0. \qquad (13)$$

By substituting in (13) the expression for Ω from (12), the following linear system of partial differential equations is obtained:

$$K_1\frac{\partial^2 u}{\partial t^2} - K_2\frac{\partial^2 u}{\partial\phi^2} - K_3\frac{\partial^2 u}{\partial\beta^2} + K_4\frac{\partial u}{\partial\phi} + K_5 u -$$

$$- K_6\frac{\partial^2 v}{\partial\phi\,\partial\beta} + K_7\frac{\partial v}{\partial\beta} + K_8\frac{\partial w}{\partial\phi} - K_9 w = 0,$$

$$L_1\frac{\partial^2 v}{\partial t^2} - L_2\frac{\partial^2 v}{\partial\phi^2} - L_3\frac{\partial^2 v}{\partial\beta^2} - L_4\frac{\partial^2 u}{\partial\phi\,\partial\beta} + L_5\frac{\partial u}{\partial\beta} - L_6\frac{\partial w}{\partial\beta} = 0,$$

$$M_1\frac{\partial^2 w}{\partial t^2} + M_2\frac{\partial^4 w}{\partial\phi^4} + M_3\frac{\partial^4 w}{\partial\beta^4} + M_4\frac{\partial^4 w}{\partial\phi^2\,\partial\beta^2} + M_5\frac{\partial^2 w}{\partial\phi^2} +$$

$$+ M_6\frac{\partial^2 w}{\partial\beta^2} - M_7 w - M_8\frac{\partial u}{\partial\phi} - M_9 u + M_{10}\frac{\partial v}{\partial\beta} = 0. \qquad (14)$$

For the case of local stability, the coefficients of (14) are known.

3. Solution of the System of Differential Equations

System (14) is solved by the Bubnov-Galerkin method. The unknown functions u, v and w are obtained in the form:

$$u = \sum_{i=1}^{\infty}\sum_{j=1}^{\infty} U_{ij}(t)\,\cos i\,(m\pi\phi/\phi_1)\,\cos jn\beta,$$

$$v = \sum_{i=1}^{\infty}\sum_{j=1}^{\infty} V_{ij}(t)\,\sin i\,(m\pi\phi/\phi_1)\,\sin jn\beta, \qquad (15)$$

$$w = \sum_{i=1}^{\infty}\sum_{j=1}^{\infty} W_{ij}(t)\,\sin i\,(m\pi\phi/\phi_1)\,\cos jn\beta,$$

where m is the number of semiwaves along the hyperbolic generators, U, V and W are unknown time-dependent functions and n is the number of full waves on the parallel circles.

The form (15) has been selected for u, v and w, so that they satisfy the boundary conditions for a simply supported shell.

Substituting u, v and w, from (15) into (14) and applying the Bubnov-Galerkin method gives the equations

$$\frac{\mathrm{d}^2 U_{ij}}{\mathrm{d}t^2} + \overline{K}_1 U_{ij} + \overline{K}_2 V_{ij} + \overline{K}_3 W_{ij} = 0,$$

$$\frac{\mathrm{d}^2 V_{ij}}{\mathrm{d}t^2} + \overline{L}_1 U_{ij} + \overline{L}_2 V_{ij} + \overline{L}_3 W_{ij} = 0,$$

$$\frac{\mathrm{d}^2 W_{ij}}{\mathrm{d}t^2} + \overline{M}_1 U_{ij} + \overline{M}_2 V_{ij} + (\overline{M}_3^{**} - \overline{M}_3^{*} \cos \theta t) W_{ij} = 0$$

$$(i, j = 1, 2, \ldots, n). \quad (16)$$

Equations (16) are a system of differential equations of the Mathieu type. For this case we will restrict ourselves to the first approximation according to Bubnov-Galerkin, i.e. $i = j = 1$.

If the meridian and parallel forces of inertia are neglected, i.e. $\mathrm{d}^2 U/\mathrm{d}t^2 = 0$ and $\mathrm{d}^2 V/\mathrm{d}t^2 = 0$, system (16) transforms into a single Mathieu equation.

$$\frac{\mathrm{d}^2 W}{\mathrm{d}t^2} + (A + B \cos 2\tau) W = 0. \quad (17)$$

Coefficients A and B of Eq. (17) depend on the geometry of the hyperbolic shell and the loading P_0 and P_t. For each pair of values of A and B we obtain a certain point which corresponds to a state of stability or instability (the darkened zones in Fig. 2), according to the well known graphic representation of the solution of the Mathieu equation (17).

The frequency of free vibration of the shell is easily derived from the basic equations (16), where the latter are considered as a first

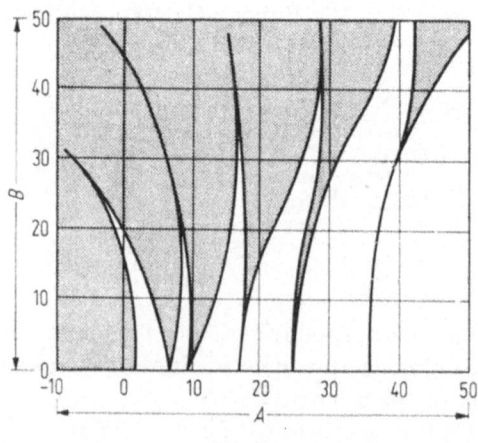

Fig. 2

approximation according to Bubnov-Galerkin $(i = j = 1)$. It is assumed

$$U = C_1 \sin \omega t^*, \quad V = C_2 \sin \omega t, \quad W = C_3 \sin \omega t. \tag{18}$$

(18) is substituted into (16). The load is assumed of the type $P = P_0$. Then it follows:

$$
\begin{aligned}
-C_1 w^2 + \bar{K}_1 C_1 + \bar{K}_2 C_2 + \bar{K}_3 C_3 &= 0, \\
-C_2 w^2 + \bar{L}_1 C_1 + \bar{L}_2 C_2 + \bar{L}_3 C_3 &= 0, \\
-C_3 w^2 + \overline{M}_1 C_1 + \overline{M}_2 C_2 + \overline{M}_3 C_3 &= 0.
\end{aligned}
\tag{19}
$$

Consequently the frequency equation assumes the form

$$
\begin{vmatrix}
\bar{K}_1 - w^2 & \bar{K}_2 & \bar{K}_3 \\
\bar{L}_1 & \bar{L}_2 - w^2 & \bar{L}_3 \\
\overline{M}_1 & \overline{M}_2 & \overline{M}_3 - w^2
\end{vmatrix} = 0.
\tag{20}
$$

If all inertia terms of the system (16) are neglected and the loading is assumed time-independent, i.e. $P_t = 0$ and $i = j = 1$, system (16) transforms into a homogeneous linear, algebraic system of 3 equations with 3 unknowns, U, V and W:

$$
\begin{aligned}
\bar{K}_1 U + \bar{K}_2 V + \bar{K}_3 W &= 0, \\
\bar{L}_1 U + \bar{L}_2 V + \bar{L}_3 W &= 0, \\
\overline{M}_1 U + \overline{M}_2 V + \overline{M}_3 W &= 0.
\end{aligned}
\tag{21}
$$

According to the assumptions for the loading, the unknowns are time-independent constants.

From the requirement for a non-zero solution of the system (21)

$$
\Delta = \begin{vmatrix}
\bar{K}_1 & \bar{K}_2 & \bar{K}_3 \\
\bar{L}_1 & \bar{L}_2 & \bar{L}_3 \\
\overline{M}_1 & \overline{M}_2 & \overline{M}_3
\end{vmatrix} = 0
\tag{22}
$$

the following formula for the critical stress of the hyperbolic shell under static loading is obtained:

$$\overline{P}_0 = (m^2 F / H) + (G / m^2 H), \tag{23}$$

where $\overline{P}_0$ is the critical stress, F, G, H are known, $k = m/n$ is a wave-generating parameter, m is the wave number along the hyperbolic generator, and n is the wave number along the parallel.

In the case of the hyperbolic shell, the wave generating parameter m/n corresponding to loss of stability could be uniquely defined, much as it has been for the case of cylindrical shells [10]. Such an approach would be more accurate, but it would result in a complicated system of nonlinear algebraic equations, which would be extremely inconvenient in this particular case.

Another approach would be to select the m/n ratio on an experimental basis. Experiments indicate that at the moment of buckling, the minute waves forming on the surface of a cylindrical shell are quadratic and rhombic ones. In the case of a hyperbolic shell, however, due to the more complicated geometry, it is assumed that the forming of such waves other the total surface is not possible. Instead, it is assumed that quadratic and rhombic waves form only in the zones of the maximum radii of curvature. The authors are not aware of any extensive experimental work.

In the numerical example it has been assumed that $k = 0.5$ ($2m = n$) because for the selected geometry $2m = n$ is a necessary condition for developing of quadratic and rhombic waves.

4. Numerical Results and Conclusions

The obtained formula (22) is similar to the classical formula for a critical stress in cylindrical shells [2]. Formula (23) can be geometrically interpreted as a summation of a parabola with a symmetry axis $m = 0$ and $\overline{P}_c = 0$.

It can be seen from Fig. 3 that $\overline{P}_c$ approaches infinity as m approaches zero or infinity. This agrees with our assumption for local loss of stability.

Hence, formula (23) is valid for $1 \ll m \ll \infty$ only. The number of semiwaves corresponding to loss of stability in the zones of maximum radii of curvature is defined by the requirement for an extremum of the

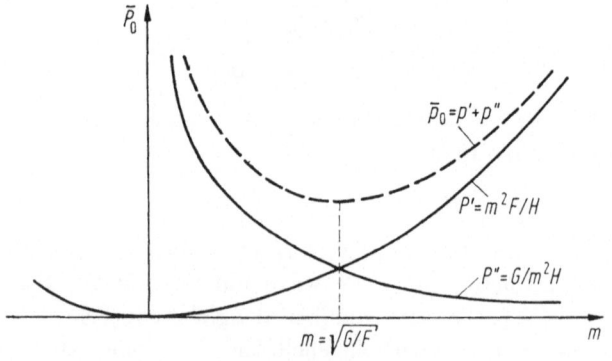

Fig. 3

function $\overline{P}_c$, i.e.

$$\frac{\mathrm{d}P_c}{\mathrm{d}m} = 0 \qquad (24)$$

or

$$m = \sqrt{G/F}. \qquad (25)$$

The critical stress for a number of different shells has been obtained by formula (23). The shells have been selected such that they are of the same length ($h = 200$ cm) and same radius of edge parallel ($r = 29$ cm). Thus, they can be considered as hyperbolic shells approaching a cylindrical shell with radius $r = 29$ cm and height $h = 200$ cm. The geometrical characteristics and critical stresses for four such shells are listed in Table 1.

Table 1

	a [cm]	b [cm]	ε^2	R_1 [cm]	R_2 [cm]	R_1/R_2	$-k\,10^{-5}$ [cm^{-2}]	m	$\overline{P}_0\,10^3$ [kp/cm^2]
I	20	100	26.0	1470	287	51	0.236	36	4.42
II	18	80	20.2	1420	294	48	0.241	36	4.31
III	15	60	17.0	1030	296	36	0.326	37	3.72
IV	13	40	10.5	860	308	28	0.414	38	3.46

The thickness of the shells is $\delta = 4.1$ cm, the elastic modulus is $E = 2.1 \cdot 10^6$ kg/cm^2, $\mu = 0.3$.

The obtained results are graphically shown in Fig. 4.

The critical stress and the corresponding number of semiwaves for the cylindrical shell are obtained according to [2] by using the formula:

$$\overline{P}_0 = (Eh/R)\,(1/\sqrt{3(1-\mu^2)}) \quad \text{corresponding to} \quad n \approx 0.91\sqrt{R/\delta}. \qquad (26)$$

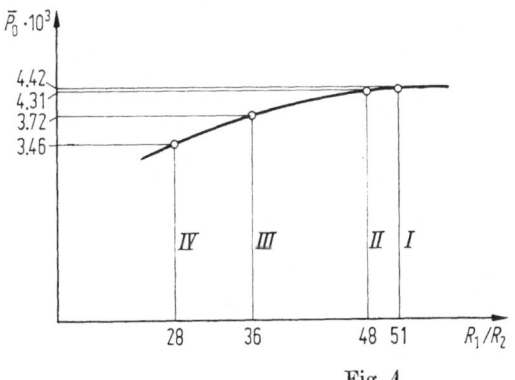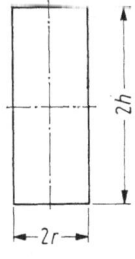

Fig. 4

Then for this particular case

$$\overline{P}_0 = 4.60 \cdot 10^3 \, \text{kp/cm}^2) \, n \approx 17 \, (m = 34). \qquad (27)$$

The obtained results indicate that increasing the R_1/R_2 ratio results in higher critical stresses, approaching the critical stress of the cylindrical shell. Simultaneously, the number of waves decreases, also approaching the one corresponding to a cylindrical shell. The obtained result enables us to compute the frequency of free vibration of the shell.

The results from Table 1 are compared to those for a cylindrical shell [9]

$$\text{hyperbolic shell:} \; \omega_{\min} = 0.103 \; 1/\text{sec},$$

$$\text{cylindrical shell:} \; \omega_{\min} = 0.082 \; 1/\text{sec}.$$

The difference of approximately 20% is due to the different geometry. However, it is also relevant that the present results are only a first approximation of the Bubnov-Galerkin solution. Furthermore, the solution is achieved through considerably more simplifying assumptions than in the case of cylindrical shells.

References

1. Bolotin, V. V.: Dynamicheskaya Ustoichivost Uprugih System. Moscow 1965.
2. Volmir, A. S.: Ustoichivost Deformiruemih System. Moscow 1961.
3. Vlasov, V. Z.: Izbranie Trudy. Moscow: "Nauka" 1964.
4. Novojilov, V. V.: Teoria Tonkih Obolochek. Moscow: Sudprongiz 1951.
5. Elsgolts, L. E.: Differentialnie Uravnenia i Variationnoe Izchislenija. Moscow: "Nauka" 1969.
6. Kozarov, M. M.: Dynamicheska Ustoichivost na Ortotropni Conichni Cherupki ot Pulsirasht Ossov Tovar. BAN, Theoretical and Applied Mechanics, Sofia, 1970, I, No. 2 (1972).
7. Kozarov, M. M.; Kolev, P. T.: Ustoichivost pri Pulzene na Ednopovurhninna Hyperbolichna Cherupka ot Ossov Tovar. BAN, Theoretical and Applied Mechanics, Sofia 1972.
8. Gontcevitch, V. S.: Sobstvennie Kolebania Plastinok i Obolochek. Kiev: Spravochnoe possobie 1964.
9. Yao, J. C.: Dynamic Stability of Cylindrical Shells under Static and Periodic Axial and Radial Loads. AIAA, J. 6, June 1963.
10. Kisliakov, S.: Kum Vuprosa za Vulnoobrazuvaneto v Teoriata na Dinamichnata Ustoichivost. Technischeska Misul. II, No. 5 (1965).

The Influence of Boundary Conditions on the Buckling of Stiffened Cylindrical Shells [1]

J. Singer, A. Rosen

Technion-Israel Institute of Technology Haifa, Israel

Abstract

Theoretical and experimental studies on the influence of boundary conditions on the buckling of stiffened cylindrical shells and their vibrations are discussed. The effect of prebuckling deformations on the buckling loads and vibrations of stiffened shells is studied and compared with that in the case of unstiffened shells. The in-plane boundary conditions are found to be of particular importance for stiffened cylindrical shells and their effect differs significantly from that in unstiffened shells. The effect of axial restraints, which are found to be of prime importance in stringer-stiffened shells, are also studied.

By correlation with the vibration tests on the same shells a method is developed for definition of the actual boundary conditions of a stiffened shell non-destructively.

The effect of eccentricity of loading on stringer-stiffened shells is studied experimentally and correlated with vibration tests on the same shells.

Preliminary results of a non-destructive experimental method for prediction of buckling loads based on vibration testing of stiffened shells are also presented.

List of Symbols

A_1, A_2	Cross section of stringer and ring, respectively
a	Ring spacing (distance between centers of rings)
A, B	Constants
b_1	Stringer spacing (distance between centers of stringers)
C1, C2, C3, C4	Notation for clamped boundary conditions
c_1	Width of stringer
d_1	Height of stringer
E	Young's modulus
e_1, e_2	Stringer or ring eccentricity, respectively (distance from shell middle surface to stiffener centroid)
$\bar{e}$	Eccentricity of loading (distance from shell middle surface to the point of application of load)
f	Frequency

[1] The research reported in this paper has been sponsored in part by the Air Force Office of Scientific Research, through the European Office of Aerospace Research, United States Air Force under Contract F44620-71-C-0116 and Grant 72-2394.

Based on T.A.E. Report 213.

h	Thickness of shell
I_{11}, I_{22}	Moment of inertia of stringer or ring stiffener cross-section about its centroidal axis
k_1	Axial elastic restraint
k_4	Rotational elastic restraint
L	Length of shell
M_x	Moment resultant in axial direction
m	Number of half longitudinal waves
N_x	Axial membrane force resultant
N_{xy}	Shear membrane force resultant
n	Circumferential wave number
P	Compressive axial load
P_{CR}	Theoretical buckling load
P_{PRE}	Theoretical buckling load with nonlinear prebuckling deformations considered
P_{mem}	Theoretical buckling load with prebuckling deformation neglected
P_{EXP}	Experimental buckling load
R	Radius to shell middle surface
SS1, SS2, SS3, SS4	Notation for simply supported boundary conditions
u, v, w	Displacements in axial, circumferential and radial directions respectively (radial direction positive inward)
x	Axial coordinate
Z	$(1 - \nu^2)^{1/2} (L/R)^2 (R/h)$, Batdorf shell parameter
ν	Poisson's ratio
ϱ_{PRE}	P_{PRE}/P_{mem}
ϱ	P_{EXP}/P_{mem}, "linearity"
η_{t1}	Torsional stiffness parameter of stringer (see [3] or [22])
ϕ	Circumferential coordinate
γ	Free parameter for exponent of frequency

Subscripts following a comma indicate differentiation

1. Introduction

The emphasis in both theoretical and experimental studies on the buckling of thin shells has in recent years been directed at the effect of geometrical imperfections, and has overshadowed the equally important problem of the influence of boundary conditions (see [1] where the studies on the influence of boundary conditions in the last decade are briefly reviewed).

For stiffened shells, and in particular closely stiffened shells (for which local panel buckling is rarely critical) the effect of geometric imperfections is less pronounced. Hence the reduction in predicted buckling loads and scatter of test results is less severe, provided the boundary conditions are adequately accounted for. The senior author has for some years now defended the earlier statement by Van der Neut [2] that linear theory is adequate for prediction of buckling loads in shells with closely spaced stiffeners. Extensive experimental evidence for integrally ring-and-stringer-stiffened cylindrical shells (reviewed in [3]), confirms the applicability of linear theory at least as a first approximation, and with at least the same reliability as for isotropic cylindrical shells under external pressure. However, the influence of the boundary conditions is found to be of prime importance in stiffened shells and sometimes even overshadows the effect of geometrical imperfections. The realization of this fact has motivated also the recent theoretical and experimental studies by the authors aimed at better understanding of the influence of boundary conditions, and their more precise assessment.

2. Effect of Prebuckling Deformations

In the extended version of the present paper [1], the earlier studies on the effect of prebuckling deformations on isotropic and stiffened cylindrical shells are reviewed and discussed. For stringer-stiffened shells under axial compression, the effect is found to depend strongly both on the shell geometry and on the stiffener geometry [1, 4 to 9]. Since only unrelated cases were studied, a partial parametric study is carried out with the BOSOR 3 program [10] and the linear theory of [11].

Figure 1 presents the variation of the influence of prebuckling deformations on the buckling loads with shell geometry for two types of stringer-stiffened shells. One with weak external stringers $A_1/b_1h = 0.25$, $I_{11}/b_1h^3 = 2.5$, $e_1/h = -2.5$ and $\eta_{t1} = 2.5$, and one with medium external stringers $A_1/b_1h = 0.5$, $I_{11}/b_1h^3 = 5$, $e_1/h = -5$ and $\eta_{t1} = 5$. The curves of ϱ_{PRE} $(= P_{PRE}/P_{mem})$ are presented for SS3 and SS4 boundary conditions. Curves for isotropic shells with the same boundary conditions, reproduced from [12] are also presented for comparison.

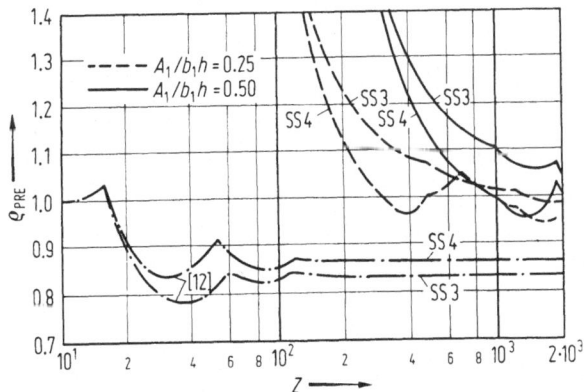

Fig. 1. Effect of prebuckling deformations on buckling of stringer-stiffened shells
—variation with shell geometry.

The general trends appear to be: (1) a shift to higher Z with increase in stiffening; (2) a large increase in buckling loads for short shells, again more pronounced the heavier the stringers; (3) even for long shells a much smaller decrease in ϱ_{PRE} than in isotropic shells, and (4) a reversal of the relative magnitudes of ϱ_{PRE} with SS3 and SS4 B.C.'s compared to those for isotropic shells.

The variation of ϱ_{PRE} with stringer geometry for 3 typical shells geometries is plotted in Fig. 2. A_1/b_1h is taken as the representative stringer geometry parameter, but the other parameters are also varied

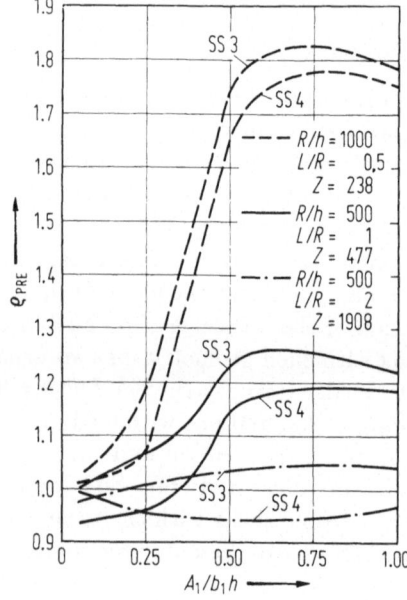

Fig. 2.
Effect of prebuckling deformations on buckling of stringer-stiffened shells — variation with stringer geometry.

proportionally (the exact relation of the other parameters to A_1/b_1h being given in Table 2 of [1]). Curves for both SS3 and SS4 B.C.'s are presented in the figure. The large increase in buckling loads for short shells is immediately evident from the curves for $Z = 238$. One can also observe the initial large range of growth in ϱ_{PRE} with stringer geometry with up to $A_1/b_1h \approx 0.5$. The growth in ϱ_{PRE} then subsides and ϱ_{PRE} even tends to decrease slightly at $A_1/b_1h > 0.75$. This behavior persists also at large Z where $\varrho_{PRE} \approx 1$.

One can summarize the salient features of this partial parametric study for stringer-stiffened shells and of the cases studied by other investigators in the following practical conclusions:

1. For short shells with medium or heavy external stringers ($Z < 250 \cdots 400$ and $A_1/b_1h > 0.3$ with corresponding other geometry

parameters in particular $-(e_1/h) > 3$, consideration of nonlinear pre-buckling deformations results in substantial increase in buckling loads. Hence linear theory will yield very conservative predictions for such shells. For similar internal stringers this increase disappears and ϱ_{PRE} has values close to those of isotropic shells. Further parametric studies for internally stiffened shells are needed.

2. For medium and long shells ($Z > 1\,000$) with external stringers $\varrho_{PRE} \approx 0.95\cdots1.05$, and the effect of nonlinear prebuckling deformations may therefore be neglected.

3. For external stringers, ϱ_{PRE} for SS4 is practically always below the value for SS3 B.C.'s, whereas in isotropic shells ϱ_{PRE} for SS4 always exceeds that for SS3.

Thus for design purposes, one can safely neglect nonlinear pre-buckling deformations for externally stringer-stiffened cylindrical shells under axial compression, except in the case of very short shells where predictions would be unduly conservative.

The influence of nonlinear prebuckling deformations on the vibra-tions of axially loaded stringer-stiffened shells has been studied in [12], as the particular case of zero load eccentricity. Very small effects have been found (see also [1]). For example, for the mode shown in Fig. 10, $\varrho_{PRE} = 0.99$ for SS4 B.C.'s at zero load, and decreases to $\varrho_{PRE} = 0.93$ near buckling. The effect for SS3 is even smaller.

3. Effect of In-Plane Boundary Conditions

Previous studies on the influence of in-plane boundary conditions for stiffened shells have been discussed in the extended version of the paper [1], in particular the recent parametric studies for ring-stiffened shells [11]. For ring-stiffened shells under axial compression the effects are similar to those in isotropic shells [14]; whereas for stringer-stiffened shells the effects differ appreciably.

Axial restraint ($u = 0$ instead of $N_x = 0$) becomes the predominant factor, whereas circumferential restraint has only a minor influence. The studies of [11] show that the effects strongly depend on the shell geometry parameter Z and on the stringer parameters A_1/b_1h and e_1/h. The effect of axial restraint is more pronounced for internal stringers but even for external ones axial restraints ($u = 0$, SS2 or SS4) may raise the predicted buckling load for moderately or heavily stiffened shells by 50% if the shell is long. The effect of axial restraint is much smaller for stringer-stiffened shells with clamped ends. For example in shells AS2 and AS3 tested in [5], $P_{C4}/P_{C3} = 1.003$ and, 1.002 respecti-vely, whereas with simple supports $P_{SS4}/P_{SS3} = 1.37$ and 1.14 respectively, where P_{C3}, P_{C4}, P_{SS3}, P_{SS4} are the theoretical buckling loads for the corresponding boundary conditions.

For better assessment of the test results, an additional parametric study of the effect of axial constraint on the buckling under axial compression of simply supported stringer-stiffened shells in the relevant geometry range has been carried out with the theory of [11] and with BOSOR 3 [10], taking into account nonlinear prebuckling in the latter. Figure 3 shows the variation of the stiffening, afforded by axial restraint, with stiffener geometry for three values of Z. The stringers are external and the relation of the other stringer geometry parameter to the area ratio $A_1/b_1 h$ is given in Table 2 of [1].

The predictions of Fig. 3 indicate that for the geometries studied, which include the range of the test shells, the largest effect occurs for medium stiffening, falling off as the stiffening increases further. By correlation with Figs. 4 and 5 of [11], one can explain this behavior as resulting from the shift to higher Z of the maximum of the axial constraint effect with increase in stringer area and other geometry parameters. Hence for a given Z the increase in $A_1/b_1 h$ yields the apparent decrease in P_{SS4}/P_{SS3}.

One may note that the influence of axial constraints on the buckling of axially compressed stringer-stiffened shells resembles that observed in isotropic shells buckling under lateral or hydrostatic pressure (see for example [15] or [16]), or vibrating freely [17]. The similarity with the influence of in-plane boundary conditions for vibrations is en-

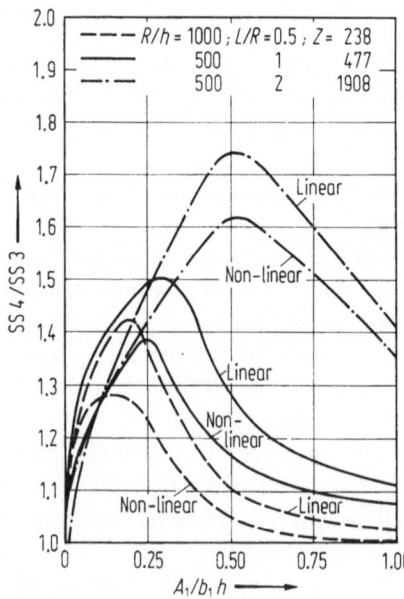

Fig. 3.
Effect of axial restraint on buckling of stringer-stiffened shells.

couraging if correlation between vibration and buckling tests is being attempted.

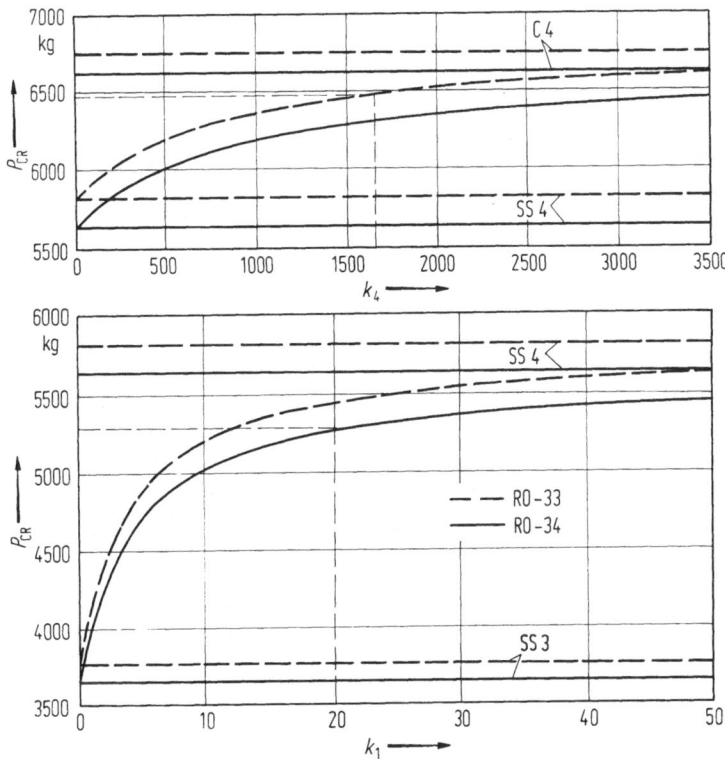

Fig. 4. Influence of elastic axial and rotational restraint on the buckling loads of shells RO-33 and RO-34.

In the studies of vibrations of stiffened shells the influence of in-plane boundary conditions has apparently not been considered. The investigators have limited themselves to consideration of rotational restraints and have analysed only SS3 or C4 boundary conditions (see for example [18] to [21]), though reference is sometimes made to the importance of restraints in isotropic shells. Hence considerable attention was given to the influence of in-plane boundary conditions in the theoretical and experimental studies of vibrations of axially loaded cylinders at the Technion. The earlier tests and calculations [22, 23]) considered only clamped edges C4. Then supports without rotational restraints, similar to those developed for buckling tests [24] or [25], were employed in tests [26] and [27] and concurrently extensive computations were carried out for different in-plane boundary conditions.

As a result of the prominence of axial constraints in the case of buckling, the emphasis in the vibration studies has also been on axial constraints [27].

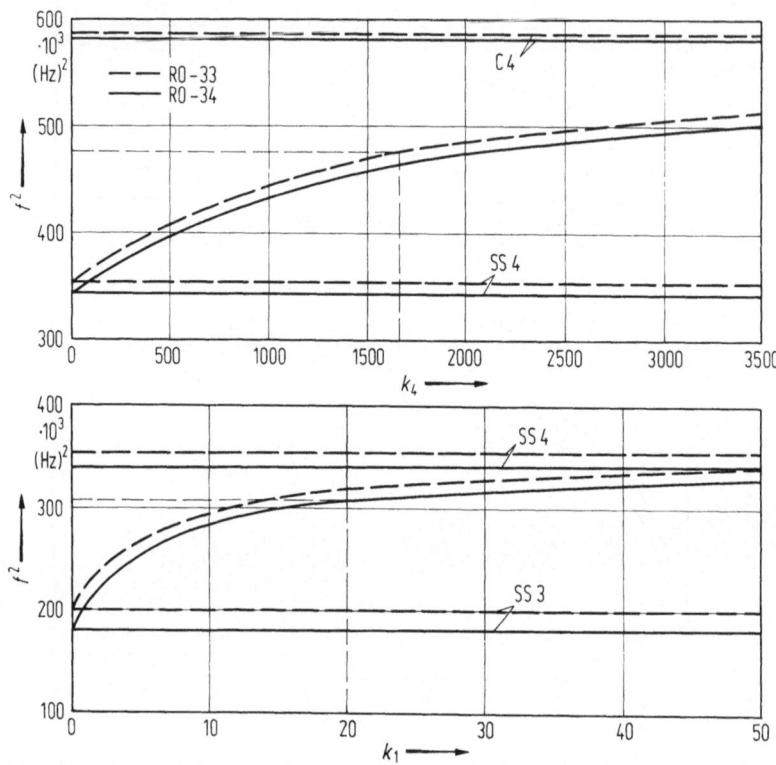

Fig. 5. Influence of elastic axial and rotational restraint on the vibrations of shells RO-33 and RO-34 ($P = 2000$ kg, $n = 11$, $m = 1$).

4. Out-Of-Plane Boundary Conditions

In the extended paper [1] the effects of out-of-plane boundary conditions on isotropic and stiffened shells are also discussed, in particular the results of [28, 29, 18, 20]. Regarding the feasibility of the inextensional buckling mode, which is the primary factor governing possible significant reductions in buckling loads [29], the behavior of stringer-stiffened shells does not differ much from that of corresponding isotropic shells. However, the radial restraints or end rings required to eliminate the inextensional mode, though larger than predicted in [28], will usually be provided by the supporting structure or laboratory fixtures. If one is concerned about a more local out-of-plane freedom, as in the tests in [9] and [13], the usual $n \geq 2$ buckling modes should be considered. The original conclusions of [28], that even small radial restraints are practically equivalent to $w = 0$ holds, as is reconfirmed by the calculations of [13].

5. Experimental Definition of Boundary Conditions by Correlation with Vibration Tests

For stringer-stiffened cylindrical shells, the definition of the boundary conditions is of prime importance. The similarity with the corresponding influence on the lower natural frequencies of vibrations of such shells indicates that correlation with vibration tests may be an appropriate tool to achieve this definition. For columns, correlation studies have been developed for assessment of the elastic restraints provided by the boundaries from vibration tests [30 to 34]. It was pointed out in [32] and [34] that the success of these techniques is restricted to supports with zero transverse displacements or strong transverse springs. Though analogous methods have apparently not been developed for shells, the similarity in behavior observed in the vibration and buckling tests of stiffened shells lead the authors to believe that such techniques could here be successfully employed.

Furthermore, if the effect of the boundary conditions on the buckling loads of stiffened shells is as important as that of the initial imperfections, or even predominant, their proper consideration should reduce experimental scatter significantly. The ratio of the experimental buckling load P_{EXP} to that predicted by linear theory P_{CR}, called in [24] and subsequently "linearity" $\varrho = P_{\mathrm{EXP}}/P_{\mathrm{CR}}$, is usually not far from unity in closely stiffened shells [3]. However, there is still a scatter of approximately $\pm 20\%$ in the collected experimental results (see Figs. 8 and 9 of [3] or Figs. 12 and 19 of [25]). This scatter can indeed be reduced by correlation with vibration tests.

Two typical shells tested recently [26], RO-33 and RO-34, are now examined as an example of the correlation technique developed. The dimensions of the specimens are given in Table 1. The shells were manufactured as twins from one blank and on one mandrel [25] and are considered as twins, though their dimensions differ very slightly. The main difference between the two specimens was in their boundary conditions. Shell RO-33 was clamped, approaching the theoretical C4 B.C.'s. Clamping for RO-33 is affected in the usual manner [23]: the end plates have circular grooves, the inner diameters of which fit the test shell tightly. Cerrobend (a low melting metal) is then poured into the external gaps. The depth of the groove was 10 mm, and the length of shell RO-33 was therefore originally 20 mm longer than RO-34. Hence the specimens were finally of identical length. Shell RO-34 was

Table 1. Geometrical properties of stringer-stiffened shells tested

Geometrical property	Shell	RO-25	RO-26	RO-27	RO-28	RO-29	RO-30	RO-31	RO-32	RO-33	RO-34
Radius to shell middle surface, R [mm]		120.1	120.1	120.1	120.1	120.1	120.1	120.1	120.1	120.1	120.1
Shell thickness, h [mm]		0.253	0.251	0.254	0.254	0.255	0.259	0.249	0.257	0.251	0.245
Shell length, L [mm]		130.0	130.0	130.0	130.0	130.0	130.0	215.0	120.0	130.0	130.0
Stiffener width c_1 [mm]		0.90	0.90	0.90	0.90	0.90	0.90	0.90	0.90	0.90	0.90
Stiffener height, d_1 [mm]		1.505	1.488	1.480	0.979	0.974	0.980	1.985	1.984	1.745	1.749
Number of stiffeners		84	84	84	84	84	84	84	84	84	84
R/h		475	479	473	473	471	463	482	467	479	490
L/R		1.08	1.08	1.08	1.08	1.08	1.08	1.79	0.999	1.08	1.08
Z		530	535	528	528	526	518	1474	445	534	548
A_1/b_1h		0.596	0.594	0.584	0.386	0.383	0.379	0.799	0.775	0.698	0.716
$-e_1/h$		3.47	3.47	3.41	2.43	2.41	2.39	4.48	4.35	3.97	4.06
I_{11}/b_1h^3		0.776	0.785	0.754	0.499	0.490	0.471	4.236	3.847	2.809	3.042
η_{t1}		6.598	6.678	6.375	3.099	3.025	2.943	10.450	9.495	8.513	9.175

7075 Aluminium alloy $E = 7500$ kp/mm^2, $\nu = 0.3$, specific gravity $= 2.80$.

mounted on simple supports, shown as edge A in Fig. 9. With these B.C.'s the shell rests in a triangular groove which is carefully fitted to the radius of each shell. The buckling loads predicted with linear theory (taking into account the small dimensional differences), the experimental buckling loads and the "linearity" values are:

RO-33 $P_{CR} = 6\,755$ kg C4,   RO-34 $P_{CR} = 3\,644$ kg SS3,
              $P_{EXP} = 4\,300$ kg, $P_{EXP} = 4\,210$ kg,

              $\varrho = 0.64$, $\varrho = 1.16$.

Hence there is an apparent scatter of about 45 percent. The experimental boundary conditions are, however, somewhere between SS3 and SS4 in the case of RO-34, and also not complete clamping for RO-33, but less than C4.

The influence of axial elastic restraint k_1 and rotational restraint k_4 on buckling and vibrations can be calculated with the theory of [22], extended to include elastic restraints at the boundaries [27]. Figure 4 shows the variation of buckling load between SS3 and SS4 and between SS4 and C4 boundary conditions for these two shells. The variation between SS3 and SS4 B.C.'s is achieved by introduction of an axial spring k_1 (the other B.C.'s are $v = 0$, $w = 0$, $M_x = 0$). The stiffness k_1 is zero for SS3 B.C.'s and infinity for SS4. Similarly, the variation between SS4 and C4 is obtained with a torsional spring (the other B.C.'s are $u = 0$, $v = 0$, $w = 0$), the stiffness of which k_4 is zero for SS4 and infinity for C4 B.C.'s. The variation of buckling load with increasing spring stiffness shows in both cases initially a steep increase and then a slow asymptotic approach to the values of SS4 and C4. It should be pointed out that the magnitudes of the springs in the two cases are different. One may also note, that in the figure the small dimensional difference in shell thickness between the two shells is taken into account and the precise results are presented for both shells. The buckling modes in all cases have here one axial half wave, whereas the circumferential wave numbers for small values of axial springs ($0 < k_1 < 4$) is $n = 10$, and for all other cases $n = 11$.

Correlation with the natural vibrations of the corresponding axially loaded shell is now employed to assess the real boundary conditions of the experiment. In Fig. 5 the influence of elastic axial and rotational restraints on the squared frequency of the mode $n = 11$, $m = 1$, at an axial load of $2\,000$ kg, is shown in the same manner in which the influence on buckling load was shown in Fig. 4. This mode of vibration was chosen, because it represents the buckling mode in most of the range of the springs. The load of $2\,000$ kg was chosen, to be well outside the range of low loads at which the shell has not yet settled in its simple supports. The behavior is indeed very similar to that shown in Fig. 4.

Hence by measuring the natural frequency at a relatively low load, one can estimate the real boundary conditions.

The experimental technique developed for vibration tests of axially loaded shells [23] is employed here. The test apparatus and procedure is described in detail in [23] and is only briefly summarized here. The test set up is shown in Fig. 6. The shell is excited by an acoustic driver

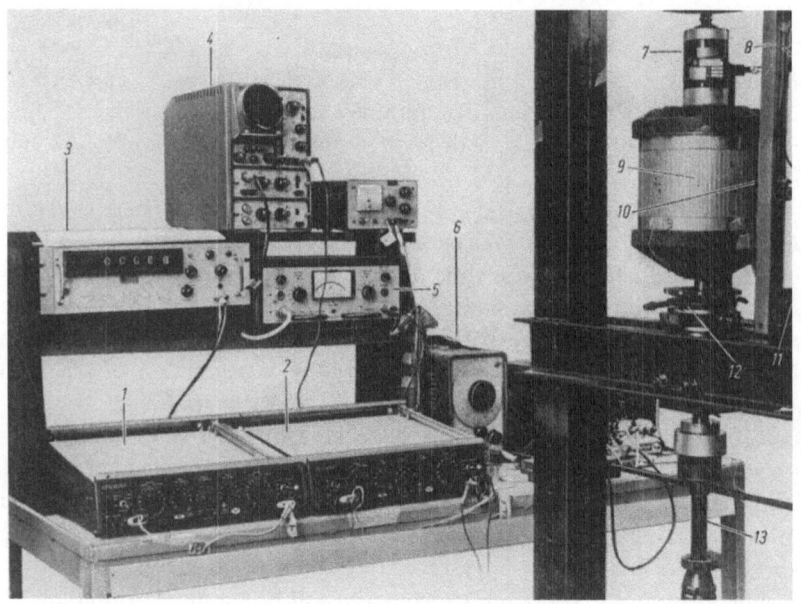

Fig. 6. Test set up.

1 Axial *x-y* recorder; *2* Circumferential *x-y* recorder; *3* Frequency counter; *4* Oscilloscope; *5* Microphone amplifier; *6* Oscillator; *7* Load cell; *8* Spanning apparatus; *9* Shell; *10* Microphone; *11* Axial potentiometer; *12* Circumferential potentiometer; *13* Loading jack.

which is inside the shell. The response of the shell is measured by a microphone outside the shell. The excitation frequency is changed and resonance is detected by the help of Lissajous figures. When a resonance frequency is detected, the mode of vibration is recorded by plotting the microphone reading versus its circumferential or axial position on *X-Y* recorders. The load is applied with a screw-jack and the load distribution is checked by an array of pairs of strain gages, which permit separation between compressive and bending strains. In the tests, the load is increased, first in relatively large increments of 400 kg, and then at progressively smaller increments of 200 kg and 100 kg. At each load, a full scan of a wide range of frequencies is performed and

the modes at the resonant frequencies are recorded. The scan is repeated with the microphone at different positions to prevent omission
of any mode due to an accidental mode-point. At buckling the drop
in load is also recorded. The experimental results include some multiple
results for the same mode shape at a certain load. These appear because
of difficulties in detection, and similar ambiguities were attributed by
Tobias [35] to imperfect axisymmetry. The multiple frequencies, however, do not cause any difficulty when they are close, as in the case
here.

Figure 7 presents the frequency squared versus axial load in shells
RO-33 and RO-34 for the mode $n = 11$, $m = 1$. For clarity, the theoretical curves are presented only for one shell, RO-34, since the results
for the twin shell RO-33 are very close (as can be seen for example in
Fig. 5).

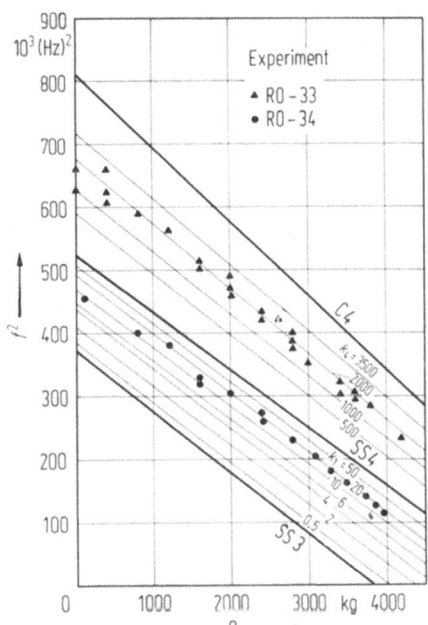

Fig. 7.
Frequency squared versus
axial load-shells RO-33 and
RO-34 for $n = 11$, $m = 1$.

The experimental results for RO-34 exhibit a clear trend to some
value of the spring k_1 and those for RO-33 to a certain value of the
spring k_4. Figure 8 is a similar plot for another mode $n = 7$, $m = 1$, and
again shows similar clear trends. The values of k_1 and k_4 from Fig. 8
are, however, found to be slightly lower than those in Fig. 7 for $n = 11$,
$m = 1$. Similar variations in the restraints with wave number were also
observed in other cases [27].

The discussion will now focus on the case $n = 11$, $m = 1$, since this is also the theoretical buckling mode. The values of frequency squared at 2 000 kg are taken from Fig. 7 and by following the dotted lines in

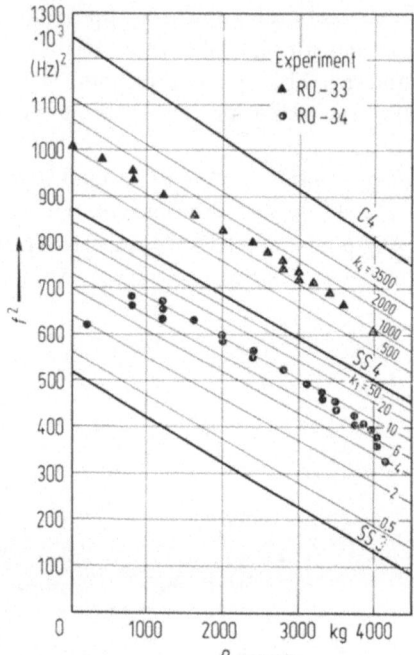

Fig. 8.
Frequency squared versus axial load-shells RO-33 and RO-34 for $n = 7$, $m = 1$.

Fig. 5, one finds that for RO-33, $k_4 = 1\,660$ and for RO-34, $k_1 = 20$. With these values of spring stiffness one turns to Fig. 4 and, by again following the dotted lines, one obtains the theoretical predicted buckling loads for these values of springs. Hence the predictions and "linearity" values for the same experimental buckling loads change to:

RO-33 P_{CR} = 6 469 kg,   elastic restraints (from vibrations)
          P_{EXP} = 4 300 kg,
          ϱ =   0.66,

RO-34 P_{CR} = 5 288 kg,   elastic restraints (from vibrations)
          P_{EXP} = 4 210 kg,
          ϱ =   0.80.

The scatter in the "linearity" reduces therefore appreciably from the previous 45% to 17%.

A similar correlation has been carried out for two other shells tested, RO-31 and RO-32. These two shells were again manufactured from one blank and on one mandrel and are twins except for different length. Their dimensions are given in Table 3. Shell RO-31 was clamped in the same manner as RO-33 and RO-32 was again supported as edge A in Fig. 9. The results of the correlation with vibrations and the reference values are:

RO-31 $P_{CR} = 6\,870$ kg C4, $P_{CR} = 6\,250$ kg elastic restraints (from
$$P_{EXP} = 4\,693 \text{ kg} \qquad \text{vibrations)}$$
$\varrho = 0.68$ $\varrho = 0.72$ after correlation

RO-32 $P_{CR} = 5\,338$ kg SS3 $P_{CR} = 6\,220$ kg elastic restraints (from
$$P_{EXP} = 4\,700 \text{ kg} \qquad \text{vibrations)}$$
$\varrho = 0.88$ $\varrho = 0.76$ after correlation.

Hence the scatter in "linearity" has been reduced from 23% to 5%.

Further correlations for other experimental results and further tests are under way to add confidence in the correlation technique.

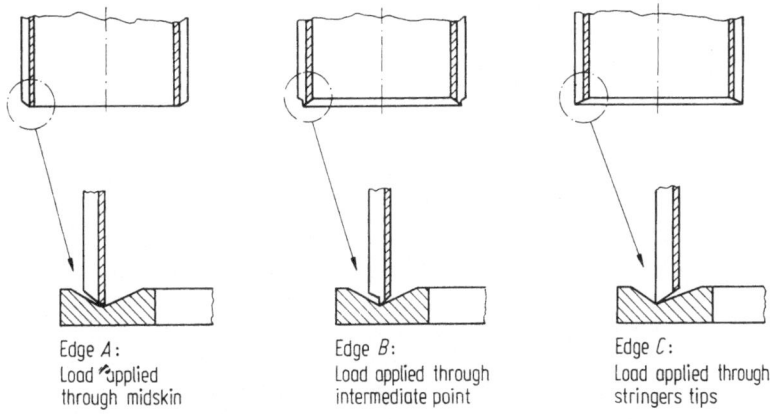

Edge *A*:
Load applied
through midskin

Edge *B*:
Load applied through
intermediate point

Edge *C*:
Load applied through
stringers tips

Fig. 9. Details of load application for different types of edges.

Edge *A*: Load applied through midskin; Edge *B*: Load applied through intermediate point; Edge *C*: Load applied through stringers tips.

6. Effect of Eccentricity of Loading

Eccentricity of loading, usually defined as the radial distance between the line of axial load application and the shell midskin, has been shown to have considerable influence on the buckling load of stringer-stiffened shells (see, for example, [6], or [9]). In the more recent study [9] the theoretical investigation was amplified by tests on integrally stringer-stiffened cylindrical shells loaded eccentrically and having different boundary conditions. These studies have now been extended to consider the influence of load eccentricity on the vibrations of axially loaded stiffened shells both theoretically and experimentally and correlate the results with those for buckling. Two families of shells are studied, one heavily stiffened and the other moderately stiffened.

The details of the extensive calculations and experiments are given in [13] and only the salient features are summarized here.

The experimental set-up and procedure is essentially identical to that shown in Fig. 6 and discussed in the previous section, (see also [23] or [27]), except for the details of load application and the specimens.

Six integrally stringer-stiffened shells were tested in the present test series. The specimens, which are similar to the shells of [23] and [25], were cut from 7075-T6 aluminium alloy extruded tubes and accurately machined by a process described in [25]. The eccentricity of loading is achieved by applying the load through the stringers. Specimens are therefore manufactured with three kinds of edges, as shown in Fig. 9. In the case of edge A, load is applied through the mid-skin of the shell, for edge B load is applied through an intermediate point along the depth of the stringers and in the case of edge C through the tip of the stringers. In all the cases special support rings (see Fig. 9) are accurately fitted to the shell edges, which restrain the radial displacement of the shell edge or stringers. The specimens were manufactured in triplets, consisting of three shells made from one blank, one with each of the 3 types of edges. Comparison was therefore between almost identical shells, as can be seen in Table 1, giving the dimensions of the shells. Specimens RO-25, 26 and 27 represent the heavily stiffened family of shells and RO-28, 29 and 30 the moderately stiffened one.

The calculations for eccentric loading were carried out with BOSOR 3 [10] and for $\bar{e} = 0$. They were compared with results obtained with the linear theory of [22]. Vibrations with one axial half wave ($m = 1$) and two axial half waves ($m = 2$), and loads up to buckling, are computed and measured. Results for a typical vibration mode ($m = 1$, $n = 7$) are shown in Fig. 10. Theoretical predictions are given for SS3 and SS4 boundary conditions and the experimental results for the 3 shells of the family are also plotted. The experimental results exhibit here a behavior similar to that predicted for SS4 B.C.'s (through the values differ): RO-25 with $\bar{e}/h = 0$ having the lowest frequencies and those corresponding to RO-27, with $\bar{e}/h = -2.84$, and RO-26 with $e/h = -6.43$ above them respectively. This is typical for high circumferential wave numbers ($n \geq 7$ for the shells with heavy stringers and $n \geq 8$ for those with medium stringers), whereas for lower n the experimental behavior is closer to that predicted for SS3 B.C.'s and even the frequencies are below the ones for SS3 (see [13]).

The influence of eccentricity of loading on the buckling loads is shown in Figs. 11 and 12. The theoretical predictions for the heavily stiffened shell (Fig. 11) show that for SS3 B.C.'s there is little effect up to an outward load eccentricity of $\bar{e}/h = -2.5$, where a shallow maximum occurs. For larger outward load eccentricities the buckling

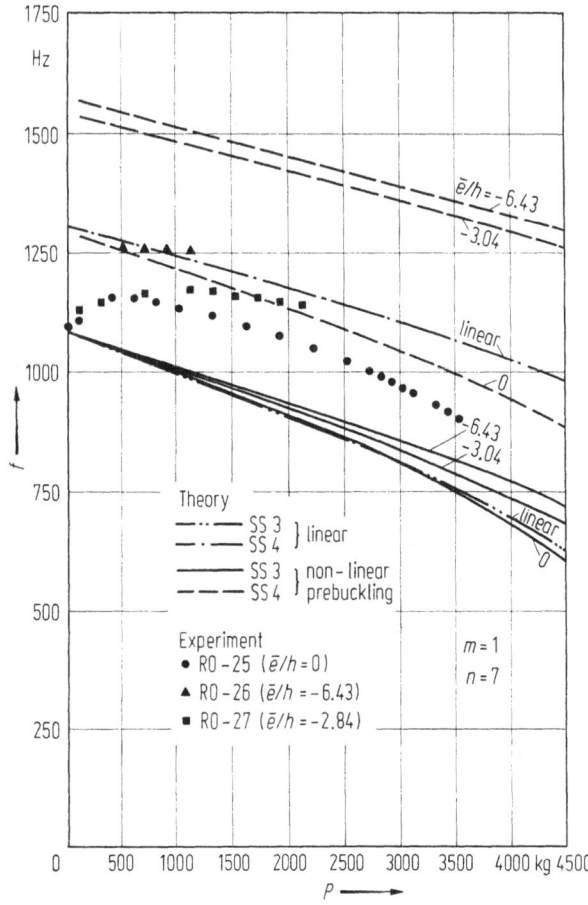

Fig. 10. Frequency versus axial load, "heavy" stringers ($A_1/b_1h = 0.59$).

load decreases and also the mode of buckling changes. For SS4 B.C.'s there is initially a steep rise in the buckling load with increase of outward eccentricity up to $\bar{e}/h = -1.5$. After that the increase is more moderate. The case of SS4 boundary conditions with load through midskin plus an equivalent moment is also plotted. This case differs from the former case of SS4 (with coinciding support and loading point) and is closer in its behavior to the SS3 case. We see that the theoretical influence of load eccentricity on the buckling load is similar to its influence on vibrations, especially for SS4 B.C.'s where also for vibra-

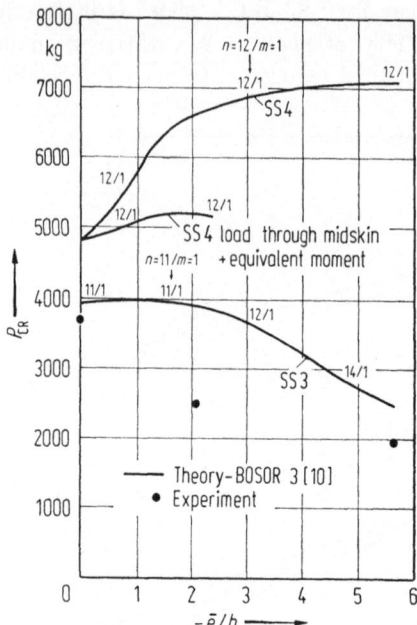

Fig. 11.
Influence of eccentricity of
loading on buckling loads,
"heavy" stringers
$(A_1/b_1h = 0.59)$.

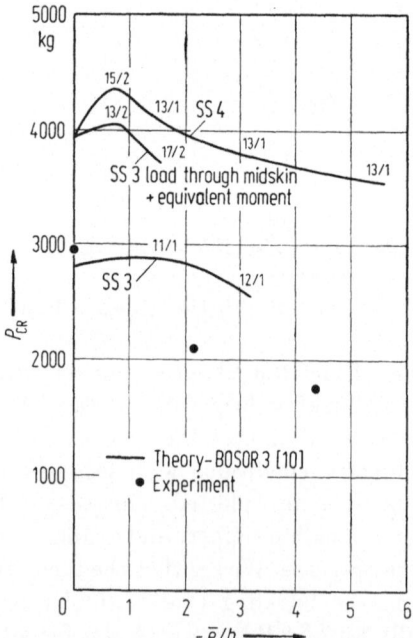

Fig. 12.
Influence of eccentricity of
loading on buckling loads,
"medium" stringers
$(A_1/b_1h = 0.38)$.

tions there is a steep rise in frequencies for small outward load eccentricity and a more moderate one for higher eccentricities (see for example, Fig. 5j of [13]). The three experimental points are also plotted in Fig. 11. The three shells buckled with $m = 1$, but exhibited different behavior near buckling. RO-25 buckled at 3 700 kg with 8 circumferential waves and the load dropped after buckling to 1 600 kg. Shell RO-26 showed noticeable bending before buckling. Buckling occured at 1 990 kg (it was not so well defined and the number of waves could not be counted). After buckling the load dropped only to 1 960 kg, the shell then continued to carry higher loads and the waves developed with increasing load (see also Figs. 9a and 9b of [13]). At 2 100 kg and 2 150 kg (the maximum load the shell carried), there were approximately 12 waves. Shell RO-27 buckled at 2 500 kg with approximately 10 waves and after buckling the load dropped to 2 240 kg. Note that for buckling calculations the load eccentricity measured after buckling is employed $\bar{e}/h = -5.66$ for RO-26 and $\bar{e}/h = -2.04$ for RO-27.

The results for the moderately stiffened shells are shown in Fig. 12. Though generally similar, Fig. 12 exhibits some significant differences compared to Fig. 11. The theoretical curve for SS4 differs having a steep maximum at a relatively low value of load eccentricity and decreasing then with $\bar{e}/h$. This behavior correlates with a similar one for the vibrations [13]. The experimental buckling loads were for RO-28: 2 960 kg with 9 waves, for RO-30: 2 100 kg with 9 waves and for RO-29: 1 740 kg with 9 waves. The buckling behavior was again more violent for RO-28 and softer for RO-30 and RO-29.

Hence the present experimental results reconfirm clearly and emphasize the behavior observed in [9] that increase in outward load eccentricity results in lower buckling loads but also in a much less violent buckling phenomenon. The agreement between the experimental results and the predictions is fair, but might possibly be improved by correlation with the vibration results. One may observe from the vibrations tests, that the test boundary conditions are nearer to SS4 than SS3, in the vibration modes close to the buckling modes. Further studies and tests are planned.

7. Prediction of Buckling Loads from Vibration Tests

The prediction of the buckling loads from vibration tests, as a basis for a nondestructive test method, has been attempted by many investigators for different structures. For columns and frames, good results have been obtained (see for example [36], [37] or [34], but applications of similar techniques to plates and shells have not yielded practical methods [36], except one successful application to spherical

246   J. Singer and A. Rosen

caps [38]. Other methods of nondestructive testing have been developed
for shells with some success (see for example [39] and [40]). The authors
believe, however, that correlation with vibration tests is a promising
direction of attack, in particular for closely stiffened shells, where the
low vibration modes observed in tests are very similar to the buckling
modes.

The prediction of buckling loads from the curves of frequency squared versus loads was already attempted by the authors in [23]. Some of the
shells tested there yielded promising results, though they were relatively weakly stiffened. The fact, that in the buckling modes for the
longer shells tested in [23] $m \geq 2$, also made the correlation more difficult. Heavier stiffening and shorter shells appeared therefore desirable
for more reliable correlation, on which the initial steps in the developments of a nondestructive test method could be based.

Linear extrapolation of the curves of frequency squared versus load
does not give satisfactory results, and the predicted buckling load
would in general be much higher than the experimental ones. The experimental curves, when continued to buckling, exhibit a steepening
at high loads and usually a rapid drop in frequency before buckling.
This behavior may be confirmed theoretically when a sophisticated
analysis which takes into account the real B.C.'s and real initial imperfections will be carried out. At present a semi-empirical curve is

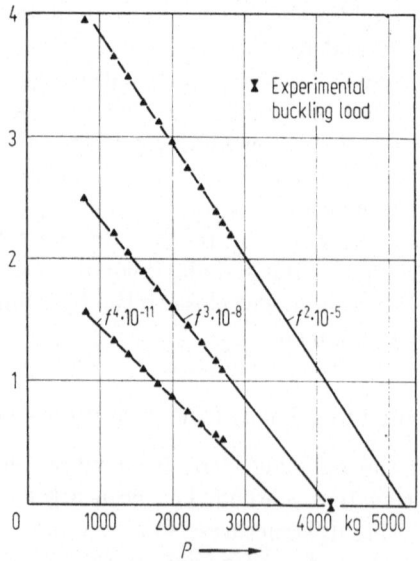

Fig. 13. Prediction of buckling load from variation of measured frequencies with
axial load—shell RO-34.

tried. One assumes that the curve is represented by a function $f^2 = \sqrt[\gamma]{A - BP}$ where $\gamma > 1$. This equation can be written in another form as

$$f^{2\gamma} = A - BP.$$

This form has the advantage that now the curve fitting is for a straight line which can be done very easily by small desk-computers. In Fig. 13, curves of experimental values of f for shell RO-34 raised to different powers are plotted. Extrapolation of f^3 yields a load very close to the experimental buckling load. If one tries to find the best fit to the last equation (taking the results up to 2700 kg) one obtains the best fit for the frequency to the power of 2.8 ($\gamma = 1.4$), which by extrapolation yields a buckling load of 4360 kg, 3% higher than the experimental value of 4220 kg. It should be pointed out that the error is larger in some shells. These results are only preliminary, and further tests and studies are in progress to obtain a more reliable correlation.

8. Conclusions

The following general conclusions can be drawn from the results obtained:

1. For externally stringer-stiffened cylindrical shells under axial compression, in the range of geometries studied, nonlinear prebuckling deformations have a relatively small effect on the buckling loads, except for very short shells where consideration of prebuckling deformation yields significantly higher buckling loads.

2. Axial restraints, and also rotational restraints, strongly affect the buckling loads of stringer-stiffened shells. This effect depends on shell and stiffener geometry. Hence a better definition of the actual boundary conditions, in particular with respect to axial and rotational restraint, will lead to more accurate predictions of buckling loads and will significantly reduce experimental scatter.

3. Axial restraints and rotational restraints also strongly affect the free vibrations of stiffened shells. This analogical behavior may be utilized as a tool for the definition of the actual boundary conditions, as was demonstrated for typical stringer-stiffened shells.

4. Eccentricity of loading has a very significant effect on the buckling load and behavior of stringer-stiffened shells, as was predicted by theory and verified by tests. The vibrations are affected in a similar manner by load eccentricity.

5. Preliminary results indicate that correlation with vibration tests, at axial loads much below the buckling loads, may yield the basis for a nondestructive test method for practical determination of buck-

ling loads of imperfect closely stiffened shells with real boundary conditions.

Acknowledgement

The authors wish to thank Mr. A. Greenwald, Mr. A. Klausner and Mr. H. Abramowits for their dedicated assistance in the experimental work and data processing. They also wish to thank Mr. S. Nachmani and the laboratory staff, R. Azulai, S. Wiesel, S. Fledel, L. Spector, for their assistance, and Miss D. Reuven for preparation of the figures.

References

1. Singer, J.; Rosen, A.: The Influence of Boundary Conditions on the Buckling of Stiffened Cylindrical Shells. TAE Report 213, Technion, Israel Institute of Technology, Dept. of Aeronautical Engineering, Haifa, Israel, June 1974.
2. van der Neut, A.: General Instability of Orthogonally Stiffened Cylindrical Shells. Collected Papers on Instability of Shell Structures—1962, NASA TN D-1510, pp. 309—319, Dec. 1962.
3. Singer, J.: Buckling of Integrally Stiffened Cylindrical Shells—A Review of Experiment and Theory. Contributions to the Theory of Aircraft Structures. Rotterdam: Delft University Press 1972, pp. 325—358.
4. Peterson, J. P.: Buckling of Stiffened Cylinders in Axial Compression and Bending—a Review of Test Data. NASA TN D-5561, December 1969.
5. Singer, J.; Arbocz, J.; Babcock, C. D.: Buckling of Imperfect Stiffened Cylindrical Shells Under Axial Compression. AIAA J. 9, No. 1, 68—75 (1971).
6. Block, D. L.: Influence of Discrete Ring Stiffeners and Prebuckling Deformations on the Buckling of Eccentrically Stiffened Orthotropic Cylinders. NASA TN D-4283, January 1968.
7. Almroth, B. O.; Bushnell, D.: Computer Analysis of Various Shells of Revolution. AIAA J. 6, No. 10, 1848—1855 (1968).
8. Card, M. F.; Jones, R. M.: Experimental and Theoretical Results for Buckling of Eccentrically Stiffened Cylinders. NASA TN D-3639, October 1966.
9. Weller, T.; Singer, J.; Batterman, S. C.: Influence of Eccentricity of Loading on Buckling of Stringer-Stiffened Cylindrical Shells. Thin-Shell Structures, Theory, Experiment and Design, Fung, Y. C. and Sechler, E. E. (ed.). Englewood-Cliffs, N.J.: Prentice-Hall 1974, pp. 305—324.
10. Bushnell, D.: Stress, Stability, and Vibration of Complex Shells of Revolution: Analysis and User's Manual for BOSOR3. Lockheed Missiles & Space Co. Report N-5J-69-1, Sept. 1969.
11. Weller, T.: Further Studies on the Effect of In-Plane Boundary Conditions on the Buckling of Stiffened Cylindrical Shells. TAE Report No. 120, Technion—Israel Institute of Technology, Dept. of Aeronautical Engineering, Haifa (Israel) January 1974.
12. Yamaki, N.; Kodama, S.: Buckling of Circular Cylindrical Shells under Compression. Rep. Inst. High Speed Mech. (Tohoku University) 23, 99—123 (1970); 24, 111—142 (1971); 25, 99—141 (1972); 27, 1—30 (1973).
13. Rosen, A.; Singer, J.: Vibrations and Buckling of Eccentrically Loaded Stiffened Cylindrical Shells. TAE Report 205, Technion—Israel Institute of Technology, Dept. of Aeronautical Engineering, Haifa, Israel, June 1974.
14. Weller, T.; Baruch, M.; Singer, J.: Influence of In-Plane Boundary Conditions on Buckling Under Axial Compression of Ring Stiffened Cylindrical

Shells. Proceedings of the Fifth Israel Annual Conference of Mechanical Engineering, Israel J. of Techn. **9**, No. 4, 397–410 (1971).
15. Sobel, L. H.: Effects of Boundary Conditions on the Stability of Cylinders Subject to Lateral and Axial Pressures. AIAA J. **2**, No. 8, 1437–1440 (1964).
16. Singer, J.: The Effect of Axial Constraint on the Instability of Thin Circula Cylindrical Shells under External Pressure. J. of Appl. Mech. **27**, No. 4, 737–739 (1960).
17. Forsberg, K.: Influence of Boundary Conditions on the Modal Response of Thin Cylindrical Shells. AIAA J. **2**, No. 12, 2150–2157 (1964).
18. Resnick, B. S.; Dujundji, J.: Effects of Orthotropicity, Boundary Conditions and Eccentricity on the Vibrations of Cylindrical Shells. AFOSR Scientific Rep. AFOSR 66-2821, ASRL TR 134-2, M.I.T. Aeroelastic and Structures Research Laboratory, Nov. 1966.
19. Penzes, L. E.: Effect of Boundary Conditions on Flexural Vibrations of Thin Orthogonally Stiffened Cylindrical Shells. J. of Acous. Soc. Am. **42**, No. 4, 901–903 (1967).
20. Sewall, J. L.; Naumann, E. C.: An Experimental and Analytical Vibration Study of Thin Cylindrical Shells With and Without Longitudinal Stiffeners. NASA TN D-4705, September 1968.
21. Patnaik, S.; Sankaran, G. V.: Vibrations of Initially Stressed Stiffened Circular Cylinders and Panels. J. of Sound and Vibration **31**, No. 3, 369–382 (1973).
22. Rosen, A.; Singer, J.: Vibrations of Axially Loaded Stiffened Cylindrical Shells: Part I—Theoretical Analysis. TAE Report No. 162, Technion—Israel Institute of Technology, Dept. of Aeronautical Engineering, Haifa, Israel, February 1974.
23. Rosen, A.; Singer, J.: Vibrations of Axially Loaded Stiffened Cylindrical Shells: Part II—Experimental Analysis. TAE Report No. 163, Technion—Israel Institute of Technology, Dept. of Aeronautical Engineering, Haifa, Israel, August 1973.
24. Singer, J.: The Influence of Stiffener Geometry and Spacing on the Buckling of Axially Compressed Cylindrical and Conical Shells. Theory of Thin Shells, Proceedings of 2nd IUTAM Symposium on Theory of Thin Shells, Copenhagen, September 1967. Berlin, Heidelberg, New York: Springer 1969, pp. 234–263.
25. Weller, T.; Singer, J.: Experimental Studies on the Buckling of 7075-T6 Aluminum Alloy Integrally Stringer-Stiffened Shells. TAE Report No. 135, Technion Research and Development Foundation, Haifa, Israel, November 1971.
26. Rosen, A.; Singer, J.: Further Experimental Studies of Vibrations of Axially Loaded Stiffened Cylindrical Shells. TAE Report 210, Technion—Israel Institute of Technology, Dept. of Aeronautical Engineering, Haifa, Israel, May 1975.
27. Rosen, A.; Singer, J.: Vibrations of Axially Loaded Stiffened Cylindrical Shells with Elastic Restraints. TAE Report 208, Technion—Israel Institute of Technology, Dept. of Aeronautical Engineering, Haifa, Israel, January 1975.
28. Almroth, B. O.: Influence of Imperfections and Edge Restraint on the Buckling of Axially Compressed Cylinders. NASA CR-432, April 1966. Also presented at the AIAA/ASME 7th Structures and Materials Conference, Cocoa Beach, Florida, April 18–20, 1966.

29. Cohen, G. A.: Buckling of Axially Compressed Cylindrical Shells with Ring-Stiffened Edges. AIAA J. **4**, No. 10, 1859—1862 (1966).
30. Lurie, H.: Effective End Restraints of Columns by Frequency Measurements. J. of the Aeronautical Sciences **18**, No. 8, 566—567 (1951).
31. Klein, B.: Determinations of Effective End Fixity of Columns with Unequal Rotational End Restraints by Means of Vibration Test Data. J. of the Royal Aeronautical Society **61**, No. 554, 131—132 (1957).
32. Jacobson, M. J.; Wenner, M. L.: Predicting Buckling Loads from Vibration Data. Experimental Mechanics **8**, No. 10, 35N—38N (1968).
33. Burgreen, D.: End Fixity Effect on Vibration and Stability. J. Engng. Mech. Div. ASCE **86**, 13—28 (1960).
34. Segall, A.: Nondestructive Dynamic Methods for the Determination of the Critical Loads of Elastic Columns. M. Sc. (Aeronautical Engineering) Thesis, Technion—Israel Institute of Technology, Haifa, Israel, June 1973 (in Hebrew).
35. Tobias, A. S.: A Theory of Imperfections for the Vibrations of Elastic Bodies of Revolution. Engng. **172**, 409—410 (1951).
36. Lurie, H.: Lateral Vibrations as Related to Structural Stability. J. of Appl. Mech. ASME **19**, 195—204 (1952).
37. Johnson, E. E.; Goldhammer, B. F.: The Determination of the Critical Load of a Column or Stiffened Panel in Compression by the Vibration Method. Proc. of the Society for Experimental Stress Analysis **11**, No. 1, 221—232 (1953).
38. Okubo, S.; Whittier, J. S.: A Note on Buckling and Vibrations of an Externally Pressurized Shallow Spherical Shell. J. of Appl. Mech. ASME **34**, 1032—1034 (1967).
39. Becker, H.: Nondestructive Testing for Structural Stability. J. of Ship Research **13**, 272—275 (1969).
40. Craig, J. I.; Duggan, M. F.: Nondestructive Shell-Stability Estimation by a Combined-loading Technique. Experimental Mechanics **13**, No. 9, 381—388 (1975).

The Effect of Shape Imperfections and Stiffening on the Buckling of Circular Cylinders

R. C. Tennyson

Institute for Aerospace Studies University of Toronto, Toronto, Ontario, Canada

Abstract

Both ring and stringer stiffened circular cylinders with and without shape imperfections have been studied under axial compression, external hyrostatic pressure and combined loading of compression—hydrostatic pressure. To provide reference data for the stiffened shell results, compressive buckling tests were performed on unreinforced cylinders containing asymmetric imperfection distributions. In addition, hydrostatic buckling of unstiffened cylinders was also investigated to assess the effects of different boundary conditions on the critical pressure for varying values of cylinder lengths.

Notation

A_r, A_s	Cross-sectional areas of ring and stringer reinforcements, respectively
c	$[3(1 - \nu^2)]^{1/2}$
d_r, d_s	Separation distance measured between centroids for ring and stringer reinforcements, respectively
D	$Et^3/12(1 - \nu^2)$
e_r, e_s	Radial distance between shell wall median surface and centroids of ring and stringer reinforcements, respectively
E	Young's modulus of elasticity
$\bar{h}$	Equivalent smeared out shell wall thickness
I_r, I_s	Bending moments of inertia about centroidal axes for ring and stringer reinforcements, respectively
K_p	$p_{cr}RL^2/D\pi^2$
K_{xi}, K_{yj}	$\pi R/q_0 l_{xi}$, $\pi R/q_0 l_{yj}$ respectively
l_{xi}, l_{yj}	Imperfection and/or buckling mode half-wavelengths in the axial and circumferential directions corresponding to the ith and j^{th} modes, respectively
L	Cylinder length
N_x, N_{xy}	Distributed normal and shear forces, respectively
p	Pressure
p_{cr}	External hydrostatic buckling pressure
$\bar{p}_{cr,0}$	Theoretical hydrostatic buckling pressure under no external axial load
P	Axial compressive load
P_{cr}	Axial compressive buckling load
$\bar{P}_{cl}$	$2\pi E\bar{h}^2/c$
$\bar{P}_{cr,0}$	Theoretical compressive buckling load at ambient pressure
q_0	$(R/t)^{1/2} [12(1 - \nu^2)]^{1/4}$

R Cylinder radius measured to the median surface of the shell wall

$t, \bar{t}$ Shell wall thickness and average measured value, respectively

U, V, W Displacements measured in the axial, circumferential and radial directions, respectively

X, Y Nondimensional orthogonal coordinate axes measured in the axial and circumferential directions, respectively

Z $(L^2/Rt)(1-\nu^2)^{1/2}$

$\alpha_{\mathrm{r}}, \alpha_{\mathrm{s}}$ $A_{\mathrm{r}}/d_{\mathrm{r}}t$, $A_{\mathrm{s}}/d_{\mathrm{s}}t$ respectively

$\beta_{\mathrm{r}}, \beta_{\mathrm{s}}$ $EI_{\mathrm{r}}/Dd_{\mathrm{r}}$, $EI_{\mathrm{s}}/Dd_{\mathrm{s}}$ respectively

$\gamma_{\mathrm{r}}, \gamma_{\mathrm{s}}$ e_{r}/t, e_{s}/t respectively

δ Imperfection amplitude

λ Ratio of the imperfect/perfect cylinder buckling loads

λ_P, λ_p Ratios of experimental/theoretical cylinder buckling loads for axial compression and external hydrostatic pressure, respectively

μ Deviation of cylinder median surface from "perfect" configuration/shell wall thickness ($\delta/2\bar{t}$ for the test models)

ν Poisson's ratio

1. Introduction

Although extensive efforts have been directed towards stability analyses and testing of stiffened cylinders (an excellent review of which can be found in [1]), it is still not resolved as to whether geometric shape imperfections actually lead to substantial buckling load reductions. Despite the availability of imperfection sensitivity calculations for stiffened cylinders (see [2] for example) indicating that these structures can indeed be susceptible to a loss in buckling strength due to the presence of imperfections, depending on the specific geometry of the shell structure, experimental evidence has been lacking particularly for prescribed imperfection distributions. It is felt that such test data are required in order to assess the structural efficiency of stiffened cylinders as compared to equivalent weight unstiffened shells. Thus a programme was initiated to investigate the buckling behaviour of both ring and stringer stiffened circular cylinders subjected to axial compression, external hydrostatic pressure and combined loading of compression-hydrostatic pressure. In order to experimentally determine the effect of shape imperfections, geometrically "near-perfect" test models were included to provide reference data for comparison purposes. The specific shape imperfection incorporated in most of the cylinders was an axisymmetric profile in which various amplitudes and wavelengths were studied. One series of stringer stiffened cylinders contained an asymmetric distribution which will be discussed in detail later. Again, because it was desirable to correlate these results with unstiffened cylinders, it was therefore necessary to undertake a study of the effect of asymmetric shape imperfections on the buckling behaviour of unreinforced circular cylinders. This was also the case for hydrostatic pressure loading since even the most recent [3] test results do not provide evidence on the difference in buckling pressures one could expect between cylinders having simply supported or clamped edges. Since all of the stiffened cylinders were tested with clamped ends, it was thus important to determine the collapse pressures of comparable geometrically "near-perfect" unstiffened cylinders.

All cylinder test models were manufactured from a photoelastic epoxy plastic using the spin-casting technique. In this way geometrically "near-perfect" specimens were readily manufactured for both unstiffened and stiffened configurations. Shape imperfections were in most cases profiled on the inner surface of either the test cylinder or the rotational mold. Thus the shell wall not only contained a prescribed deviation in its median surface but also a thickness variation. This latter effect has been shown to have a negligible influence on the predicted buckling load based on the average shell wall thickness for unstiffened cylinders, although this may not be the case for stiffened shells. However, the thickness variation was not considered in the numerical analysis.

It should also be noted that the epoxy plastic used in the fabrication of the cylinders has been well defined in terms of its material properties. In particular, it exhibits a linear stress-strain response virtually up to its fracture stress, with little "ductility". Consequently, nonlinear stress-strain behaviour during testing was not a problem, although in several cases the cylinders did fracture after buckling. In most instances however, each buckling test was performed several times and the repeatability of the test results was excellent. Thus each cylinder was used for a wide variety of loading conditions and this provided a good basis for comparing relative buckling loads.

Because of the diversity of the test programme, this report is subdivided into three major sections dealing with unstiffened cylinders under external hydrostatic pressure, the effect of asymmetric shape imperfections on the axial compressive buckling load of unstiffened cylinders and lastly, buckling of stiffened cylinders. For more details, reference should be made to the three UTIAS reports cited in the text.

2. Unstiffened Cylinders under External Hydrostatic Pressure

An investigation[1] of the buckling behaviour of unstiffened circular cylinders subjected to hydrostatic pressure loading was undertaken using geometrically "near-perfect" test models. Because of the con-

---

[1] The analysis and experiments reported in this section were performed by Mr. K. H. Chan and Mr. C. E. Seibel, respectively (Research Assistants, UTIAS).

siderable scatter in previously published data (except for the results of [3]), it was of interest to determine the effect of varying boundary conditions on the buckling loads as a function of the Z parameter. In particular, two sets of boundary conditions were employed: the clamped case in which[1]

$$W = W_{,x} = U_{,y} = V = 0$$

and simple support where

$$W = W_{,xx} = N_x = N_{xy} = 0$$

at the shell ends. In our nomenclature, the clamped and simple support cases will be referred to as CUV and SNT, respectively. It has been shown in [4] that the buckling condition $U_{,y} = 0$ also corresponds to $U = 0$. Furthermore, Yamaki [3] has demonstrated that for various clamped and simple support boundary conditions, two asymptotic buckling solutions result which can be described by

$$K_p = \begin{cases} 1.04Z^{1/2} \\ 1.56Z^{1/2} \end{cases} \quad \text{for} \quad Z > 100. \tag{1}$$

Essentially, the boundary conditions corresponding to these two cases can be classed into two different groups depending on whether $N_x = 0$ or $U = 0$, respectively, at the inception of buckling.

Based on the particular CUV and SNT boundary conditions employed in this study, buckling load solutions were obtained using the coupled nonlinear Donnell equations [4]. In this analysis, axisymmetric prebuckling solutions were first derived satisfying the given boundary conditions. Subsequently, the linearized buckling equations in the form of two coupled ordinary differential equations with variable coefficients were solved numerically by a forward integration scheme, again applying the appropriate boundary conditions.

In total, thirty-six cylinder configurations were tested, a description of which can be found in Table 1. The clamped edge constraint was achieved rather easily by bonding the cylinder ends to aluminum plates containing an annular groove. However, to obtain the simple support condition (SNT) experimentally required several attempts at providing minimal axial, circumferential and rotational restraints. As a result, the final configuration selected was composed of an aluminum end plate with a small outer lip, the inside diameter of which provided a close fit with the outside surface of the shell to eliminate outward radial end displacements yet permit relatively free edge rotation. The shell was then placed on a lubricated rubber gasket attached to the end

---

[1] A comma indicates differentiation with respect to the subscript variables indicated.

plate. In this way, circumferential shear forces were maintained small (but not necessarily zero) and free axial contraction occurred alt hough some elastic reaction was encountered in the expansion direction.

Table 1. Description of cylinders subjected to external hydrostatic pressure

Shell No.[1]	R [in]	$\bar{t}$ [in]	L [in]	Z	Experimental	
					p_{cr} (psi)	K_p
1S	3.90	0.0186	12.0	1817	0.23	52.2
2S	3.90	0.0245	12.0	1381	0.45	43.6
3S	3.90	0.0292	12.0	1159	0.71	41.0
4S	3.90	0.0341	12.0	993	0.99	35.6
5S	3.89	0.0404	12.0	839	1.57	34.0
6S	3.89	0.0482	12.0	704	2.43	31.0
7S	3.90	0.0187	8.31	867	0.34	35.7
8S	3.90	0.0244	8.23	652	0.65	30.1
9S	3.90	0.0292	8.22	543	1.03	27.8
10S	3.90	0.0340	8.43	492	1.50	27.0
11S	3.89	0.0403	8.63	434	2.27	25.7
12S	3.90	0.0484	8.66	364	3.45	22.6
13S	3.90	0.0187	4.48	251	0.63	19.3
14S	3.90	0.0244	4.49	194	1.14	15.8
15S	3.90	0.0292	4.13	137	2.01	13.7
16S	3.90	0.0340	4.10	116	3.21	13.6
17S	3.89	0.0403	3.74	81	5.06	10.7
18S	3.89	0.0484	3.49	59	8.00	8.6
1C	3.90	0.0186	11.44	1652	0.33	67.3
2C	3.90	0.0245	11.44	1255	0.64	56.9
3C	3.90	0.0292	11.50	1064	1.01	53.5
4C	3.90	0.0341	11.00	834	1.54	47.0
5C	3.89	0.0404	11.50	770	2.30	45.9
6C	3.89	0.0482	11.44	639	3.52	40.9
7C	3.90	0.0187	7.75	754	0.48	43.7
8C	3.90	0.0244	7.71	571	0.93	37.8
9C	3.90	0.0292	7.71	478	1.44	34.3
10C	3.90	0.0340	7.71	422	2.09	32.4
11C	3.89	0.0403	8.13	385	3.07	30.8
12C	3.89	0.0484	8.06	316	4.75	27.1
13C	3.90	0.0187	3.88	188	0.89	20.5
14C	3.90	0.0244	3.85	142	1.80	18.4
15C	3.90	0.0292	3.50	98	3.06	15.0
16C	3.90	0.0340	3.33	76	4.92	13.8
17C	3.89	0.0403	3.11	56	7.94	11.7
18C	3.89	0.0484	2.93	42	12.72	9.5

[1] S, C refer to simply supported (SNT) and clamped (CUV) boundary conditions, respectively.

$E = 4 \cdot 10^5$ psi, $\nu = 0.4$.

External hydrostatic pressure was applied to the shells by eva-
cuating the inside of the cylinder from a reservoir connected to a
vacuum pump. Thus controlled pressure loading was obtained, as
measured by an LVDT type pressure transducer in combination with
an $X-Y$ plotter. From these pressure-time curves, it was possible to
determine the critical buckling load which represented the peak value
obtained prior to a sudden pressure drop and collapse of the cylinder.
A summary of the buckling pressures and coefficients K_p can also be
found in Table 1. From the data shown plotted in Fig. 1, it is clear

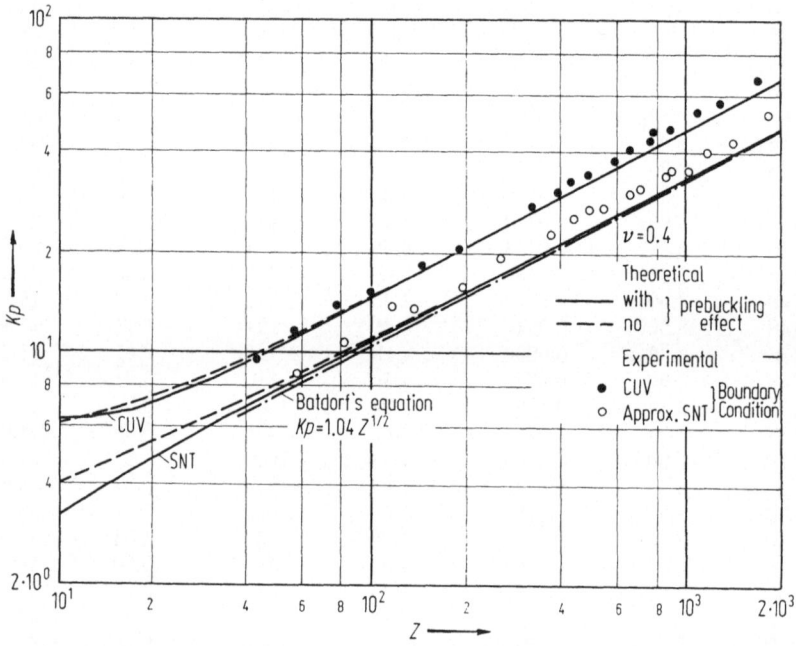

Fig. 1. Comparison of experimental and theoretical buckling results for external
hydrostatic pressure.

that a significant difference exists in the buckling pressures for the cylinders with clamped (CUV) and simple support (SNT) boundary conditions, as predicted theoretically. Furthermore, it is evident that the
simple support edge constraint achieved was not quite ideal, as expected,
but still rather close to the assumed condition. For both cases, consistent
buckling behaviour was observed, with very little scatter in the data, and
in excellent agreement with theory. It is also of interest, to note that
prebuckling effects only become significant for $Z < 100$.

3. Unstiffened Cylinders with Asymmetric Shape Imperfections

The effect of asymmetric shape imperfections on the axial compressive buckling behaviour of unstiffened circular cylindrical shells has been investigated[1] both analytically and experimentally. In particular, imperfection distributions described by

$$W_0(X, Y) = \begin{cases} -\mu_{x0} \cos K_{x0}X + \mu_1 \cos K_{x1}X \cos K_{y1}Y, & (2) \\ -\mu_{y0} \cos K_{y0}Y, & (3) \end{cases}$$

were studied for the specific case when $K_{x0} = 2K_{x1}$. From the coupled nonlinear Donnell equations with initial shape imperfection terms included, buckling load solutions were obtained using the Galerkin procedure for both imperfection cases, neglecting the effect of end constraints. Theoretical results have already been presented for imperfections of the form of Eq. (2) in [5, 6 and 7]. However, for the pure circumferential shape imperfection [Eq. (3)], the corresponding buckling load can be determined from

$$\lambda = \frac{(K_1^2 + K_2^2/4)^2}{2K_1^2} + \frac{K_1^2}{2(K_1^2 + K_2^2/4)^2} - \frac{CK_1^2K_2^2\mu_2}{(K_1^2 + K_2^2/4)^2} +$$

$$+ \frac{C^2K_1^2K_2^4\mu_2^2}{2}\left(\frac{1}{(K_1^2 + K_2^2/4)^2} + \frac{1}{(K_1^2 + 9K_2^2/4)^2}\right), \quad (4)$$

where $\mu_2 = \mu_{y0}$, $K_2 = K_{y0}$ and K_1 is varied to yield a minimum value for λ.

For comparison purposes, experiments were performed using test models having clamped end constraints and the following imperfection distributions:

$$W_0 = \begin{cases} -\mu_2 \cos K_2Y \\ -\mu_1 \cos 2K_1X - \mu_2 \cos K_2Y. \\ \mu_2 \cos K_1X \cos K_2Y \end{cases} \quad (5)$$

It should be noted that in general, the shape imperfection is present only on one surface of the shell wall, thus resulting in a median surface having the same shape but half the amplitude. However, for the imperfection distribution comprised of a sum of two functions, each surface had to be profiled. A total of 15 cylinders was tested, the results of which are summarized in Table 2 together with a description of their properties. From Figs. 2 and 3 it can be seen that for the case of the pure circumferential imperfection, there is a relatively small reduction in the buckling load compared to that which one would obtain for an equivalent axisymmetric shape imperfection. Of some interest is the

---

[1] The analysis and experiments reported in this section were performed by Mr. M. Booton and Mr. D. Zimcik, respectively (Research Assistants, UTIAS).

Table 2. Description of imperfect unstiffened cylinders

Shell No.	R [in]	$\bar{t}$ [in]	L [in]	μ_1	μ_2	K_1	K_2	λ (EXP'T)	λ_P
1	3.86	0.0259	8.5	—	0.0342	—	0.551	0.865	0.88
2	3.89	0.0393	11.0	—	0.0662	—	0.670	0.868	0.91
3	3.89	0.0262	11.0	—	0.1050	—	0.551	0.823	0.87
4	3.90	0.0258	11.0	—	0.0816	—	1.090	0.805	0.90
5	3.89	0.0202	11.0	—	0.1035	—	0.975	0.771	0.91
6	3.86	0.0162	10.2	0.0285	0.0525	0.477	0.437	0.630	0.88
7	3.90	0.0253	10.2	0.0403	0.0326	0.703	0.542	0.784	0.97
8	3.89	0.0228	11.0	0.0471	0.1210	1.045	0.500	0.748	0.91
9	3.89	0.0253	11.0	0.0472	0.1100	0.198	0.535	0.703	0.88
10	3.89	0.0266	11.0	0.0406	0.1050	0.824	0.555	0.686	0.87
11[1]	3.09	0.0177	8.6	—	0.0844	0.412	0.425	0.569	0.80
12	3.08	0.0289	10.8	—	0.0495	0.554	0.544	0.806	1.07
13	3.08	0.0223	10.8	—	0.0628	0.461	0.476	0.730	1.03
14	3.08	0.0250	11.5	—	0.0566	0.978	0.960	0.833	0.83
15	3.09	0.0174	11.5	—	0.0813	0.817	0.801	0.787	0.79

[1] Data based on local end region where buckling occurred.

$E = 4 \cdot 10^5$ psi, $\nu = 0.4$.

presence of a critical circumferential imperfection wavelength leading to a minimum buckling load at $K_2 = 1$. This corresponds to a wavelength equal to the classical axisymmetric buckling mode value.

The second set of imperfect cylinders contained an imperfection distribution defined by the sum of an axial and circumferential periodic function. The purpose of these experiments was to determine if the presence of the circumferential imperfection would substantially affect

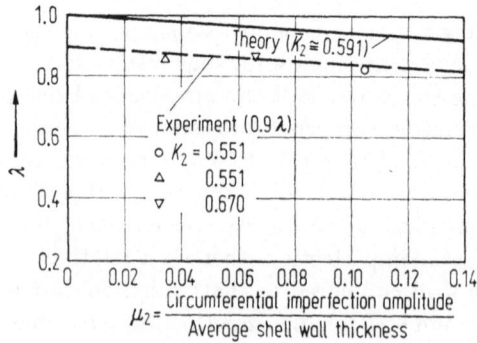

Fig. 2.
Buckling loads for circular cylinder containing a circumferential imperfection

the loads associated with the pure axisymmetric imperfection. As can be seen in Fig. 4, this particular asymmetric distribution did not result in buckling loads significantly below the predicted values based on axisymmetric imperfection theory, including the clamped edge constraint.

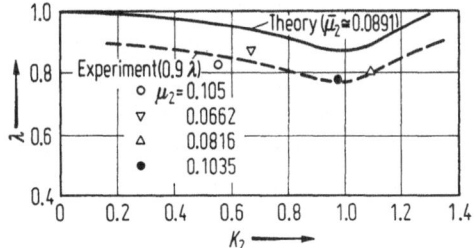

Fig. 3. Effect of circumferential imperfection wavelength on buckling loads.

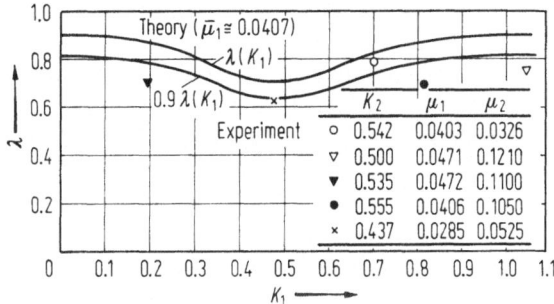

Fig. 4. Comparison of asymmetric imperfect shell buckling loads with theory for varying axial imperfection wavelengths.

The most difficult experimental models to fabricate were those containing the asymmetric imperfection having the form $\mu_2 \cos K_1 X \cos K_2 Y$ Considerable development went into devising a method of cutting this profile with a tracer tool apparatus operating in conjunction with gears and templates mounted on a lathe. Details of this manufacturing process will be published in [8]. Using the analysis described in [8], it was possible to calculate buckling loads, neglecting end conditions. As demonstrated previously [7], a particular combination ($K_1 = K_2 = 1/2$) corresponding to the classical asymmetric buckling mode yields a maximum load reduction. It can also be shown that these loads do not differ significantly from those values predicted for an equivalent axisymmetric imperfection and thus the effect of the circumferential component is small. From the experimental results which are presented in Table 2, agreement with theory is reasonably good for the variety of imperfection amplitudes and wavelengths studied.

4. Buckling of Stiffened Circular Cylindrical Shells

An extensive investigation[1] of the buckling of stiffened circular cylindrical shells was undertaken in which both circumferentially (ring) and axially (stringer) reinforced configurations were studied. The effects of axisymmetric shape imperfections were also considered for a variety of loading modes. In most cases, theoretical buckling loads were calculated based on a computer programme supplied by Hutchinson [9], although for axisymmetric imperfect models, the analysis in [2] was extended to incorporate arbitrary wavelengths. However, this latter analysis did not include prebuckling edge constraint effects which were taken into account in the more general programme [9]. In all of the stiffened cylinder studies, a clamped edge condition corresponding to the CUV state was considered.

4.1. Ring Stiffened Cylinders

A series of tests were conducted on both inner and outer surface ring stiffened cylinders, including several having a prescribed axisymmetric imperfection distribution. A description of these test models is given in Table 3 and Fig. 5. The various loading modes applied to the shells included axial compression, external hydrostatic pressure and combinations of compression-internal pressure and compression-hydrostatic pressure.

For the case of pure axial compression, theoretical buckling loads were calculated based on the nonlinear Donnell-type equations taking into account the prebuckling deformations associated with the clamped edge constraint. It was found that a mean curve could be fitted through these solutions such that for the range of shell parameters considered, the predicted buckling loads were all within $\pm 3.4\%$ of this curve. It was also observed that the classical axisymmetric solution (for $\gamma_r > 0$) coincided quite closely with this curve. However, the classical asymmetric solution was found to be consistently lower by an amount which increased with increasing values of α_r. Figure 6 shows a comparison between the mean predicted behaviour based on nonlinear theory and the experimental data for perfect cylinders for $Z \geqq 480$. For smaller values of the Z parameter, it was obsverved that the cylinder buckling loads increased, as can be seen in Table 3. It is evident that for the degree of stiffening achieved, no significant difference could be attributed to eccentricity effects (i.e., $\gamma_r \lessgtr 0$). On the other hand, the presence of small axisymmetric shape imperfections led to load reductions from 10% to 20% below the perfect shell values. In comparing these buck-

---

[1] The analysis and experiments reported in this section were pe rformed by Mr. K. H. Chan and Mr. E. Boros, respectively (Research Assistants, UTIAS).

Table 3. Description of ring stiffened cylinders

Shell No.[1]	α_r	β_r	γ_r	$\bar{t}$ [in]	$\bar{h}$ [in]	R [in]	Z	μ	K	Experiment P_{cr} [lbs]	Experiment p_{cr} [psi]	λ_P	λ_p
1	0.213	0.394	−1.170	0.0165	0.0198	3.93	1635	—	—	458	—	0.99	—
1A	0.213	0.394	−1.170	0.0165	0.0198	3.93	354	—	—	486	—	1.02	—
2	0.585	2.805	−1.634	0.0312	0.0483	3.92	479	—	—	1978	—	1.04	—
2A	0.585	2.805	−1.634	0.0312	0.0483	3.92	187	—	—	2032	—	1.07	—
3	0.206	0.207	−1.009	0.0204	0.0244	3.93	1261	—	—	747	—	1.05	—
3A	0.206	0.207	−1.009	0.0204	0.0244	3.93	183	—	—	806	—	1.13	—
4	0.416	1.152	−1.336	0.0173	0.0245	3.93	1631	—	—	570	—	1.03	—
4A	0.416	1.152	−1.336	0.0173	0.0245	3.93	195	—	—	609	—	1.08	—
4B	0.416	1.152	−1.336	0.0173	0.0245	3.93	311	—	—	616	—	1.10	—
5	0.283	0.503	−1.166	0.0199	0.0256	3.93	1420	—	—	728	−1.25	1.04	1.10
6	0.523	1.563	−1.576	0.0181	0.0276	3.93	1290	—	—	647	−1.95	1.01	1.15
7	0.344	0.520	−1.114	0.0216	0.0290	3.93	898	—	—	925	−1.92	1.10	1.13
11	0.240	0.371	1.122	0.0167	0.0207	3.89	794	—	—	460	−0.92	0.95	1.03
12	0.452	1.054	1.276	0.0173	0.0251	3.93	1634	—	—	538	−1.04	0.96	0.95
13	0.667	2.694	1.511	0.0165	0.0273	3.93	1604	—	—	543	−1.58	0.99	1.02
14	0.215	0.189	0.967	0.0276	0.0335	3.93	542	—	—	1362	—	1.04	—
21	0.332	0.687	1.221	0.0185	0.0246	3.89	1580	0.0506	0.593	480	−1.03	0.94	—
22	0.295	0.484	1.143	0.0202	0.0263	3.89	1430	0.0186	0.519	597	−1.10	0.91	—
23	0.310	0.575	1.182	0.0188	0.0244	3.89	1538	0.0719	0.164	515	−1.02	—	—
24	0.265	0.357	1.082	0.0223	0.0283	3.89	843	0.0437	0.752	771	−1.50	0.97	—
31	0.326	0.626	−1.194	0.0192	0.0256	3.91	1316	0.0625	0.609	574	−1.28	—	—
32	0.537	1.441	−1.329	0.0201	0.0310	3.91	1026	0.0597	0.620	672	−2.52	—	—
33	0.361	0.877	−1.276	0.0184	0.0253	3.90	1368	0.0653	0.684	547	−1.33	—	—
34	0.475	1.313	−1.337	0.0187	0.0274	3.90	1548	0.0646	0.686	573	−1.62	—	—

[1] Ring cross-section and imperfection profile are shown in Fig. 5.
$E = 4 \cdot 10^5$ psi, $\nu = 0.4$.

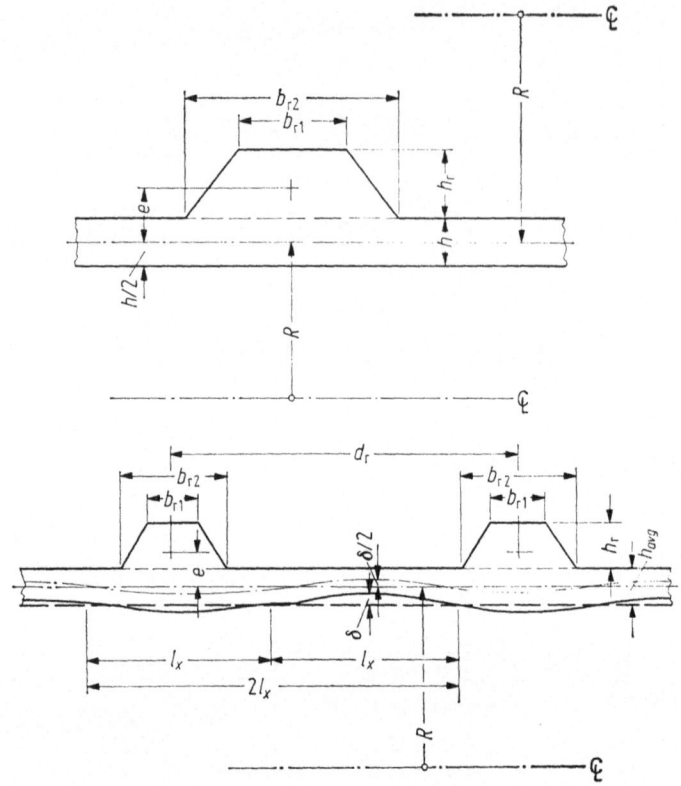

Fig. 5. Detailed ring geometry. $\gamma_r > 0$ Outside rings. $\gamma_r < 0$ Inside rings.

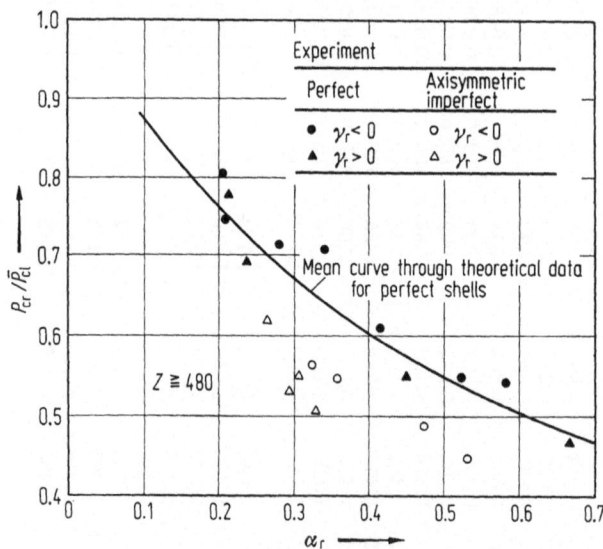

Fig. 6. Ring stiffened cylinders under axial compression.

ling loads with axisymmetric imperfection theory, it was found that for $\gamma_r > 0$, the experimental results were within 10% of theory whereas for $\gamma_r < 0$, the experiments were about 25% higher than predicted. No explanation for this latter discrepancy has yet been determined. It is of interest to also compare these loads with those for equivalent weight unstiffened cylinders under axial compression. Based on the work of [10], it was found that the ring stiffened axisymmetric imperfect cylinder buckling loads ranged from 26% to 61% lower than their equivalent weight unstiffened shells. In other words, both perfect and imperfect ring stiffened cylinders are substantially less efficient in axial compression than equivalent weight unstiffened cylinders.

The cylindrical shells were also tested under axial compression with the addition of internal pressure. Figure 7 illustrates the buckling results obtained for varying values of internal pressure. For comparison

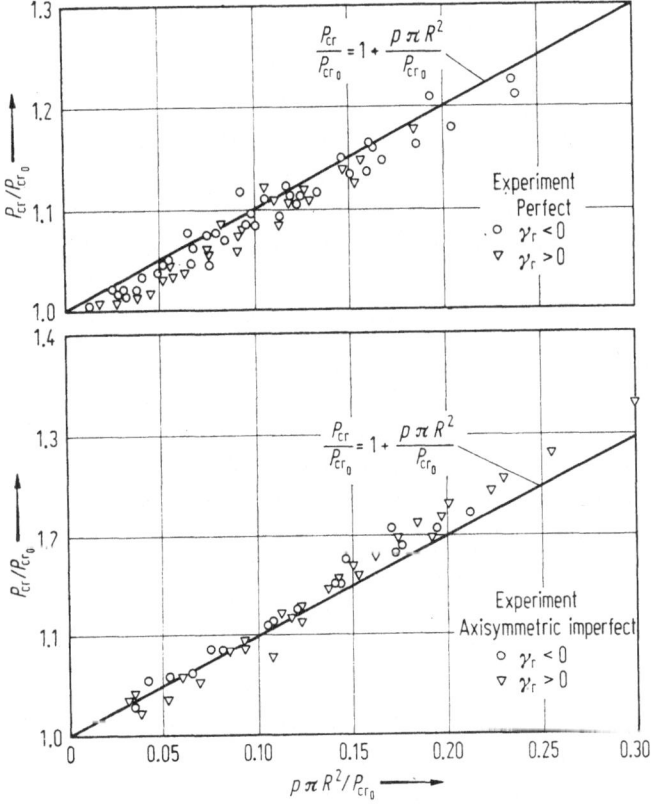

Fig. 7. The effect of internal pressure on the axial compressive buckling load of ring stiffened cylinders.

purposes, the simple formula

$$P_{cr}/P_{cr0} = 1 + (p\pi R^2/P_{cr0}) \tag{6}$$

was employed to determine to what extent the effect of internal pressure could be taken into account in increasing the critical compressive loads. It can readily be seen that Eq. (6) provides accurate estimates of the buckling loads for both inner and outer ring stiffened cylinders. Furthermore, axisymmetric imperfect shells behaved similarly, although Eq. (6) did consistently underestimate the buckling stresses for $p\pi R^2/P_{cr0} > 0.13$. The relatively larger increases in the buckling loads for the imperfect cylinders probably result from the reduced influence of the initial shape imperfections as the internal pressure is increased. This has also been shown to be the case for unstiffened shells [11]. In general, it can be concluded that for the range of cylinder parameters considered, compressive buckling loads can be predicted rather well using Eq. (6).

Combined loading tests involving axial compression and external hydrostatic pressure were also conducted on the ring stiffened cylinders. Although the geometry varied between cylinder test models, it was possible to construct mean interaction curves through the theoretical solutions within $\pm 3\%$ deviation for both inner and outer stiffened shells. A comparison of the experimental results with these mean interaction curves for the geometrically "near-perfect" ring stiffened cylinders is shown in Fig. 8. In general, the test data lie within $\pm 15\%$ of the theory, and demonstrate essentially the predicted interaction behaviour. It is also of interest to note that in most cases, the external hydrostatic buckling pressure observed experimentally was higher (by as much as 15%) than predicted theoretically. From a design viewpoint, the equivalent weight unstiffened cylinders having the same end constraints (CUV) were found to have buckling pressures ranging from 50% to 100% less than the ring stiffened configurations. Clearly, in terms of hydrostatic buckling efficiency, the ring reinforced cylinders are much superior to the unstiffened shells. As far as imperfection sensitivity is concerned, the major contribution appears to be in the axial compression loading mode with far less influence evident under the action of external pressure, (for the range of shell parameters considered) as can be seen in Fig. 9. The hydrostatic buckling pressures for the axisymmetric imperfect cylinders agree quite well with the predicted values for the geometrically perfect configurations and the predominant load reductions can be associated with the axial buckling loads. The axisymmetric imperfection distributions considered (refer to Table 3) included wavelengths corresponding to the critical axisymmetric buckling mode defined by

$$l_{xcr} = \pi(Rt)^{1/2} \left[12(1 - \nu^2)\,(1 + \alpha_r)\right]^{-1/4}. \tag{7}$$

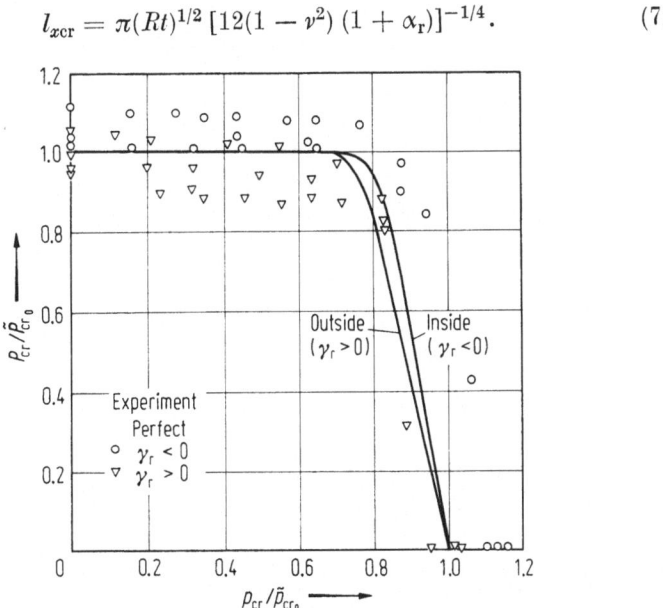

Fig. 8. Comparison of theory with experiment for ring stiffened cylinders subjected to combined loading of axial compression and external hydrostatic pressure.

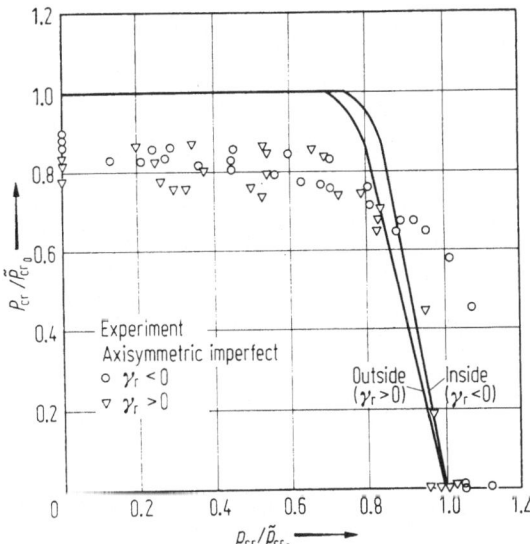

Fig. 9. The effect of axisymmetric shape imperfections on the buckling of ring stiffened cylinders subjected to combined loading of axial compression and external hydrostatic pressure.

It is of interest to note that for the α_r parameters of these shells, l_{xcr} is only slightly different from the classical axisymmetric buckling mode for the corresponding unstiffened cylinders.

4.2. Stringer Stiffened Cylinders

A study of stringer stiffened circular cylinders under axial compression, external hydrostatic pressure and combined loading of compression-hydrostatic pressure was undertaken to determine the effects of varying shell and reinforcement parameters on their buckling behaviour. In addition, the presence of geometric shape imperfections in the form of axisymmetric and asymmetric distributions were investigated to evaluate the imperfection sensitivity of axially stiffened cylinders. A description of the test models used is given in Tables 4, 5 and in Fig. 10. It should also be emphasized that only outer surface stiffening ($\gamma_s > 0$) was considered. As in the case of the ring stiffened cylinders, the shells were all manufactured using the spin-casting technique and thus the reinforcements were integrally part of the shell wall. For comparison with the experimental data, buckling loads were computed taking into account the clamped edge constraint (CUV condition) and the associated prebuckling deformations.

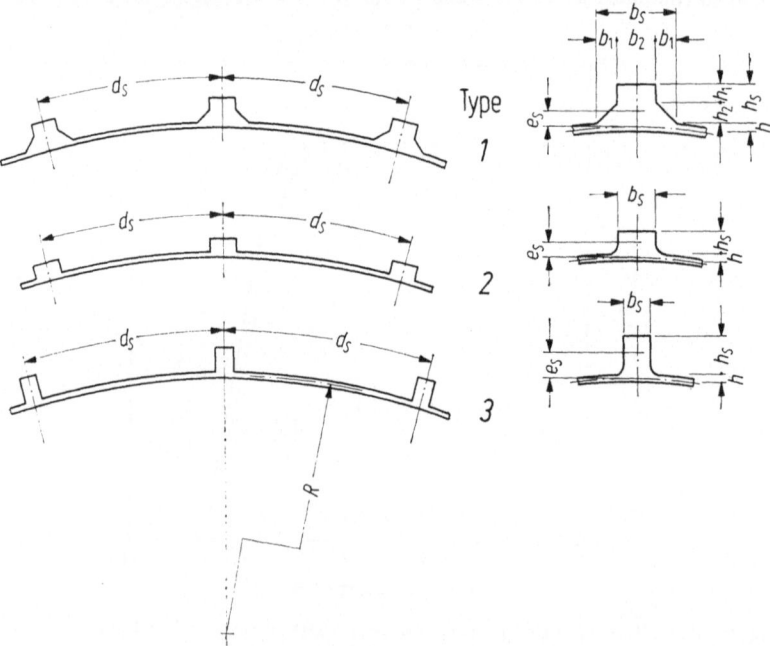

Fig. 10. Detailed stringer geometry.

Table 4. Description of geometrically "near-perfect" stringer stiffened cylinders

Shell No.[1]	α_s	β_s	γ_s	$\bar{t}$ [in]	$\bar{h}$ [in]	R [in]	Z	Experiment $\bar{P}_{cr}$ [lbs]	Experiment p_{cr} [psi]	λ_P	λ_p
101	1.150	17.00	2.26	0.0349	0.0752	3.69	843	7026	—	1.02	—
102	1.540	40.50	2.85	0.0261	0.0664	3.70	1126	5390	−2.40	0.99	1.09
103	0.800	4.47	1.72	0.0503	0.0907	3.69	583	10855	−8.01	1.05	1.22
104	0.990	17.86	2.02	0.0403	0.0806	3.69	731	8173	−4.40	1.01	1.00
105	2.680	99.00	3.67	0.0194	0.0597	3.70	1512	3980	−1.65	0.84	1.05
106	1.840	79.60	3.32	0.0218	0.0620	3.70	1345	4355	−1.85	0.89	1.05
111	0.964	15.12	2.76	0.0189	0.0373	3.76	1560	1877	—	0.95	—
112	0.721	6.30	2.11	0.0253	0.0437	3.76	1167	2556	—	0.92	—
113	0.606	3.74	1.86	0.0301	0.0486	3.76	970	3096	−1.72	0.93	1.07
114	0.877	11.34	2.47	0.0208	0.0392	3.76	1418	1989	−0.85	0.99	1.12
114A	0.877	11.34	2.47	0.0208	0.0392	3.76	143	3285	—	0.94	!
114B	0.877	11.34	2.47	0.0208	0.0392	3.76	300	2393	—	0.91	—
114C	0.877	11.34	2.47	0.0208	0.0392	3.76	66	4294	—	1.16	—
115	0.926	13.36	2.58	0.0197	0.0382	3.76	1525	1995	−0.64	0.98	0.92
115A	0.926	13.36	2.58	0.0197	0.0382	3.76	61	4210	—	1.19	—
115B	0.926	13.36	2.58	0.0197	0.0382	3.76	98	3682	—	1.07	—
115C	0.926	13.36	2.58	0.0197	0.0382	3.76	68	4160	—	1.17	—
116	1.675	78.60	4.25	0.0109	0.0294	3.77	2640	949	−0.26	0.73	0.84
117	0.666	4.94	1.99	0.0274	0.0459	3.76	1080	2990	—	1.00	—
118	0.970	15.37	2.67	0.0188	0.0373	3.76	1542	1975	−0.73	1.01	1.12
119	0.710	6.00	2.10	0.0257	0.0442	3.76	1130	2630	−1.23	0.93	1.06
151	0.673	24.45	3.81	0.0232	0.0388	3.74	948	2720	−1.14	0.93	0.87
152	0.517	11.07	3.04	0.0302	0.0459	3.74	1020	3398	—	0.89	—
153	0.903	58.97	4.96	0.0173	0.0330	3.74	1668	1815	−0.68	0.84	0.72
154	0.592	16.60	3.39	0.0264	0.0421	3.74	1124	2596	−1.29	0.80	0.97

[1] Stringer cross-sections are shown in Fig. 10. Shell Nos. 101—106, 111—119 and 151—154 are described by types (1), (2) and (3), respectively.

$E = 4 \cdot 10^5$ psi, $\nu = 0.4$.

Table 5. Description of stringer stiffened cylinders with shape imperfections

Shell No.[1]	α_s	β_s	γ_s	$\bar{t}$ [in]	$\bar{h}$ [in]	R [in]	Z	μ	K_1	K_2	Experiment P_{cr} [lbs]	Experiment p_{cr} [psi]
121	0.804	8.714	2.230	0.0227	0.0411	3.76	1324	0.0460	0.645	—	2374	−0.99
121A	0.804	8.714	2.230	0.0227	0.0411	3.76	123	0.0460	0.645	—	3390	—
121B	0.804	8.714	2.230	0.0227	0.0411	3.76	276	0.0460	0.645	—	2640	—
121C	0.804	8.714	2.230	0.0227	0.0411	3.76	52	0.0460	0.645	—	4244	—
121D	0.804	8.714	2.230	0.0227	0.0411	3.76	105	0.0460	0.645	—	3390	—
122	0.654	4.690	1.965	0.0279	0.0464	3.76	1058	0.0291	0.715	—	2944	—
123	0.511	2.242	1.645	0.0357	0.0542	3.75	819	0.0230	0.809	—	4182	−2.64
123A	0.511	2.242	1.645	0.0357	0.0542	3.75	99	0.0230	0.809	—	5140	—
123B	0.511	2.242	1.645	0.0357	0.0542	3.75	60	0.0230	0.809	—	6560	—
124	1.012	17.50	2.770	0.0180	0.0365	3.76	1589	0.0500	0.574	—	1802	−0.63
124A	1.012	17.50	2.770	0.0180	0.0365	3.76	61	0.0500	0.574	—	3360	—
124B	1.012	17.50	2.770	0.0180	0.0365	3.76	900	0.0500	0.574	—	1949	—
125	0.970	15.37	2.230	0.0188	0.0373	3.76	1530	0.0400	0.684	—	1939	−0.73
125A	0.970	15.37	2.230	0.0188	0.0373	3.76	60	0.0400	0.684	—	3758	—
125B	0.970	15.37	2.230	0.0188	0.0373	3.76	113	0.0400	0.684	—	3234	—
125C	0.970	15.37	2.230	0.0188	0.0373	3.76	283	0.0400	0.684	—	2500	—
126	0.793	8.380	1.820	0.0230	0.0415	3.76	1282	0.0455	0.748	—	2450	−1.00
126A	0.793	8.380	1.820	0.0230	0.0415	3.76	60	0.0455	0.748	—	4514	—
126B	0.793	8.380	1.820	0.0230	0.0415	3.76	99	0.0455	0.748	—	4280	—
126C	0.793	8.380	1.820	0.0230	0.0415	3.76	210	0.0455	0.748	—	2870	—
127	1.340	40.50	3.503	0.0136	0.0321	3.76	2166	0.0662	0.579	—	1305	−0.40
161	0.531	12.00	3.090	0.0294	0.0451	3.74	1045	0.0313	0.026	0.299	3170	—
162	0.528	11.75	3.070	0.0296	0.0453	3.74	1038	0.0777	0.026	0.300	2780	−1.55
163	0.676	24.73	3.790	0.0231	0.0390	3.74	1341	0.0390	0.023	0.265	2060	−0.98
164	0.668	23.79	3.750	0.0234	0.0391	3.74	1312	0.1136	0.023	0.266	2067	−0.96

[1] Stringer cross-sections are shown in Fig. 10. Shell Nos. 121—127 and 161—164 are described by types (2), and (3) respectively.

$E = 4 \cdot 10^5$ psi, $\nu = 0.4$.

Under pure axial compression, it was found that most of the geometrically "near-perfect" cylinders buckled at loads less than the classical values for the equivalent weight unstiffened shells ($\overline{Pcl}$). Only in one series of tests (shell nos. 151—154) were the observed buckling loads consistently higher than the predicted $\overline{Pcl}$ values. Significant increases in the buckling loads also occurred when the Z parameter was varied as can be seen in Fig. 11 where the data for one series of tests

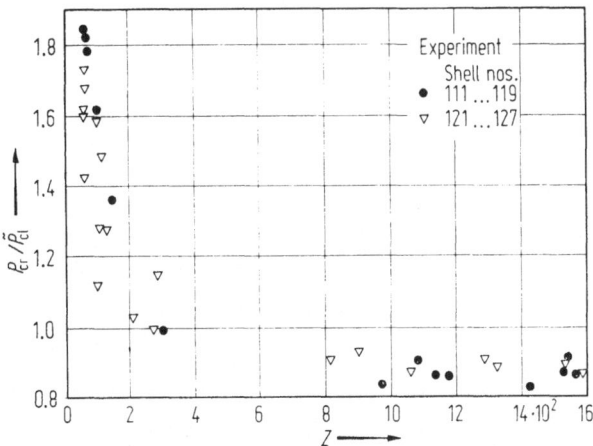

Fig. 11. The effect of varying Z and axisymmetric shape imperfections on the compressive buckling load of stringer stiffened cylinders.

(shell nos. 111—119) are plotted, as a function of Z. It would appear that for the range of parameters considered, this effect becomes substantial for $Z < 200$. At this point it is worth commenting that in the range of $Z < 150$, the axisymmetric solutions provided the best agreement with the experimental results. For corresponding cylinders having axisymmetric imperfection distributions (shell nos. 121—127), similar increases in buckling loads due to varying Z were also observed. However, the imperfect cylinder buckling loads were less than those for the "perfect" cylinder configurations for $Z < 300$ even though there appears to be no measurable imperfection effect for $Z > 300$. This is further illustrated in Fig. 12 where a comparison between the "perfect" and imperfect cylinder buckling loads is given as a function of β_s. It should be noted that λ_p, the ratio of (experiment/theory), was based on the computed buckling load for a geometrically perfect shell. Again it is evident that the effect of the axisymmetric shape imperfections studied ($\mu < 1$) resulted in no singificant load reductions for $Z > 200$. On the other hand, the stiffened shells (nos. 161—164) containing an asymmetric imperfection distribution of the form $\mu_2 = \cos K_1 X \cos K_2 Y$ (with the axial half-wavelength equal to the cylinder length) were found to buckle under axial compression at loads considerably less than the "perfect" shell values. However, for this particular cylinder configura-

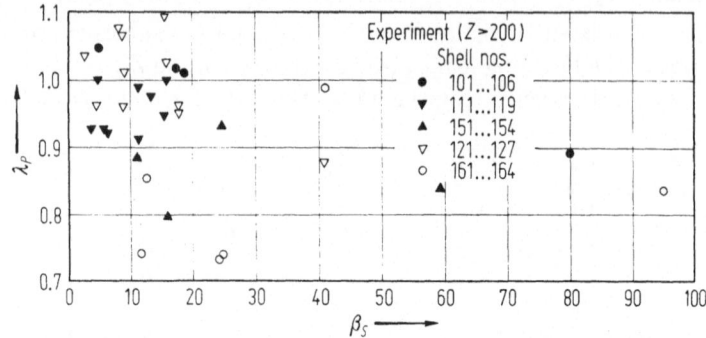

Fig. 12. Comparison of experimental data for both perfect and imperfect stringer stiffened cylinders with theory based on perfect shell configuration.

tion, these loads were still higher than those which one would obtain on equivalent weight unstiffened shells having the same imperfection distribution. Consequently, although stringer stiffened cylinders do not exhibit the same degree of imperfection sensitivity as is found in unstiffened circular cylinders, from a design point of view it is necessary to choose very carefully the reinforcement parameters to ensure that the buckling efficiency of the stringer stiffened cylinders exceeds that for equivalent weight unreinforced cylinders due to the presence of inevitable geometric shape imperfections. Perhaps the simplest criterion to impose is the requirement that the buckling load of a stringer stiffened cylinder $\geq \overline{Pcl}$.

The same cylinder test models were also subjected to external hydrostatic pressure and combined loading of compression and external pressure. A plot of the experimental results obtained is shown in Figs. 13, 14 together with a mean interaction curve fitted through the theoretical solutions. It was found that none of the predicted values deviated from this curve by more than $\pm 3\%$. In general, the agreement between theory and experiment is reasonable with most of the interaction data lying within $\pm 17\%$, although larger deviations were encountered in some pure compression and external pressure buckling loads. Of particular interest is the comparison between the axisymmetric imperfect cylinder buckling data with the mean theoretical interaction curve calculated for the "perfect" shell configurations. It is evident from Fig. 14 once again that for the range of axisymmetric imperfection amplitudes considered, no measurable reductions in buck-

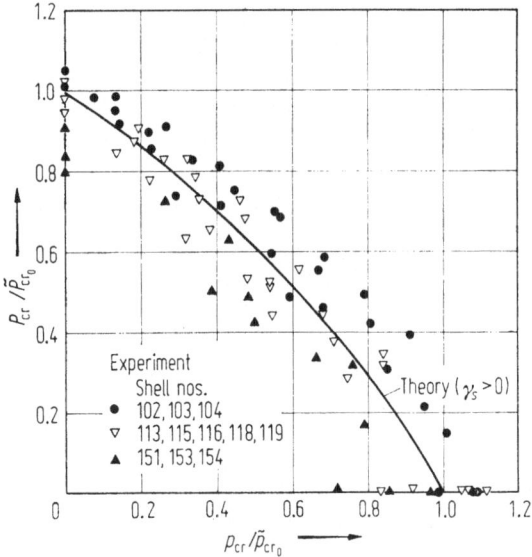

Fig. 13. Comparison of theory with experiment for stringer stiffened cylinders subjected to combined loading of axial compression and external hydrostatic pressure.

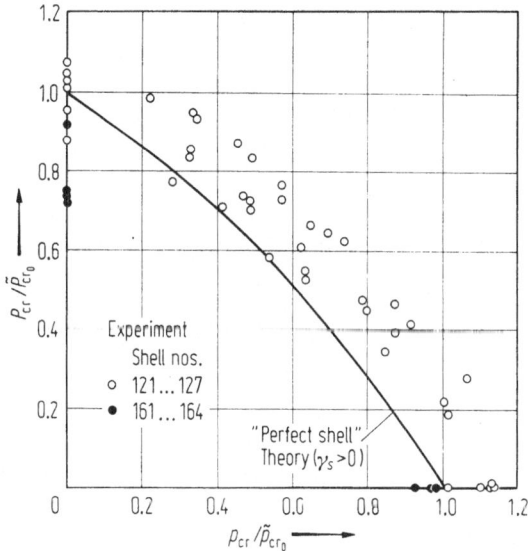

Fig. 14. The effect of shape imperfections on the buckling of stringer stiffened cylinders subjected to combined loading of axial compression and external hydrostatic pressure.

ling strengths occurred. Interaction data for the asymmetric imperfect cylinders has not yet been obtained and thus it is difficult to assess the effect of this type of imperfection on the combined loading buckling behaviour. For the case of external hydrostatic pressure acting alone on the stringer stiffened cylinders it can be seen that the imperfect cylinders were not affected, as indicated by the close agreement between "perfect" shell theory and experiment. Although there was considerable scatter in the critical external pressure results shown in Fig. 13, most of the data fell within $\pm 13\%$ of the predicted values. Based on these pressures, it was estimated from Fig. 1 that the equivalent weight unstiffened cylinders had hydrostatic buckling pressures ranging from two to four times higher. Thus it would appear for the range of shell parameters considered that axially stiffened cylinders are substantially less efficient in terms of buckling pressure than equivalent weight unstiffened cylinders.

Conclusions

Based on the results presented, the following conclusions can be made:

1. The effect of clamped boundary conditions leads to hydrostatic buckling pressures substantially larger than those obtained for simply supported edges, as predicted by theory.

2. For unstiffened cylinders under axial compression, asymmetric shape imperfections do not lead to load reductions in excess of those obtained for equivalent axisymmetric imperfections.

3. Ring stiffened cylinders with and without shape imperfections were found to be less efficient in terms of buckling strength under axial compression than equivalent weight unstiffened circular cylinders, but considerably stronger for hydrostatic pressure. Under combined loading of compression-hydrostatic pressure, the effect of the external pressure was relatively small until it approached its critical value. Furthermore, the buckling pressure did not appear to be particularly sensitive to the presence of axisymmetric shape imperfections. It was also found that the increase in axial buckling strength due to the addition of internal pressure could easily be taken into account by means of the tensile stress correction factor.

4. The axial compressive buckling loads of stringer stiffened cylinders did not always exceed the corresponding strength of equivalent weight unstiffened cylinders. Moreover, it was found that asymmetric shape imperfections were far more severe than axisymmetric distributions and resulted in considerable load reductions. For hydrostatic pressure loading, the stringer stiffened shells did not exhibit significant sensitivity to either axisymmetric or asymmetric imperfections, at

least for the range of parameters considered. Under combined loading of compression-hydrostatic pressure, the interaction was quite strong as compared to the behaviour of ring stiffened cylinders.

Acknowledgements

The author wishes to gratefully acknowledge the contributions made to this paper by his several graduate research assistants; Mr. M. Booton, Mr. E. Boros, Mr. K. Chan, Mr. C. Seibel and Mr. D. Zimcik. Furthermore, the generosity of Prof. J. W. Hutchinson in supplying us with a computer programme to permit theoretical comparisons to be made with our multitude of experimental data is also very much appreciated. In addition, the financial support by the National Research Council of Canada (NRC Grant No. A-2783) has allowed us to continue our work in shell mechanics.

References

1. Singer, J.: Buckling of Integrally Stiffened Cylindrical Shells—A Review of Experiment and Theory. Contributions to the Theory of Aircraft Structures, Delft University Press 1972.
2. Hutchinson, J. W.; Amazigo, J. C.: Imperfection—Sensitivity of Eccentrically Stiffened Cylindrical Shells. AIAA J. 5 (1967).
3. Yamaki, N.; Otomo, K.: Experiments on the Postbuckling Behaviour of Circular Cylindrical Shells Under Hydrostatic Pressure. Experimental Mechanics. SESA, July 1973.
4. Seibel, C. E.; Chan, K. H.; Tennyson, R. C.: Buckling of Circular Cylinders Under Hydrostatic Pressure (to be published).
5. Hutchinson, J. W.: Axial Buckling of Pressurized Imperfect Cylindrical Shells. AIAA J. 3, No. 8 (1965).
6. Arbocz, J.; Babcock, C. D.: The Effect of General Imperfections on the Buckling of Cylindrical Shells. J. Appl. Mech. March 1969.
7. Tennyson, R. C.; Muggeridge, D. B.; Caswell, R. D.: New Design Criteria for Predicting Buckling of Cylindrical Shells under Axial Compression. J. Spacecraft and Rockets, AIAA, 8, No. 8 (1971).
8. Booton, M.; Zimcik, D.; Tennyson, R. C.: Buckling of Circular Cylindrical Shells with Asymmetric Shape Imperfections (to be published).
9. Hutchinson, J. W.: Private Communication Concerning Computer Programme to Analyze Stiffened Circular Cylinders. Harvard University 1973.
10. Tennyson, R. C.; Muggeridge, D. B.: Buckling of Axisymmetric Imperfect Circular Cylindrical Shells under Axial Compression. AIAA J. 7, No. 11 (1969).
11. Hutchinson, J. W.: Axial Buckling of Pressurized Imperfect Cylindrical Shells. AIAA J. 3, No. 8 (1965).

Calculated Postbuckling Loads as Lower Limits for the Buckling Loads of Thin-Walled Circular Cylinders [1]

M. Esslinger, B. Geier

Deutsche Forschungs- und Versuchsanstalt für Luft- und Raumfahrt, Braunschweig, Germany

1. Introduction

It is well known that the experimental buckling loads of thin-walled circular cylinders are lower than the bifurcation loads, since thin-walled cylinders are sensitive to initial imperfections. This imperfection sensitivity depends on the radius, the wall thickness and the length of the cylinders as well as on the shape and arrangement of stiffeners. Therefore in searching for the most effective stiffening, it is not sufficient to compare bifurcation loads; one has to take into account the imperfection sensitivity too.

More than 30 years ago v. Kármán found that for thin-walled shells, post-buckling equilibrium states exist with loads far below the bifurcation load. In those days the idea came into fashion, that it should be possible to find postbuckling loads which could be taken as lower limits to the scatter region of the buckling loads. These postbuckling loads would be reasonable index values for optimization.

The experimental and theoretical search for these postbuckling loads has first been performed for isotropic cylinders, for which the experiments are cheap and the analysis is simple. In this paper we shall report on our results found for isotropic cylinders. Moreover, it will be indicated how the investigation methods, having worked successfully for isotropic cylinders, can be applied to orthotropic cylinders, too.

2. Cylinders Subjected to External Pressure

For cylinders subjected to external pressure the buckling and post-buckling patterns as well as the buckling and postbuckling loads do depend on the cylinder length (Fig. 1a).

A plot of the theoretical postbuckling curves [1] computed for different circumferential wave numbers shows, that one of the curves attains

---

[1] This work was supported by the Deutsche Forschungs-Gemeinschaft. This aid is gratefully acknowledged.

the lowest postbuckling load (Fig. 1 b). We denote this curve, the corresponding postbuckling pattern and the lowest postbuckling load as "characteristic" for the cylinder. The characteristic postbuckling pattern appears in buckling tests after the snap-through.

Figure 1 c shows two characteristic postbuckling curves, one calculated and the other one recorded in an experiment.

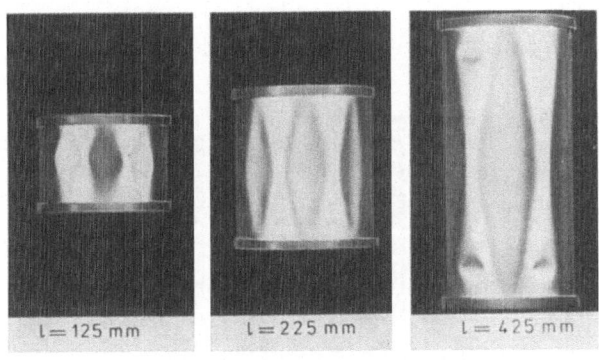

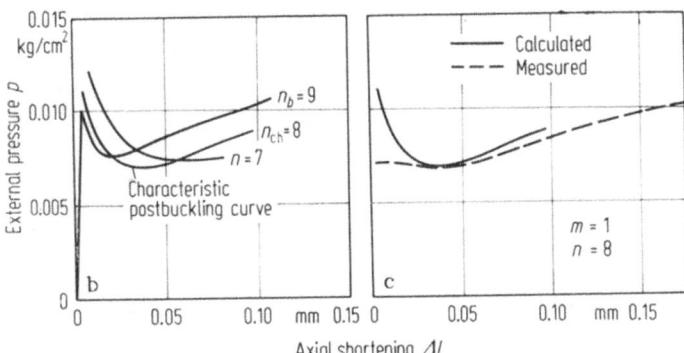

Fig. 1 a—c. Cylinder under hydrostatic external pressure ($r = 100$ mm, $t = 0.254$ mm).
a) Postbuckling patterns; b) Selection of the characteristic postbuckling curve $l = 240$ mm; c) Theoretical and experimental characteristic postbuckling curves.

The characteristic postbuckling load is the lower limit to the scatter region of the buckling loads and a reasonable index value for optimization. For isotropic cylinders it amounts to approximately 65% of the bifurcation load. The definition and calculation of this index load can easily be applied to orthotropic cylinders.

3. Axially Loaded Cylinders with Test Boundary Conditions

3.1. Controlled Shortening

A cylinder loaded by controlled axial shortening exhibits a postbuckling pattern consisting of two staggered rows of buckles (Fig. 2a). The circumferential wave number stable after the snap-through, and the corresponding postbuckling loads depend on the cylinder length.

A plot of the theoretical postbuckling curves [2] computed for different circumferential wave numbers shows that one of the curves

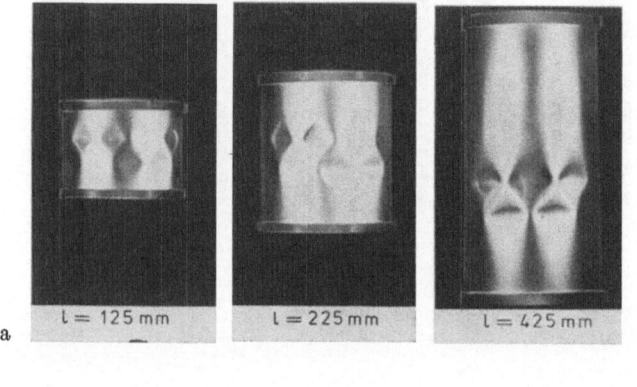

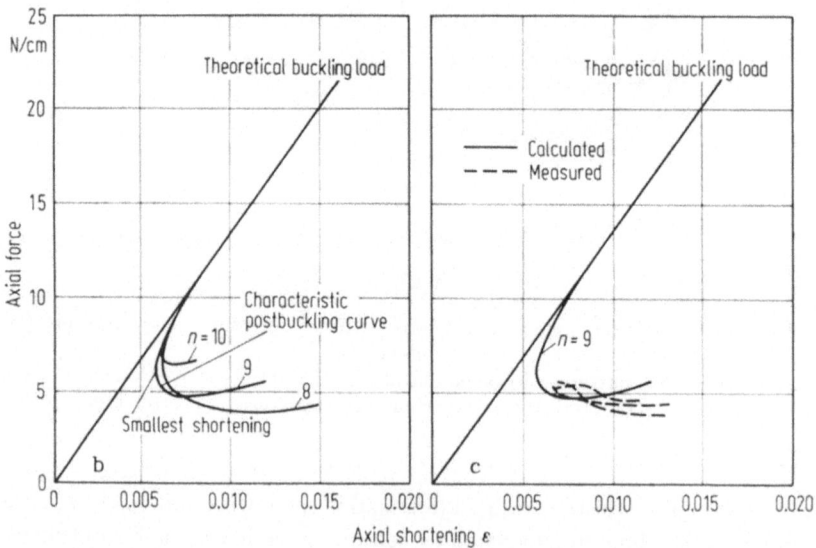

Fig. 2a—c. Unpressurized axially loaded cylinders ($r = 100$ mm, $t = 0.254$ mm).
a) Postbuckling patterns; b) Selection of the characteristic postbuckling curve
$l = 300$ mm; c) Theoretical and experimental characteristic postbuckling curves.

attains the smallest postbuckling shortening (Fig. 2b). We denote this curve, the corresponding postbuckling pattern, the smallest shortening and the corresponding load as "characteristic" for the cylinder. The characteristic postbuckling pattern appears in buckling tests after the snap-through.

Figure 2c shows four characteristic postbuckling curves, one calculated and the other three recorded in an experiment.

A cylinder loaded by controlled shortening will not buckle as long as its shortening is less than the smallest postbuckling shortening (Fig. 3a). When this shortening is attained on the prebuckling line

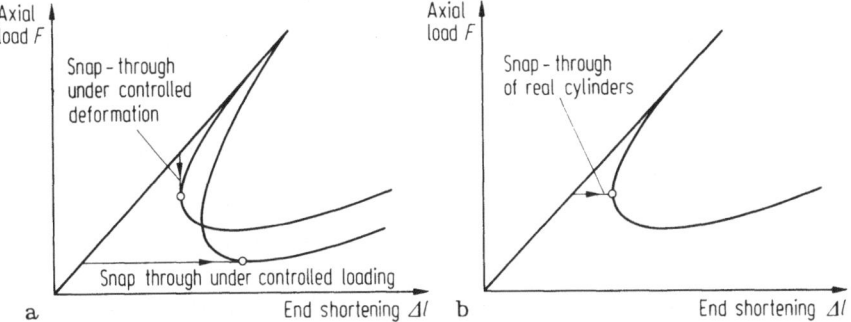

Fig. 3a and b. Lower limits of the buckling load.
a) Theoretical; b) Practical.

the load carried by the cylinder is somewhat higher than the characteristic postbuckling load. If one wants to take regard of irregularities not accounted for in the computation, one may define the characteristic postbuckling load as lower limit to the scatter region of the experimental buckling loads (Fig. 3b).

In order to find out whether it is a reliable lower limit, we stimulated the cylinders to premature buckling by short-time application of local radial point loads, and we found that all buckling loads of the disturbed cylinders were higher than the characteristic postbuckling load.

3.2. Dead Weight Loading

Buckling under controlled *shortening* is a laboratory test. For practical applications buckling under controlled *load* is more important.

Cylinders subjected to controlled shortening will not buckle before the smallest postbuckling shortening is reached, whereas cylinders subjected to controlled load can snap-through as soon as the level of the lowest-postbuckling load is exceeded (Fig. 3a).

In our tests on cylinders under dead weight loading a radial disturbance was applied with the aid of a device by which the indentation depth and force could be measured. Figure 4 presents the axial buckling

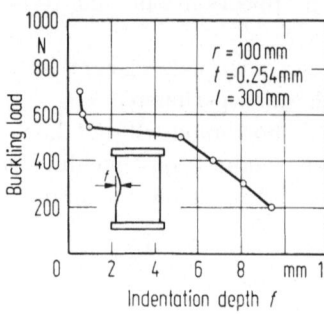

Fig. 4.
Buckling load as function
of indentation depth.

loads as function of the indentation depth [2], which was necessary to stimulate premature buckling. There are indeed lower buckling loads than at the cylinder subjected to controlled shortening. However, these very low buckling loads are connected with large indentation depths. Disturbances of this size need not be considered as unavoidable but may be introduced, if necessary, into the stability proof as a defined external load.

For the evaluation of the load carrying capacity of slightly imperfect cylinders it is important to know that a distinct load exists at which the indentation depth causing the cylinder to buckle increases rapidly. Only above this limit can buckling be induced by small initial imperfections. This limit load is somewhat higher than the characteristic postbuckling load.

3.3. Summary

Figure 5 shows the experimental buckling loads of slightly disturbed cylinders and the characteristic postbuckling loads, as functions of the cylinder length. It can be seen, that the characteristic postbuckling load is a lower limit to the buckling loads of the disturbed cylinder with exception of three outside points at a very small length. We shall return to this exception in a later chapter.—The characteristc postbuckling load decreases with increasing cylinder length.

Actually we are looking for the buckling loads of undisturbed cylinders. These, however, are independent of the cylinder length. Therefore the characteristic postbuckling load cannot be regarded as the desired lower limit.

However, the finding that the buckling behaviour of the axially loaded cylinder is characterized by the smallest postbuckling *shortening*, shall be kept in mind.

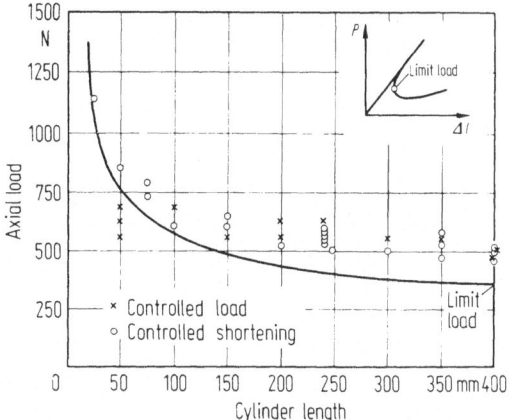

Fig. 5. Buckling loads of disturbed cylinders ($r = 100$ mm, $t = 0.254$ mm).

4. Axially Loaded Cylinders without Boundary Conditions

4.1. Preliminary Remarks

In the meantime it has been found by high-speed motion pictures that immediately after buckling an unstable Kármán type diamond pattern appears which is independent of the cylinder length. This leads to the fascinating task of defining a postbuckling load which can be regarded as a lower limit to the scatter region of the experimental buckling loads and as a reasonable index value for optimization studies, on the basis of

the unstable diamond-shaped postbuckling pattern appearing immediately after buckling, and

the finding that the buckling behaviour is characterized by the smallest postbuckling shortening.

4.2. Experimental Investigations

We begin with a thorough intrepretation of the experimental results. The pictures compiled in Fig. 6 show the buckling process of a Mylar cylinder. The buckling cylinder was filmed with 5 000 fold time dilatation.

The buckling process begins with the snapping-in of a single buckle having nearly equal dimensions in longitudinal and circumferential direction. Buckles of that kind will be called square in the following. With some benevolent intention one may state that the size of the first buckle corresponds to the circumferential wave number $n = 18$. This is the same wave number as resulting from the classical buckling formula under the condition of square buckles.

Around the first buckle additional buckles appear until after 2 ms a field of buckles, extending over a large part of the cylinder surface, has developed. The buckles of this field are still square but larger than the initial buckle. Their dimensions correspond approximately to the circumferential wave number $n = 13$.

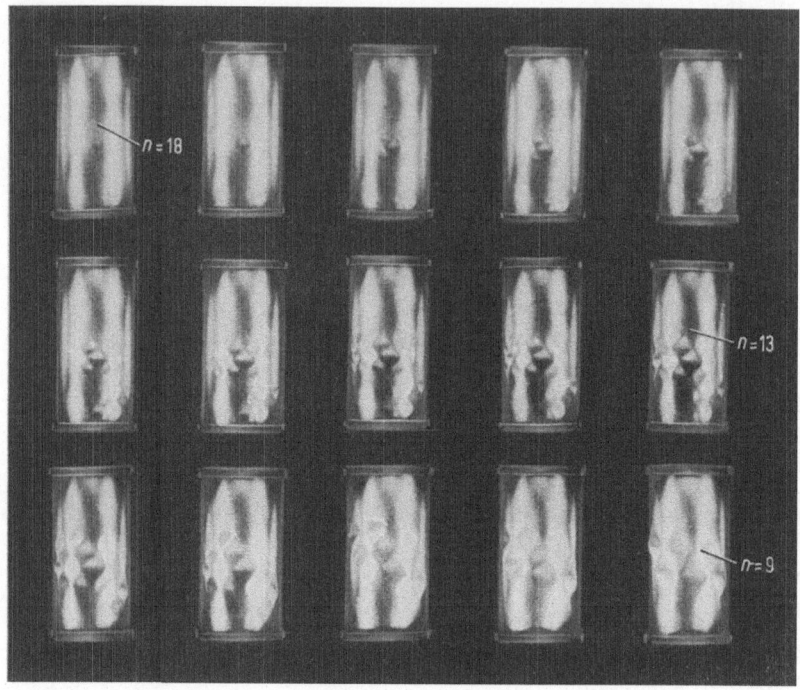

Fig. 6. Buckling of a long axially loaded cylinder.

After these 2 ms the intrinsic buckling process has come to an end, i.e. no further buckles supervene. Henceforth the square shape of the buckles is abandoned. The buckles grow more elongated in axial direction until finally the stable two-tier postbuckling pattern has developed, which is well known from numerous experimental postbuckling investigations on axially loaded thin-walled cylinders.

4.3. Theoretical Investigations

In order to get a theoretical view of the unstable postbuckling behaviour we calculated the minimum values of axial shortening attained with diamond shaped postbuckling patterns of different circumferential wave numbers and aspect ratios of the buckles. Figure 7 shows the results of these computations: The minimum values of axial shortening as function of the wave number and the aspect ratio.

On this surface the path, run through by the cylinder, has been traced. The buckling process begins with square buckles the circumferential wave length of which corresponds to $n = 18$. Then the circumferential

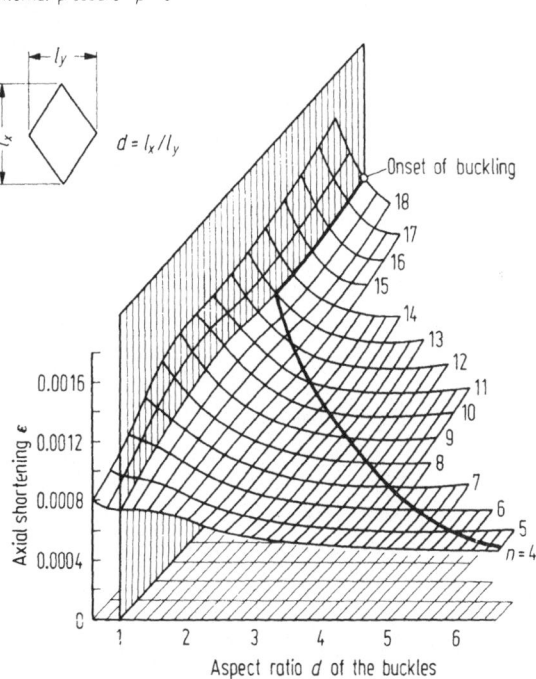

Fig. 7. Smallest values of axial shortening.

wave number decreases, the square shape of the buckles being maintained. In this direction which is emphasized in the diagram by a plane with vertical hatching a minimum of the shortening is reached at $n = 13$. Here the intrinsic buckling process terminates. From then on the path follows a new line. The buckles grow more elongated in axial direction; the circumferential wave number decreases further. In our test cylinders the buckling process was stopped since the cylinder edges prevented further elongation of the buckles.

It is not immediately convincing that the cylinder loaded by controlled shortening changes its shortening continuously during the buckling process. But the high speed pictures are so clear and the analysis is so transparent, that their validity cannot be doubted. There are two arguments, able to prove that the seemingly confused logics are not violated in reality.

At the beginning of the snap-through the cylinder buckles only within a narrow region. It is quite probable, that this narrow buckled region is shortened more than on the average (Fig. 8). During the snap-

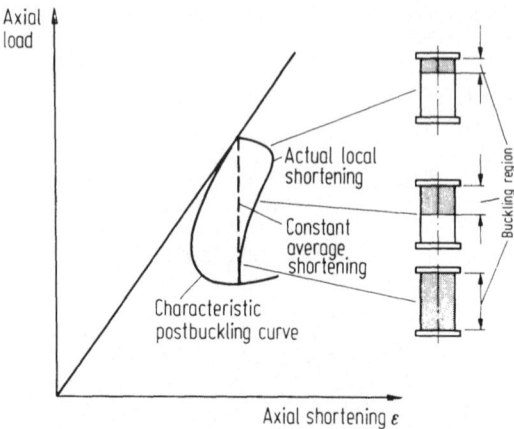

Fig. 8. Comparison of local and average axial shortening.

through the buckled region grows larger and its shortening gets smaller. When the buckling process has ended, the buckled region covers the whole length of the cylinder and hence, its shortening has decreased to the average shortening again.

By the way, Fig. 8 makes plausible that the buckling behaviour of imperfect and/or disturbed cylinders is equal under controlled load and controlled shortening. This figure confirms, that it was reasonable to define, as the lower limit to the scatter region of the buckling loads the load corresponding to the smallest postbuckling shortening, rather than the load corresponding to the same shortening on the prebuckling line (Fig. 3).

For a cylinder subjected to pure axial load in a deformation controlled test device, the area under the load shortening curve represents the potential energy. Therefore, the equilibrium states connected with the smallest axial shortening are likewise the equilibrium states connected with the lowest potential energy. The surface of the smallest potential energies, shown in Fig. 9, has in the main the same shape as the surface of smallest shortenings, shown in Fig. 7. The minimum values of the potential energy may be interpreted as index values for the energy level of the different postbuckling patterns, in the region of small shortening. Thus the trace shown in Fig. 9 only indicates the sequence of the post-buckling patterns during the snap-through; but it does not imply that the postbuckling deformations exactly follow the smallest shortenings.

The only distinguished point in the diagrams (Figs. 7 and 9) is the minimum on the curve of square buckles at $n = 13$, where the intrinsic buckling process terminates. The corresponding axial load for the iso-

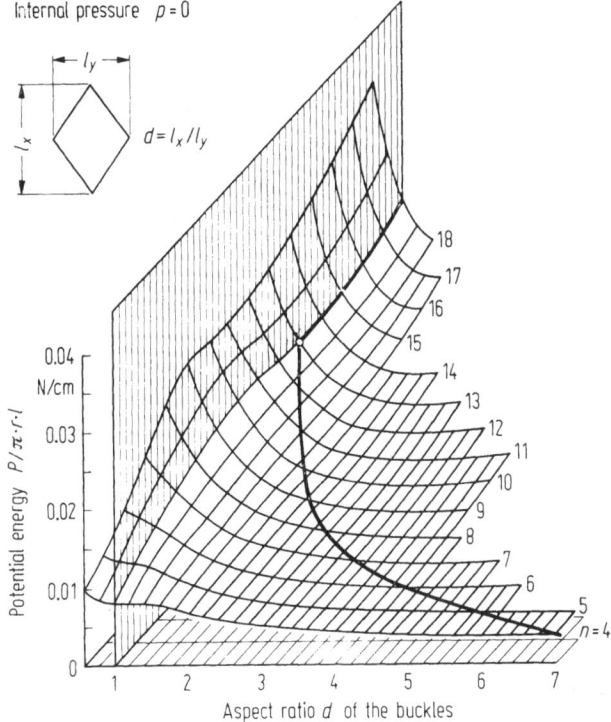

Fig. 9. Smallest values of potential energy.

tropic cylinder amounts to 50%, say, of the classical buckling load. We named it Kármán's load, because it has been calculated with the aid of Kármán's diamond shaped postbuckling pattern.

Since the snapping-in of buckles terminates as soon as the load carried by the cylinder has decreased down to Kármán's limit load, the hypothesis is advanced, that Kármán's load is the lower limit to the scatter region of the local buckling loads of undisturbed cylinders.

The above hypothesis implies that the local buckling stress of isotropic cylinders can be reduced by geometric imperfections not more than 50%. One may object that theoretical investigations [3] under the assumption of an extraordinarily unfavourable imperfection pattern yielded a much greater reduction. But this objection is not convincing since it is improbable that such an unfavourable imperfection pattern is produced by chance.

Kármán's load was taken as a lower limit to the local buckling loads. It is obvious that the establishment of a lower limit for the load carrying capacity of an axially loaded cylinder must rely on local stress, since it is well known from numerous buckling and postbuckling tests that cylinders with extremely low buckling loads always buckle locally.

Thin-walled and short cylinders are particularly sensitive to local disturbances by out-of-planeness of the edge support, because their critical end shortening is particularly small

$$\Delta l = \varepsilon l = 0,6(t \cdot l/r) \, .$$

In Fig. 5 three outside points have remained unexplained. Probably at these points the limit load was exceeded locally, whereas on large parts of the circumference the stress level was lower. The outside points shall now be used to lodge the remark, that for extremely short cylinders the concept of the resulting buckling load is not meaningful.

The hypothesis that the local buckling stress of axially loaded cylinders is reduced by random deviations from the perfect shape, not more than down to 50% of the classical buckling load, shall be checked by measuring the stress in the cylinder before buckling. Convinced that this verification of the hypothesis will be successful, we extended the search for Kármán's limit load to cylinders with internal pressure.

5. Pressurized Axially Loaded Cylinders without Boundary Conditions

5.1. Experimental Investigations

On axially loaded cylinders with internal pressure the buckling process, as on the cylinder without internal pressure, begins with the snapping-in of a single buckle near one edge (Fig. 10, left hand frame). This single buckle is the source of a narrow postbuckling pattern consisting of two or three rows of buckles arranged around the circumference near an

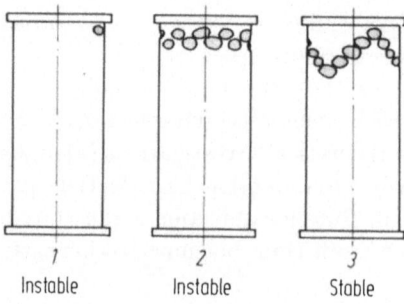

Fig. 10.
Postbuckling patterns of an axially loaded cylinder with medium internal pressure.

edge (Fig. 10, middle frame). At medium values of internal pressure these tiers of buckles are unstable. Buckles are pushed towards the top and the bottom such that in long cylinders a regular zig-zag pattern

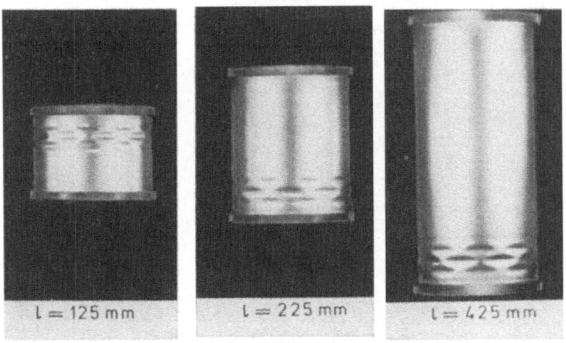

Fig. 11. Pressurized axially loaded cylinders ($r = 100$ mm, $t = 0.254$ mm, $p = 0.35$ kp/cm^2 = 3.5 N/cm^2).

is formed (Fig. 10, right hand frame). At short cylinders the zig-zag pattern is generally not so strongly marked, since the postbuckling pattern gets stable, before the transformation has terminated.

At high values of internal pressure the three-tiered pattern appearing immediately after buckling is stable and independent of the cylinder length (Fig. 11).

Figure 12 gives a survey of the stable postbuckling patterns of axially loaded cylinders subjected to different values of internal pressure.

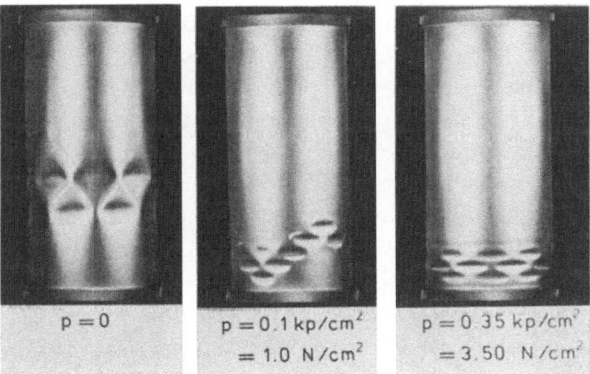

Fig. 12. Pressurized axially loaded cylinders ($r = 100$ mm, $t = 0.254$ mm, $l = 425$ mm).

Figure 13 presents load-shortening-curves recorded in tests on pressurized axially loaded cylinders. The horizontal line above the diagram is the classical buckling load, equally valid for cylinders with

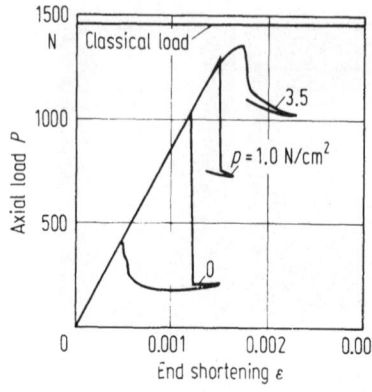

Fig. 13.
Load shortening curves of pressurized axially loaded cylinders ($r = 100$ mm, $t = 0.254$ mm, $l = 225$ mm).

and without internal pressure. From this diagram two statements can be deduced:

The internal pressures raises the buckling and postbuckling loads.

At high internal pressure all postbuckling equilibrium states are stable.

The latter statement is in agreement with the fact that at high internal pressure the postbuckling patterns have been found to be stable; at medium pressure the tiers of buckles existing at the beginning of the buckling process turn into zig-zag pattern.

5.2. Theoretical Investigations

For the calculation of Kármán's load it is most important to know the aspect ratio of the first buckle. On the motion pictures it seems to be discernible that the buckles snapping in are square at low internal pressure, and are longer in circumferential than in axial direction at high internal pressure.

The perfect pressurized cylinder would buckle with an axisymmetric pattern. From this, one may deduce that the axisymmetric imperfections are amplified most strongly. Koiter [3] stated that in cylinders with axisymmetric imperfections the axial wave length of the periodic buckling pattern is twice that of the imperfection pattern. If one combines this statement with the experimental result that in cylinders with medium internal pressure, the buckles snapping in are square, one is led to the circumferential wave number $n = 18$, which is the same as for the unpressurized perfect cylinder.

For the pressurized cylinder we performed principally the same computations as carried out before for the unpressurized cylinder, i.e. we computed the smallest potential energy for different circumferential wave numbers and aspects ratios of the buckles. In this case the equilibrium states of smallest shortening and smallest potential energy do not coincide, since the radial pressure contributes to the potential energy.

Figure 14 shows results of the computations for a cylinder with small internal pressure.

On the curve $d = 1$, emphasized in the diagram by a verticllay hatched plane, there is again a distinct minimum, indicating the end of the intrinsic buckling process. However, compared with unpressurized cylinders this minimum is situated at a larger circumferential wave number and corresponds to a higher axial load.

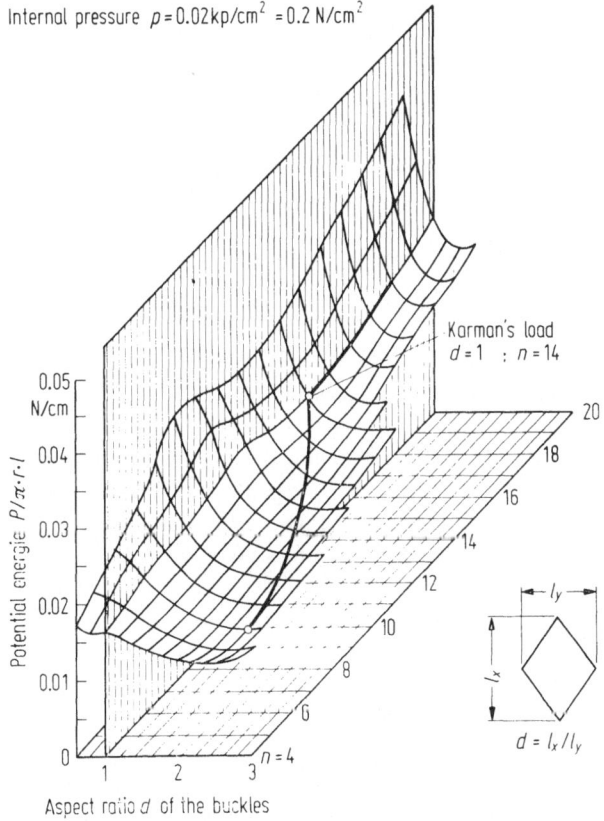

Fig. 14. Smallest values of potential energy.

It is remarkable that for the cylinder with internal pressure an absolute minimum of the potential energy exists. For small internal pressure this minimum is connected with such a large axial wave length that it could not be attained by our test cylinders.

For higher values of internal pressure the minima on the curve $d = 1$ tend towards higher circumferential wave numbers. Kármán's load increases further.

The loads corresponding to the absolute minima of potential energy approach the Kármán load. The existence of these minima does not imply that the corresponding equilibrium states are stable, since the computations were restricted to diamond-shaped postbuckling patterns; at medium internal pressure the cylinder escapes into a zig-zag post-buckling pattern not contained in the analysis.

Figure 15 presents a schematic view of the calculated buckling paths leading downhill at the various potential energy mountains. All paths run to smaller circumferential wave numbers. At small internal pressure

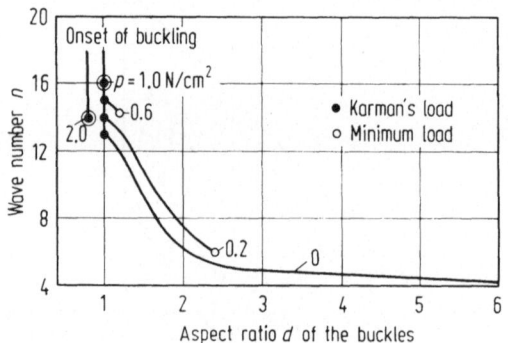

Fig. 15. Schematic representation of buckling paths.

the deepest valley lies at an aspect ratio which is appreciably greater than $d = 1$; the stable postbuckling pattern consists of two staggered rows of buckles in the middle of the cylinder (Fig. 12). At medium internal pressure the deepest canyon is near Kármán's load; the stable postbuckling pattern is zig-zag shaped. At high internal pressure the bottom of the valley lies at an aspect ratio smaller than $d = 1$; the stable postbuckling pattern consists of three tiers of low diamond shaped buckles.

Figure 16 gives a dimensionless presentation of Kármán's loads as function of the internal pressure. This curve, determined by computations for perfect cylinders, is compared with an experiment-based curve, taken from a report by Weingarten, Seide and Morgan [4]. For small internal pressure the agreement is good.

For high internal pressure the computations yield no reduction of the classical buckling load, whereas the measured buckling loads are lower. This discrepancy may be explained by the fact that for cylinders buckling at loads close to the classical load, the edge effects must no longer be neglected in the computations.

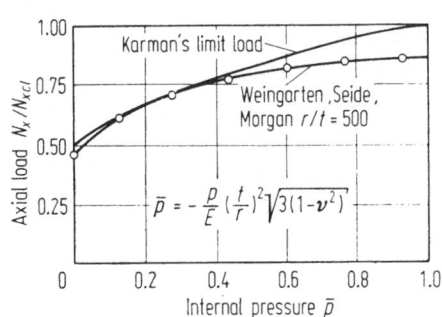

Fig. 16.
Calculated and measured lower limits for the axial buckling loads of isotropic cylinders.

6. Outlook on Stiffened Shells

The investigations reported so far aimed at providing the tools for the optimization of stiffened shells. This aim has been attained for cylinders with zero and small internal pressure. Therefore, the optimization studies for orthotropic cylinders could be attacked:

For longitudinally stiffened cylinders buckling with one wave over the cylinder length one calculates and compares the characteristic postbuckling loads for cylinders of finite length. Figure 17 shows the standard example [5] for successful theoretical comprehension of the sensitiveness to initial imperfections. For the externally stiffened cylinder the bifurcation load is considerably higher than for the inter-

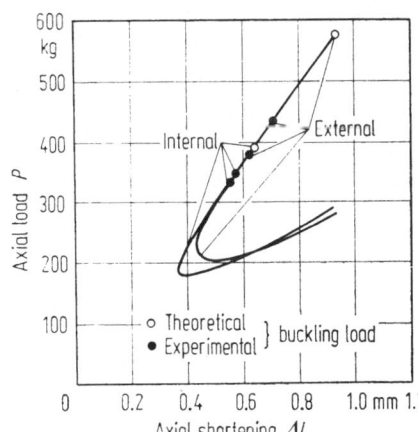

Fig. 17.
Internally and externally stiffened cylinders.

nally stiffened one. But the experimental buckling loads, as well as the characteristic postbuckling loads of the two cylinders, are much closer.

For cylinders with chess-board buckling patterns the definition and computation of Kármán's load are easy to perform. The circumferential wave number and the aspect ratio of the first buckle can be assumed to be those of the theoretical buckling pattern. On travelling down the potential mountains maintaining the aspect ratio of the first buckle, a minimum of the potential energy is reached. The load corresponding to this minimun will be defined as Kármán's load. It seems to be a reasonable lower limit to the scatter region of the buckling loads and a reasonable index value for optimization. The experimental verification has not been performed yet.

For cylinders with an axisymmetric theoretical buckling pattern we do not know the aspect ratio of the first buckle. We hope to find it with the help of high-speed motion pictures in the future.

References

1. Esslinger, M. E.; Geier, B. M.: On the Buckling and Postbuckling Behavior of Thin-Walled Circular Cylinders. RILEM Int. Symposium III, Instituto Nacional de Technologia Industrial, Buenos Aires, Argentinien, 1971, pp. 97—124.
2. Esslinger, M.; Geier, B.: Gerechnete Nachbeullasten als untere Grenze der experimentellen axialen Beullasten von Kreiszylindern (in German). Der Stahlbau 41, No. 12, 353—360 (1972).
3. Koiter, W. T.: The Effect of Axisymmetric Imperfections on the Buckling of Cylindrical Shells under Axial Compression. Lockheed Missiles & Space Comp., Sunnyvale, Calif., U.S.A., Techn. Rep. 6-90-93-86, Aug. 1963, III.
4. Weingarten, V. I.; Seide, P.; Morgan, E. J.: Elastic Stability of Thin-Walled Cylindrical and Conical Shells under Combined Internal Pressure and Axial Compression. AIAA J. **3**, 1118—1125 (1965).
5. Esslinger, M.: Beulen und Nachbeulen exzentrisch versteifter dünnwandiger Kreiszylinder unter axialsymmetrischer Belastung (in German). DLR FB 70-48 (Sept. 1970).

Prediction of Buckling Loads Based on Experimentally Measured Initial Imperfections [1]

J. Arbocz, Ch. D. Babcock, Jr.

California Institute of Technology, Pasadena, California, U.S.A.

Abstract

Correlation studies between experimental buckling loads and analytical predictions based on experimentally measured initial imperfections were carried out for axially compressed isotropic and stiffened cylindrical shells. By expanding the response of a cylindrical shell in truncated Fourier series, the nonlinear Donnell type shell equations for imperfect stiffened shells were reduced to a set of linear equations in the correction terms by Newton's method of quasilinearization. Solutions were obtained for isotropic and for ring and stringer stiffened shells. The amplitudes of the initial imperfections used in the analysis were calculated from the corresponding Imbert-Donnell imperfection models. The free parameters in this imperfection model were obtained by least square fitting the harmonics of the experimentally measured initial imperfections. It was possible in all cases to achieve satisfactory correlation using only a few suitably chosen deflection and imperfection modes.

1. Introduction

The buckling behavior of axially compressed thin walled stiffened or unstiffened cylindrical shells has been a major concern to practicing structural engineers for many years. Initial geometrical imperfections have been accepted qualitatively as the explanation for the discrepancy between the analytical predictions and the experimental values and for the frequently large scatter of the experimental results. This acceptance is mainly due to the work of a few investigators [1, 2, 3] who, using specialized imperfections, have demonstrated the sensitivity of the buckling load to initial imperfections.

However, despite the accepted theoretical explanation of the buckling behavior of axially compressed shells, the incorporation of the idea of imperfection sensitivity into engineering practice has not been accomplished except in an empirical manner. The engineers who design actual shell structures against buckling do it by the method

---

[1] This work was supported by the National Science Foundation under Grant GK 16934. This aid is gratefully acknowledged.

developed in the early 40's [4, 5] by using an empirical "knockdown factor" applied to the results of the classical small deflection theory. The "knockdown factor" is chosen so that its product with the classical buckling load leads to a lower bound to all the experimental data for the shell-loading configuration under consideration.

This apparent reluctance of the practicing structural engineers to accept and assimilate the findings of the theoreticians is based on two often overlooked but very important facts. In the first place, with the exception of a few papers by Hutchinson [6], Thurston and Freeland [7] and Arbocz and Babcock [8], the bulk of the imperfection studies deal with idealized axisymmetric shapes which are seldom, if ever realized in the actual shell structures. The other, and from the designer's standpoint probably the more important reason is the complete absence of measurement of imperfections for full scale structures.

For laboratory scale shell structures, imperfections have been measured by the Galcit Group for several years [8, 9]. Buckling load predictions based on these imperfections have been published previously [8, 10]. These imperfection correlation studies were based on approximate solutions of the Donnell type imperfect shell equations. These solutions did not satisfy the correct experimental boundary conditions. In the present work special attention has been paid to account for the effects of the nonlinear prebuckling deformations due to the edge constraints and the correct experimental boundary conditions.

2. Effect of General Imperfections-Multimode Solution

The correlation studies reported in this paper were carried out using an analytical solution of the imperfect shell equations that can incorporate general imperfection shapes. Initially the nonlinear Donnell type shell equations are reduced to a set of linear partial differential equations by Newton's method of quasilinearization. (This type of solution was first applied to the imperfect shell problem by Thurston and Freeland [7].) A combination of Fourier expansion and Galerkin's procedure then results in a set of algebraic equations in terms of the harmonic components of the correction terms. Finally, the system of algebraic equations is solved by a standard iterative procedure.

2.1. Stiffened Shell Equations

Using the sign convention defined in Fig. 1 the Donnell type equations for imperfect stiffened cylindrical shells can be written [11]

$$L_H(F) - L_Q(W) = -W_{,xx}/R - L_{NL}(W, W + 2\overline{W})/2, \qquad (1)$$

$$L_Q(F) + L_D(W) = F_{,xx}/R + L_{NL}(F, W + \overline{W}), \qquad (2)$$

where the linear operators are

$$L_D(\;) = D_{xx}(\;)_{,xxxx} + D_{xy}(\;)_{,xxyy} + D_{yy}(\;)_{,yyyy}, \qquad (3)$$

$$L_H(\;) = H_{xx}(\;)_{,xxxx} + H_{xy}(\;)_{,xxyy} + H_{yy}(\;)_{,yyyy}, \qquad (4)$$

$$L_Q(\;) = Q_{xx}(\;)_{,xxxx} + Q_{xy}(\;)_{,xxyy} + Q_{yy}(\;)_{,yyyy}, \qquad (5)$$

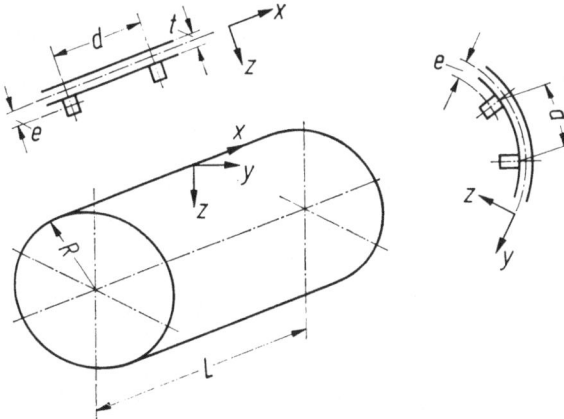

Fig. 1. Notation and sign convention.

and the nonlinear operator is

$$L_{\mathrm{NL}}(S, T) = S_{,xx}T_{,yy} - 2S_{,xy}T_{,xy} + S_{,yy}T_{,xx}. \qquad (6)$$

Commas in the subscripts denote repeated partial differentiation with respect to the independent variables following the comma. The stiffener properties have been "smeared out" to arrive at effective bending, stretching and torsional stiffnesses. The stiffener parameters D_{xx}, H_{xx}, Q_{xx}, D_{xy}, ... etc., are defined in [11]. These equations were first derived by Geier [12], however in the present notation they are due to Singer [13] and Hutchinson and Amazigo [14]. Here $\overline{W}$ is the initial radial imperfection, W is the component of displacement normal to the shell midsurface and F is the Airy stress function.

2.2. Newton's Method of Quasilinearization

Let us represent the $(m+1)^{\mathrm{th}}$ approximation to a solution of Eqs. (1) and (2) by

$$\begin{aligned} W_{m+1} &= W_m + \delta W_m, \\ F_{m+1} &= F_m + \delta F_m \end{aligned} \qquad (7)$$

where

$$W_m, F_m = m\text{-th approximation to the solution,}$$

$$\delta W_m, \delta F_m = \text{correction to the } m\text{-th approximation.}$$

Substituting into Eqs. (1) and (2) and neglecting squares of the correction quantities yields the following set of linear equations for determining the correction terms

$$L_H(\delta F_m) - L_Q(\delta W_m) + \delta W_{m,xx}/R + L_{NL}(W_m + \overline{W}, \delta W_m) = -E_m^{(1)}, \quad (8)$$

$$L_Q(\delta F_m) + L_D(\delta W_m) - \delta F_{m,xx}/R - L_{NL}(F_m, \delta W_m) - $$
$$- L_{NL}(W_m + \overline{W}, \delta F_m) = -E_m^{(2)}, \quad (9)$$

where

$$E_m^{(1)} = L_H(F_m) - L_Q(W_m) + W_{m,xx}/R + L_{NL}(W_m, W_m + 2\overline{W})/2, \quad (10)$$
$$E_m^{(2)} = L_Q(F_m) + L_D(W_m) - F_{m,xx}/R - L_{NL}(F_m, W_m + \overline{W}). \quad (11)$$

2.3. Reduction to a Set of Algebraic Equations

If we represent the initial imperfections by

$$\overline{W} = t \sum_{i=1}^{N_1} \overline{W}_{i0} \cos i\bar{x} + t \sum_{k,l=1}^{N_2} \overline{W}_{kl} \sin k\bar{x} \cos l\bar{y} + $$
$$+ t \sum_{k,l=1}^{N_3} \overline{W}'_{kl} \sin k\bar{x} \sin l\bar{y}, \quad (12)$$

where

$$\bar{x} = \pi x/L, \quad \bar{y} = y/R,$$

then Eqs. (1) and (2) admit separable solutions of the form

$$\left\{ \begin{matrix} W_m \\ \delta W_m \end{matrix} \right\} = t \left\{ \begin{matrix} W_\nu \\ 0 \end{matrix} \right\} + t \sum_{i=1}^{N_1} \left\{ \begin{matrix} W_{i0} \\ \delta W_{i0} \end{matrix} \right\} \cos i\bar{x} + t \sum_{k,l=1}^{N_2} \left\{ \begin{matrix} W_{kl} \\ \delta W_{kl} \end{matrix} \right\} \sin k\bar{x} \cos l\bar{y} + $$
$$(13)$$
$$+ t \sum_{k,l=1}^{N_3} \left\{ \begin{matrix} W'_{kl} \\ \delta W'_{kl} \end{matrix} \right\} \sin k\bar{x} \sin l\bar{y},$$

$$\left\{ \begin{matrix} F_m \\ \delta F_m \end{matrix} \right\} = \frac{E R t^2}{c} \left\{ \begin{matrix} -\lambda^2 \bar{y}^2/2 \\ 0 \end{matrix} \right\} + \frac{E R t^2}{c} \sum_{i=1}^{N_1} \left\{ \begin{matrix} F_{i0} \\ \delta F_{i0} \end{matrix} \right\} \cos i\bar{x} + $$

$$+ \frac{E R t^2}{c} \sum_{k,l=1}^{N_2} \left\{ \begin{matrix} F_{kl} \\ \delta F_{kl} \end{matrix} \right\} \sin k\bar{x} \cos l\bar{y} + \frac{E R t^2}{c} \sum_{k,l=1}^{N_3} \left\{ \begin{matrix} F'_{kl} \\ \delta F'_{kl} \end{matrix} \right\} \sin k\bar{x} \sin l\bar{y}, \quad (14)$$

where

$$W_\nu = -\frac{\nu}{c} \frac{\overline{H}_{xx}}{1 + \mu_1} \lambda, \quad c = \sqrt{3(1 - \nu^2)}.$$

The unknown coefficients are determined by Galerkin's procedure yielding a set of linear algebraic equations in terms of the unknown correction terms. In matrix notation

$$[A]\{\delta F\} + [B]\{\delta W\} = -\{E^{(1)}\}, \quad (15)$$
$$[C]\{\delta F\} + [D]\{\delta W\} = -\{E^{(2)}\}. \quad (16)$$

To obtain the buckling load for a given imperfect cylindrical shell one begins by making an initial guess for $\{W\}$ and $\{F\}$ at a small initial load level λ. Iteration is then carried out until the correction vectors are smaller than some preselected value. The converged solutions then are used as the initial guess at the next higher axial load level $\lambda + \Delta\lambda$. The entire process is repeated for increasing values of the axial load parameter λ. The nonlinear analysis then will locate the limit point of the prebuckling states. By definition the value of the loading parameter λ corresponding to the limit point will be the theoretical buckling load.

It is shown in [15] that the solution satisfies the circumferential periodicity condition. It also contains details of the coefficient matrices A, B, C and D and the error vectors $E^{(1)}$ and $E^{(2)}$.

3. Comparison with other Solutions

In order to assess the accuracy of the multimode solutions comparisons with Koiter's asymptotic theory and with a numerical solution, which include rigorous satisfaction of the experimental boundary conditions, were carried out.

3.1. Koiter's Asymptotic Theory

It was pointed out by Koiter [16] that if for an isotropic shell the initial imperfections are represented by the following 3 modes

$$\overline{W} = t\bar{\xi}\{\cos i_{\text{cl}}\bar{x} + \sqrt{2}\, i_{\text{cl}}/k_1 \sin k_1\bar{x} \cos l\bar{y}$$

$$- \sqrt{2}\, i_{\text{cl}}/k_2 \sin k_2\bar{x} \cos l\bar{y}\}, \tag{17}$$

where

$$i_{\text{cl}} = (L/\pi)\sqrt{2c/Rt}, \tag{18}$$

and k_1 and k_2 are the two roots of the quadratic equation

$$k^2 \frac{Rt}{2c}\left(\frac{\pi}{L}\right)^2 + l^2 \frac{Rt}{2c}\left(\frac{1}{R}\right)^2 - k\sqrt{\frac{Rt}{2c}}\frac{\pi}{L} = 0, \tag{19}$$

then buckling occurs by reaching a limit point at an axial load level λ_{s} given by the equation

$$(\lambda_{\text{s}} - 1)^2 = -6c\bar{\xi}\lambda_{\text{s}}. \tag{20}$$

This result and the results using the multimode solution are given in Fig. 2. As can be seen the multimode solution agrees with Koiter's asymptotic formula for sufficiently small values of the imperfection amplitude $\bar{\xi}$. However, for increasing values of $\bar{\xi}$ the multimode solution, which includes higher order terms in the approximate solution, predicts lower buckling loads. It should also be noticed that the imperfection represented by a suitable combination of the buckling modes is far more adverse than a single axisymmetric imperfection [2].

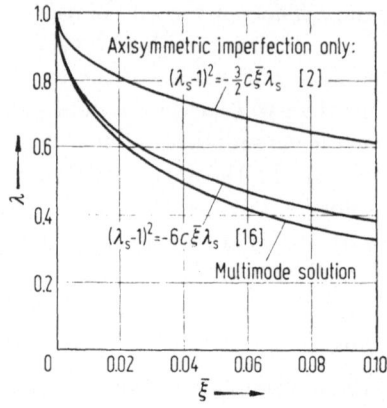

Fig. 2.
Comparison of buckling load predictions for imperfect shells.

3.2. The Effects of Boundary Conditions and Nonlinear Prebuckling Deformations

The effect of experimental boundary conditions has been extensively discussed by Almroth [17], Hoff and Soong [18] and Weller et al. [19]. The effects can be separated into two major items that will be discussed separately. These are the effect of nonlinear prebuckling deformation caused by the end constraint of the test shell and the effect of end fixity on the buckling deformation (eigenfunction) and its associated buckling load (eigenvalue).

The multimode analysis neglects completely the effect of the nonlinear prebuckling deformation caused by the edge constraint. The boundary conditions satisfied for the eigenvalue problem are the classical simple support conditions SS-3 ($w = M_x = v = N_x = 0$). In order to assess the difference in buckling load due to these effects the BOSOR [20] and SRA [21] computer codes were used. Two isotropic, two stringer stiffened and two ring stiffened shells were used in this study. The properties of these shells are given in Table 1. Initially the perfect shell behavior will be discussed.

Our study recovered the results of previous investigators [22, 23, 24], who showed that the buckling load of a moderate length perfect isotropic shell with a membrane prebuckling state is insensitive to boundary conditions, provided the out-of-plane deflection w and the in-plane circumferential displacement v are suppressed. The buckling load (expressed in lb/in) is always very close to the classical value:

$$N_{cl} = Et^2/[R\sqrt{3(1 - v^2)}].$$ (21)

Table 1. Geometric and material properties of the shells

Shell	$t \cdot 10^3$	R/t	L/R	A/dt	$-e/t$	I/dt^3	d	L	$E \cdot 10^{-6}$ [psi]	No. of Stiffeners
X-1	4.00	1000	1.00	—	—	—	—	1.0	15.2	—
A-8	4.64	862	2.00	—	—	—	—	8.0	15.2	—
B-1	8.07	496	1.94	—	—	—	—	7.75	15.45	—
XR-1	9.29	431	1.00	0.205	1.13	0.0288	0.250	4.0	10.0	14
AR-1	9.29	431	1.31	0.205	1.13	0.0288	0.250	5.25	10.0	20
XS-1	7.74	517	1.00	0.506	1.71	0.2466	0.316	4.0	10.0	80
AS-2	7.74	517	1.38	0.506	1.71	0.2466	0.316	5.5	10.0	80

All dimensions in inches, unless otherwise indicated. $\nu = 0.3$

If, however, one includes the nonlinear prebuckling deformations caused by the end constraints in the analysis, then the buckling load depends on the boundary conditions of the problem. For the SS-3 condition ($w = M_x = v = N_x = 0$) the load is reduced by about 16% for the C-3 condition ($w = w_{,x} = v = N_x = 0$) by about 7%. These results are given in detail in Table 2.

Table 2. Buckling loads of the perfect shells

Shell	N_{xM} SS-3	C-3	C-4	N_{xNL} SS-3	C-3	C-4
X-1	37.0 (13)	37.3 (27)	36.8 $(l)^1$	31.0 (26)	33.4 (26)	34.0 (27)
A-8	49.6 (9)	50.0 (9)	49.5 $(l)^1$	41.3 (24)	44.8 (24)	45.7 $(l)^1$
B-1	152.7 (8)	154.1 (8)	152.2 $(l)^1$	127.2 (18)	137.7 (18)	140.6 $(l)^1$
XR-1	142.4 (0)	143.8 (0)	143.8 (0)	135.3 (12)	142.2 (9)	143.1 (13)
AR-1	142.5 (0)	143.2 (0)	143.1 (0)	135.0 (13)	141.8 (7)	142.6 (11)
XS-1	141.6 (11)	161.6 (12)	204.0 (16)	134.4 (11)	160.2 (12)	197.1 (16)
AS-2	131.2 (10)	146.7 (10)	183.2 (14)	127.9 (10)	146.4 (10)	180.9 (14)

Numbers in parentheses are the number of circumferential waves, l,

N_{xM} Buckling load from BOSOR [20] or SRA [21] using membrane prebuckling analysis (lb/in),

N_{xNL} Buckling load from BOSOR [20] or SRA [21] using nonlinear prebuckling analysis (lb/in),

SS-3 $w = M_x = v = N_x = 0$,

C-3 $w = w_{,x} = v = N_x = 0$,

C-4 $w = w_{,x} = v = u = 0$.

[1] Values taken from Reference [24].

For the lightly ring stiffened shells studied in this paper the buckling load with membrane prebuckling, like for the isotropic shells, varies only slightly with the different boundary conditions. The inclusion of the nonlinear prebuckling deformations caused by the end constraints in the analysis results in an 8% decrease in the buckling load for the weak (SS-3) boundary condition; however, this effect is considerably less (about 1%) for the stiffer (C-3 and C-4) boundary conditions. These results are also given in Table 2.

Finally, for the stringer stiffened shells studied it is found that contrary to the behavior of the isotropic and the lightly ring stiffened shells the buckling load with membrane prebuckling depends strongly on the boundary conditions specified. Stiffening the boundary condition raises the buckling load by about 12% for C-3, by about 39% for the C-4 boundary condition. On the other hand, the inclusion of the nonlinear prebuckling deformation (with the shell loaded through the skin midsurface) has an insignificant effect.

From the results displayed in Table 2 it appears that for perfect shells the nonlinear prebuckling deformation is important only for the isotropic shell and for the lightly ring stiffened shell with the weak (SS-3) boundary condition. On the other hand, boundary conditions (with membrane prebuckling) will only have a singificant effect for the stringer stiffened shell.

In order to investigate the behavior of shells with initial imperfections, an analysis developed by Arbocz [25] which takes into account both effects will be used. In this analysis the nonlinear partial differential equations are reduced to a set of nonlinear ordinary differential equations by a two mode circumferential expansion and a Galerkin procedure. The resulting nonlinear ordinary differential equations are solved numerically by a parallel shooting technique. This analysis, in which arbitrary imperfections (with appropriate circumferential dependence) and arbitrary boundary conditions can be accounted for, will be referred to as the "extended" analysis. In order to reduce the computer cost with the numerical analysis, the test shells were slightly shortened in length for this investigation. The properties of the shells X-1, XR-1 and XS-1 are listed in Table 1.

First we shall examine the effect of an idealized imperfection, consisting of an axisymmetric and an asymmetric mode, on the buckling load of an isotropic shell. Thus, for

$$\overline{W} = -0.5t \cos 2\bar{x} + 0.05t \sin \bar{x} \cos 13\bar{y} \tag{22}$$

it was found that the multimode and the extended analysis with C-3 boundary conditions gave virtually the same buckling load [26]. The results are displayed in Fig. 3, where in addition to a load-displacement relationship the deformations near the maximum load level as computed by the two analyses are displayed. It is clear from this figure that despite the complete difference in edge constraint (and hence in the deformation near the shell edges) the imperfection is dominant and it controls the behavior of the shell. This implies that for dominant imperfections the nonlinear prebuckling deformation due to end constraint can be neglected.

Investigating the effect of an idealized imperfection consisting of one axisymmetric and two asymmetric modes on the buckling load of a lightly ring stiffened shell, good agreement was found between the results of the multimode and the extended analysis with C-3 boundary conditions [27]. As can be seen from Table 3 the difference in the critical load predicted by the two analyses is approximately 1%.

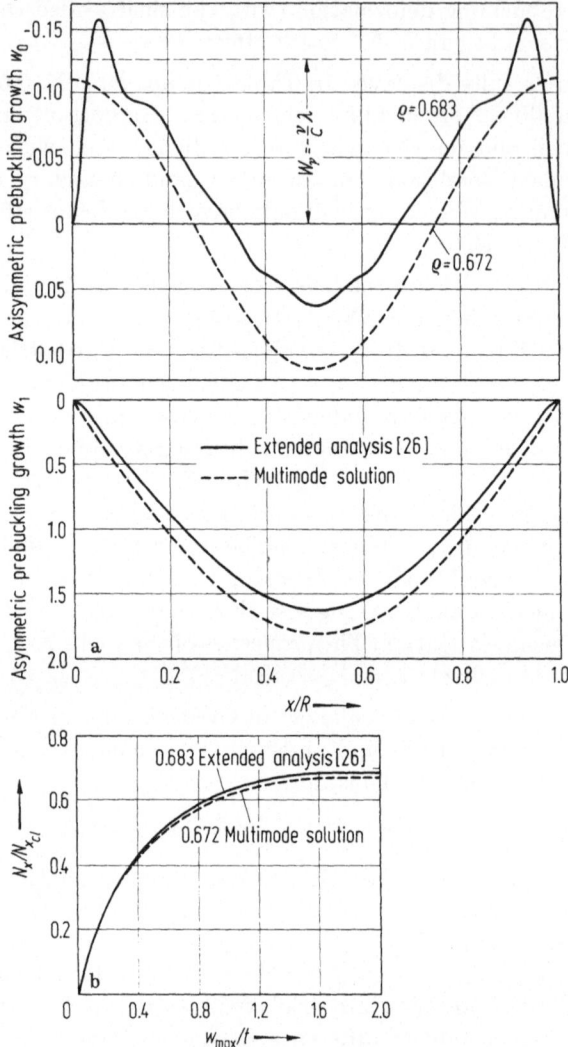

Fig. 3a and b. Comparison between multimode and extended analyses.
a) Axial dependence of the prebuckling growth for shell X-1; b) Load-displacement relationship for shell X-1.

When working with the stringer stiffened shell XS-1 one must remember that for the perfect shell the buckling load depends strongly on the specified boundary conditions. Thus, using an idealized imperfection consisting of one axisymmetric and one asymmetric imperfection it was found that the multimode and the extended analysis with SS-3 boundary conditions gave virtually the same buckling load [27].

Table 3. Buckling loads of the imperfect shells

	X-1		XR-1		XS-1	
	SS-3	C-3	SS-3	C-3	SS-3	C-3
N_{xM}	37.0 (13)	37.3 (27)	142.4 (0)	143.8 (0)	141.6 (11)	161.6 (12)
N_{sMM}	24.8 (13)	—	123.4 (9)	—	88.3 (11)	—
N_{sEXT}	—	24.6 (13)	—	119.2 (9)	87.7 (11)	102.5 (11)
N_{sMM}/N_{xM}	0.67	—	0.87	—	0.62	—
N_{sEXT}/N_{xM}	—	0.66	—	0.83	—	0.63

Numbers in parentheses are the number of circumferential waves, l,

N_{xM} Buckling load from BOSOR [20] or SRA [21] using membrane prebuckling analysis [lb/in],

N_{sMM} Imperfect shell buckling load by the multimode analysis,

N_{sEXT} Imperfect shell buckling load by the extended [26, 27] analysis.

Notice also from the results displayed in Table 3 that the imperfection resulted in a 38% decrease in the buckling load when compared with the analysis using a membrane prebuckling state and the same SS-3 boundary conditions. Next we calculated the buckling load of the same shell with the same initial imperfections but with C-3 boundary conditions. This resulted in a 37% decrease in the buckling load when compared with the analysis using a membrane prebuckling state and the same C-3 boundary conditions. This similarity in the reduction of the buckling load for the two boundary conditions motivated the decision to take boundary conditions into account (when they are important) by comparing the imperfect shell buckling load to the corresponding perfect shell buckling load using in both cases the same boundary conditions. This procedure will be followed in the correlation study with experimentally measured initial imperfections. It should be noted that in this study this will only be necessary for the stringer stiffened shell.

4. The Averaged Imperfection Model

The correlation studies to be carried out using the previously described multimode analysis require the measured initial imperfections in the form of double Fourier expansions. The measurements were taken by equipment specifically designed for this purpose at GALCIT and the results are reported in the literature [8, 9, 28]. The amplitudes of the Fourier coefficients are calculated from the measured data in a standard manner. When observing such data displayed on a log-log basis [28] it is evident that the imperfection amplitude coefficient can be approximated by straight lines as follows:

$$\overline{W}_{i0} = \overline{X}_A/i^q, \qquad \overline{W}_{kl} = \overline{X}/k^r l^s, \qquad (25)$$

where

$\overline{W}_{i0}$ is the amplitude of the i-th axisymmetric Fourier harmonic,
$\overline{W}_{kl}$ is the amplitude of the k, l-th asymmetric Fourier harmonic, and
$\overline{X}_A$, $\overline{X}$, q, r, s are coefficients determined by least-square
fitting the measured data.

This model, introduced by Imbert [29] following an idea by Donnell
and Wan [1] was extensively utilized by Arbocz and Babcock [10] in
previous correlation studies. The same imperfection model will be used
in this study for the following reasons.

In the first place, the correlation studies carried out in this paper
required in some cases imperfection amplitudes at wave numbers that
were not measured. This was due to the fact that in the early experimental work the experimental data spacing was not sufficiently close
to resolve all the harmonic amplitudes of interest. Therefore, the imperfection model was fitted over the wave numbers actually measured and
then the amplitudes of the harmonics of interest could be obtained by
extrapolation. The accuracy of this procedure is unknown. Secondly,
the imperfection model fitting is a numerical smoothing operation of
the experimental data. It was felt that such an operation was desirable
due to the experimental scatter experienced in obtaining the imperfection measurements. Thirdly, it is highly desirable to have an imperfection model that represents a class of shells manufactured by a given
process. The utilization of the imperfection sensitivity calculations will
undoubtedly occur before the detailed shell imperfections are available.
Therefore it is necessary to make predictions based on an imperfection
model of this type. The parameters of the imperfection model used in
this study are given in Table 4.

Table 4. Imperfection model summary

	Cosine Representation Axisymmetric		Sine Representation Asymmetric		
	$\overline{X}_A$	q	$\overline{X}$	r	s
A-8	0.1280	1.18	1.630	1.01	1.33
B-1	0.1780	1.43	0.960	1.13	1.18
AR-1	0.0208	1.50	0.206	1.18	1.22
AS-2	0.0068	0.25	0.786	1.12	1.23

5. Mode Selection

The number of modes of deformation included in the analysis is limited
by practical considerations, like the available core size and the time
required for obtaining the solution. Thus, since the shell buckling load

will be determined from the governing equations by using a particular set of modes, an attempt at optimizing the selection of these modes must be made. That is, we want to locate those modes which dominate the prebuckling and buckling behavior of the shell.

Examples of attempts to locate "critical modes", defined as that combination of axisymmetric and asymmetric modes which would yield the lowest buckling load, have been reported by Arbocz and Babcock [8, 10, 11] and Imbert [29]. These imperfection studies have shown that in order to yield a decrease from the buckling load of the perfect shell the initial imperfection harmonics used must include at least one mode with a significant initial amplitude and an associated eigenvalue that is close to the buckling load of the perfect shell. Hence, in this investigation, the selection of modes will be based on the following three considerations:

1. the eigenvalue of the mode,

2. the initial amplitude of the mode,

3. the coupling of the mode selected with other modes of the solution.

Based on the models selected the theoretical buckling load will be calculated. Different combinations will be used in order to investigate the influence of wave numbers and initial amplitudes on the calculated buckling load. Finally the lowest buckling load calculated will be compared with the experimentally determined buckling loads.

5.1. Eigenvalue Maps and Initial Imperfection Amplitudes

Figures 4, 5 and 6 show maps of the classical buckling loads for three of the experimental shells used in this investigation. The buckling loads were calculated for a perfect using membrane prebuckling and classical simply supported boundary conditions ($w = M_x = v = N_x = 0$).

For the isotropic shell A-8, the lowest buckling load ($\varrho = 1.0$) is a multiple eigenvalue occurring for a family of modes. We know from previous correlation studies [8, 10] that at least one of these modes should be included in the assumed solution. Also, as can be seen from Fig. 4 there are many modes whose eigenvalue is only slightly higher than the lowest eigenvalue $\varrho = 1.0$.

For the ring stiffened shell AR-1, the lowest eigenvalue ($\varrho = 1.0$) is single valued and is associated wiht a short wavelength axisymmetric mode. Once again there are several modes whose eigenvalue is only slightly higher than the lowest eigenvalue. However, it is clear from the results of Fig. 5 that the ring stiffened shell AR-1 has only a few modes with eigenvalues less than or equal to 1.01 (within 1% of the lowest eigenvalue $\varrho = 1.0$) and that all these modes have short wavelength in the axial direction. It should also be noticed that the first asymmetric mode with one half wave in the axial direction has an associated eigenvalue of 1.5 (50% higher than the lowest eigenvalue $\varrho = 1.0$).

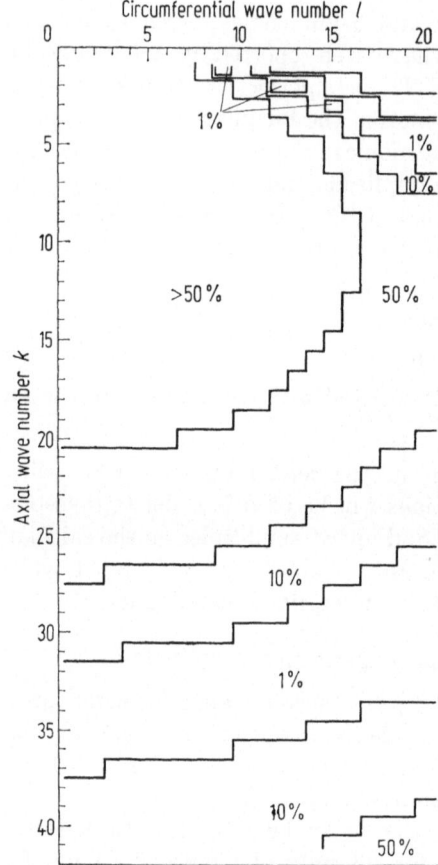

Fig. 4.
Buckling loads from linear theory for shell A-8 (SS-3 boundary condition $-w = M_x = v = N_x = 0$).

The lowest eigenvalue ($\varrho = 1.0$) of the stringer stiffened shell AS-2 is also single valued but it is associated with an asymmetric mode that has one half wave in the axial direction. Also, as can be seen from Fig. 6, there are only three modes with eigenvalues less than 1.10 (within 10% of the lowest eigenvalue $\varrho = 1.0$). As a matter of fact there are only a few modes with eigenvalues less than 1.50.

The amplitudes of the initial imperfections used in the correlation studies are computed from the averaged imperfection model. This analytical imperfection model, which expresses the variation of the imperfection amplitudes with axial and circumferential wave number, combined with the eigenvalue maps provide the investigator with a very useful guide when selecting the modes to be included in the analysis. A relatively large initial amplitude, combined with a mode shape whose eigenvalue is close to the lowest value, will always result in a buckling load which is lower than the classical value.

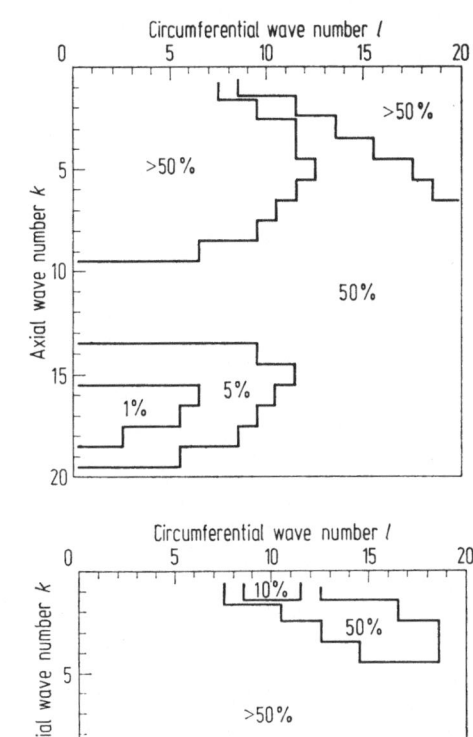

Fig. 5.
Buckling loads from linear theory for shell AR-1 (SS-3 boundary condition $-w = M_x = v = N_x = 0$).

Fig. 6.
Buckling loads from linear theory for shell AS-2 (SS-3 boundary condition $-w = M_x = v = N_x = 0$).

5.2. Coupling of Modes

By definition one or more modes are coupled, if their inclusion in the analysis results in nonzero off-diagonal terms in the matrices B, C and D of Eqs. (15) and (16). (Matrix A is a diagonal matrix.) It has been shown in [15] that coupling between one axisymmetric mode with wave numbers (i, o) and two asymmetric modes with wave numbers (k, l) and (m, n) will occur, if the relation $i = |k \pm m|$ and $l = n$ are satisfied. For the degenerate case of one axisymmetric (i, o) and one asymmetric mode (k, l) the coupling conditions reduce to the single relation $i = 2k$ [8]. Further it has been found that coupling between three asymmetric modes with wavenumbers (k, l), (m, n) and (p, q) will occur if the relations $k + m + p = $ odd integer and $q = |l \pm n|$

are satisfied. If these coupling conditions are satisfied, then the resulting buckling load of the shell is generally lower than the buckling load when each mode is considered separately.

6. Imperfection Correlation Studies

Next the experimental buckling loads of the shells A-8, B-1, AR-1 and AS-2 were predicted by using some of the mode shape combinations found previously to have important coupling properties. The experimentally determined averaged imperfection model was used in this study.

The buckling load calculations for shell A-8 are summarized in Table 5. In this table the notation (2, 0) denotes an axisymmetric mode with two half waves in the axial direction, whereas (1, 9) stands

Table 5. Buckling loads calculated by the multimode analysis (shell A-8)

```
2-modes                    ϱs              10-modes                       ϱs
(2,0) + (1,9)                = 0.901   (1,11) (1,7) (33,11) (33,7)
                                           \   /         \   /
                                           (1,2)         (33,2)
                                              \            /
4-modes                                  (2,0) + (1,9) + (33,9) + (34,0)    = 0.699
(2,0) + (1,9) + (33,9) + (34,0) = 0.846

6-modes                                  15-modes
              (33,7)                         (1,11) (1,7) (33,11) (33,7)
                |                                \   /          \   /
              (33,2)                             (1,2)          (33,2)
                |                                   \             /
(2,0) + (1,9) + (33,9) + (34,0) = 0.763  (2,0) + (1,9) + (33,9) + (34,0)
                                                        |          |
                                                      (2,4)      (32,4)
                                                        |          |
8-modes                                  (4,0) + (2,13) + (32,13) + (34,0) = 0.693
      (1,7)     (33,7)
        |         |
      (1,2)     (33,2)
        |         |
(2,0) + (1,9) + (33,9) + (34,0) = 0.744
```

for an asymmetric mode with a single half wave in the axial direction and nine full waves in the circumferential direction. The fundamental role that the basic combination (2,0) + (1, 9) + (33,9) + (34, 0) plays becomes evident as more and more modes are included in the solution. The importance of the basic combination is explained by the fact that as shown in Fig. 7 the three modes (1, 9), (33,9) and (34, 0) lie on the Koiter circle associated with the lowest eigenvalue $\varrho = 1.0$.

The results of Table 5 indicate that in going from a 6-mode to a 10-mode solution the addition of four new modes resulted in a 9% decrease in the predicted buckling load, whereas the inclusion of five additional modes (15-mode solution) resulted only in a further 1/2% decrease.

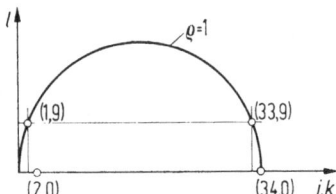

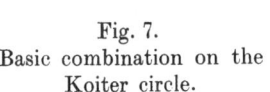

Fig. 7.
Basic combination on the
Koiter circle.

This behavior suggests that there is a point beyond which the addition of more modes will not necessarily result in further significant decrease of the predicted buckling load. The buckling load calculations for the isotropic shell B-1, summarized in Table 6, followed the same general pattern.

Table 6. Buckling loads calculated by the multimode analysis (shell B-1)

2-modes	ϱ_s	15-modes
$(2,0) + (1,8)$	$= 0.837$	$(1,10)\ (1,6)\quad (25,10)\ (25,6)$
		$\searrow\diagup\qquad\searrow\diagup$
		$(1,2)\qquad\quad (25,3)$
4-modes		$\searrow\qquad\qquad\diagup$
		$(2,0) + (1,8)\ + (25,8) + (26,0)$
$(2,0) + (1,8) + (25,8) + (26,0) = 0.803$		$\quad\big|\qquad\qquad\quad\big|$
		$(2.3)\qquad\ (24.3)$
		$\quad\big|\qquad\qquad\quad\big|$
		$(4,0) + (2,11) + (24,11) + (26,0) = 0.663$

The buckling load calculations for the ring stiffened shell AR-1 are summarized in Table 7. The selection of the modes follows the same pattern as for the isotropic shells A-8 and B-1. However, in this case the inclusion of additional modes results in only relatively small decreases in the predicted buckling loads. The reason for this behavior becomes evident if one considers the distribution of the eigenvalues for the shell AR-1 shown in Fig. 5. All those modes whose associated eigenvalues are close to the lowest eigenvalue have many half waves in the axial direction. Thus their initial amplitudes, as given by Eq. (25) of the averaged imperfection model are very small. Conversely, the associated eigenvalues of the modes with significant initial amplitudes (modes with few half waves in the axial direction) are considerably higher than the lowest eigenvalue. Thus their contribution to the lowering of the calculated buckling loads is ineffective.

Table 7. Buckling loads calculated by the multimode analysis (shell AR-1)

2-modes

(16,0) + (8,16)   = 0.995

4-modes

(2,0) + (1,8) + (15,8) + (16,0) = 0.985

6-modes

```
          (15,6)
            |
          (15,2)
            |
(2,0) + (1,8) + (15,8) + (16,0) = 0.965
```

8-modes

```
  (1,6)   (15,6)
    |        |
  (1,2)   (15,2)
    |        |
(2,0) + (1,8) + (15,8) + (16,0) = 0.959
```

12-modes

```
(1,6)   (1,5)   (15,6)   (15,5)
  |       |        |        |
(1,2)   (1,3)   (15,2)   (15,3)
    \     |        |      /
(2,0) + (1,8) + (15,8) + (16,0)   = 0.946
```

20-modes

```
        (1,13)  (15,13)
          |        |
        (1,3)   (15,3)
          |        |
        (1,10)  (15,10)
          |        |
        (1,2)   (15,2)
          |        |
(2,0) + (1,8) + (15,8) + (16,0)
          |        |
        (1,15)  (15,15)
       /  |        |     \
(1,9)  (1,7)    (15,7)   (15,9)
  |                         |
(1,6)                    (15,6)   = 0.932
```

ϱ_s

The buckling load calculations for the stringer stiffened shell AS-2 are summarized in Table 8. Comparing the result of the 2-mode solution with that of the 4-mode solution it is evident that the additional short wavelength modes have only an insignificant effect. The reason for this becomes immediately evident if one considers the distribution of the eigenvalues for the shell AS-2 shown in Fig. 6. Only the eigenvalues of a few asymmetric modes with long wavelength in the axial direction are close to the lowest eigenvalue, which in this case is asymmetric. Coupling of these modes resulted in a significant decrease in the predicted buckling load. The insignificant effect of the short wave-length axial modes is further illustrated by the fact that the elimination of these modes from the 14-mode solution resulted in a buckling load of $\varrho = 0.828$, only slightly higher than the value of $\varrho = 0.824$ predicted by the 14-mode solution itself.

Finally in Table 9, the lowest buckling loads predicted by the multimode solutions are compared with the corresponding experimental values. The agreement is very good for the isotropic shells A-8 and B-1 and satisfactory for the ring and stringer stiffened shells AR-1 and AS-2.

Table 8. Buckling loads calculated by the multimode analysis (shell AS-2)

2-modes	ϱ_s	14-modes		ϱ_s
(2,0) + (1,10)	= 0.904	(1,19)	(9,19)	
		│	│	
		(1,9)	(9,9)	
4-modes		│	│	
		(2,0) + (1,10) + (9,10) + (10,0)		
(2,0) + (1,10) + (9,10) + (10,0) = 0.903		│	│	
		(1,11)	(9,11)	
		│	│	
		(1,21)	(9,21)	
		│	│	
		(1,2)	(9,2)	= 0.824

Table 9. Comparison between theory and experiments

Shell	ϱ_s	ϱ_{EXP}	$\Delta\varrho$
A-8	0.69	0.66	0.03
B-1	0.66	0.60	0.06
AR-1	0.93	0.81	0.12
AS-2	0.82	0.71	0.11

$\varrho_s = N_{sMM}/N_{xSS-3};$ $\varrho_{EXP} = N_{xEXP}/N_{xC-4}$
N_{xSS-3}, N_{xC-4} Perfect shell buckling loads using membrane pre buckling

7. Conclusions

The main conclusion that can be drawn from this investigation is that, with the proposed multimode analysis, it is possible to predict reasonably well the buckling load of axially compressed isotropic and stiffened shells from measured initial imperfections.

Further, from the detailed imperfection correlation studies one must conclude that, if realistic variation of the imperfection amplitudes with wavelength is taken into account then suitable combinations of axisymmetric and asymmetric modes are always more damaging than either a single axisymmtric or a single asymmetric mode.

The proposed multimode analysis does not include the effect of the prebuckling deformations caused by the edge constraints. However, comparisons with the so-called "extended" analysis that includes such effects have shown that prebuckling deformations are unimportant in the presence of reasonably sized imperfections which tend to dominate the shell response, if the resulting deformations are symmetric with respect to the mid-plane of the shell. It was also found that the importance of the different boundary conditions can be properly assessed by the linear theory using membranes prebuckling solutions, and whenever the boundary conditions are important they can be taken into account by proper normalizing. These conclusions concerning the influence of the edge condition are based upon a limited investigation but appear to substantiate standard engineering practice.

Finally, it is also abundantly clear that the further incorporation of imperfection sensitivity ideas into engineering practice will have to await the measurements of imperfections of full scale structures and subsequent correlation studies.

Acknowledgement

The authors wish to thank Prof. E. E. Sechler for his contributions to this work. Not only did he gladly share his technical expertise, but he also calmly received innumerable requests for additional computer money.

References

1. Donnell, L. M.; Wan, C. C.: Effect of Imperfections on Buckling of Thin Cylinders and Columns under Axial Compression. J. Appl. Mech. **17**, 73 (1950).
2. Koiter, W. T.: The Effect of Axisymmetric Imperfections on the Buckling of Cylindrical Shells under Axial Compression. Koninkl. Ned. Akad. Weterschap Proc. **B. 66**, 265—279 (1963).
3. Budiansky, B.; Hutchinson, J. W.: Dynamic Buckling of Imperfection-Sensitive Structures. Proc. XI Intern. Congr. Appl. Mech. Edited by H. Görtler. Berlin, Göttingen, Heidelberg, New York: Springer 1964, pp. 636—651.
4. Kanemitsu, S.; Nojima, N. M.: Axial Compression Tests of Thin Circular Cylinders. M. S. Thesis, California Institute of Technology, 1939.
5. NASA Space Vehicle Design Criteria (Structures), NASA SP-8007, Revised, August 1968.
6. Hutchinson, J. W.: Axial Buckling of Pressurized Imperfect Cylindrical Shells. AIAA J. **3**, No. 8, 1461—1466 (1965).
7. Thurston, G. A.; Freeland, M. A : Buckling of Imperfect Cylinders under Axial Compression. NASA CR-541, July 1966.
8. Arbocz, J.; Babcock, C. D.: The Effect of General Imperfections on the Buckling of Cylindrical Shells. J. Appl. Mech. **36**, No. 1, 28—38 (1969).
9. Singer, J.; Arbocz, J.; Babcock, C. D.: Buckling of Imperfect Stiffened Cylindrical Shells under Axial Compression. AIAA J. **9**, No. 1, 68—75 (1971).
10. Arbocz, J.; Babcock, C. D.: On the Role of Imperfections in Shell Buckling. (Paper presented at the 13th IUTAM Congress in Moscow, Aug. 21—26, 1972.)
11. Arbocz, J.: The Effect of Initial Imperfections on Shell Stability. In: Thin-Shell Structures. Edited by Y. C. Fung; E. E. Sechler. Prentice-Hall 1974.
12. Geier, B.: Das Beulverhalten versteifter Zylinderschalen, Teil 1: Differential-Gleichungen. Z. Flugwiss. **14**, 306—323 (1966).
13. Singer, J.: Personal Communication, 1968.
14. Hutchinson, J. W.; Amazigo, J. C.: Imperfection Sensitivity of Eccentrically Stiffened Cylindrical Shells. AIAA J. **5**, No. 3, 392—401 (1967).
15. Arbocz, J.; Babcock, C. D.: A Multimode Analysis for Calculating Buckling Loads of Imperfect Cylindrical Shells. GALCIT Report SM 74-4, California Institute of Technology, Pasadena, June 1974.

16. Koiter, W. T.: Personal Communication, California Institute of Technology, April 1974.
17. Almroth, B. O.: Influence of Edge Conditions on the Stability of Axially Compressed Cylindrical Shells. NASA CR-161, February 1965.
18. Hoff, N. J.; Soong, T. E.: Buckling of Circular Cylindrical Shells in Axial Compression. SUDAER No. 204, Stanford, California, August 1964.
19. Weller, T.; Singer, J.; Batterman, S. C.: Influence of Eccentricity of Loading on Buckling of Stringer-Stiffened Cylindrical Shells. Proc. of the Thin Shell Structures Symposium, Prentice-Hall, 1974, pp. 305—324.
20. Bushnell, D.; Almroth, B. O.; Sobel, L. H.: Buckling of Shells of Revolution with Various Wall Construction. Vol. 3. User's Manual for BOSOR, NASA CR-1051, 1968.
21. Cohen, G. A.: Buckling of Axially Compressed Cylindrical Shells with Ring Stiffened Edges. AIAA J. 4, No. 10, 1859—1862 (1966).
22. Ohira, H.: Local Buckling Theory of Axially Compressed Cylinders. Proc. of the Eleventh Japan National Congress for Applied Mechanics, 1961.
23. Hoff, N. J.: Buckling of Thin Shells. Proc. of an Aerospace Symposium of Distinguished Lecturers in Honor of Dr. Theodore von Kármán on his 80th Anniversary. The Institute of Aerospace Sciences, New York, 1961.
24. Yamaki, N.; Kodama, S.: Buckling of Circular Cylindrical Shells under Compression/Report 2, Solutions Based on the Flügge Equations Neglecting Prebuckling Edge Rotations. Rep. Inst. High Speed Mech., Japan, Vol. 24 (1971).
25. Arbocz, J.: Buckling of Axially Compressed Imperfect Cylindrical Shells. Proc. of the AIAA/ASME 12th Structures, Structural Dynamics and Materials Conference, Anaheim, California, April 19—21, 1971.
26. Arbocz, J.; Sechler, E. E.: On the Buckling of Axially Compressed Imperfect Cylindrical Shells. J. Appl. Mech. 41, 737—743 (1974).
27. Arbocz, J.; Sechler, E. E.: On the Buckling of Axially Compressed Ring- and Stringer-Stiffened Imperfect Cylindrical Shells. GALCIT Report SM 73-10, California Institute of Technology, December 1973.
28. Babcock, C. D.: Experiments in Shell Buckling. In: Thin-Shell Structures. Edited by Y. C. Fung; E. E. Sechler, Prentice-Hall, 1974.
29. Imbert, K.: The Effect of Imperfections on the Buckling of Cylindrical Shells. Aeronautical Engineer Thesis, California Institute of Technology, June 1971.

Experiments on the Postbuckling Behavior of Circular Cylindrical Shells under Torsion

N. Yamaki

Tohoku University, Sendai, Japan

1. Introduction

It is one of the basic problems on the stability of structures, to clarify the buckling and postbuckling behaviors of circular cylindrical shells under torsion. Hence, numerous researches have been made on these subjects. Concerning the buckling problem, approximate solutions were obtained by Donnell [1] and Batdorf [2] under two different boundary conditions, which were ascertained to be in fair agreement with experimental results [1, 3]. Later, based on both Donnell and Flügge equations, accurate solutions under eight sets of boundary conditions were obtained by the author [4] for a wide range of shell geometries. Hence, the buckling problem seems to be almost fully investigated.

As for the postbuckling problem, solutions of rough approximate nature were obtained by Loo [5], Nash [6] and Hayashi [7], and the corresponding experimental researches were performed by Nash [8] and Weingarten [9]. However, it seems that the agreement between theory and experiment has not been achieved. On the other hand, the so-called initial postbuckling problem was solved by Budiansky [10] to estimate the initial imperfection sensitivity.

Recently, the author conducted precise experimental studies on the postbuckling behavior of clamped cylinders under external pressure [11] as well as edge compression [12], the results of which were confirmed to be in reasonable agreement with the corresponding theoretical results [13, 14]. As a continuation of these studies, the present research aims to clarify experimentally the whole aspect of the postbuckling behaviour of cylindrical shells under torsion.

Test cylinders and test apparatus are almost the same as those reported previously except for the loading system, for which great care is taken that the cylinder is subjected to purely torsional loads without the effect of bending and axial forces. Although a slight axial constraint is observed to remain, the critical load and the corresponding wave number for each test cylinder are found to agree well with those theoretically predicted [4], and the present results are considered to provide fundamental data for elaborate theoretical analyses in the future.

2. Test Specimen and Test Apparatus

As stated in the previous paper [11], test cylinders with radius $R = 100$ mm and thickness $h = 0.247$ mm are made of polyester film with Young's modulus $E = 5.55 \cdot 10^3$ MPa ($= 567$ kg/mm^2) and Poisson's ratio $\nu = 0.30$, by lapjointing along the longitudinal seam and attaching duralumin cover plates along the ends. Only the thickness of the cover plate was changed from 11 mm to 18 mm. The lengths L of the cylinder are chosen so that nominal values of the geometric parameter $Z = \sqrt{1 - \nu^2} L^2 / Rh$ are 20, 50, 100, 200, 500 and 1 000. Actual dimensions of test specimens are listed in Table 1. Preliminary experiments are carried out using the previous specimens, with almost the same results for the postbuckling behavior as here presented.

Table 1. Dimensions of test specimens
($R = 100$ mm, $h = 0.247$ mm; $E = 5.55 \cdot 10^3$ MPa, $\nu = 0.3$)

No.	20-5	50-4	100-4	200-4	500-4	1000-4
L [mm]	22.9	35.9	51.0	71.2	113.8	160.8
Z	20.2	49.8	100.3	195.8	500.2	998.6

A schematic diagram of the test setup is shown in Fig. 1. In the figure, 1 is the table of a tension test machine, 2 is a base plate, 3 is a test cylinder and 4 is a stiffening steel plate attached to the cover plate, to which four equidistant cylindrical rods 5 are fixed. To counter-balance the weight acting on the cylinder (about 6.1 kg = 60 N), two sets of the weight 6 are hung with a thin wire 7 through a pulley 8.

The loading system comprises from 9 to 23, in which 9 is the cross-head of the test machine, 10 is a reduction gear box and 11 is the lever to rotate the shaft 12. To avoid the effect of eccentricity and axial constraint, an universal joint 13 and spline shaft 14 are used, to which a torque meter 15 with capacity 200 Nm is connected. The loading head 16 has a circular disc with four equally spaced holes and the twisting moment is transmitted to the test cylinder through the ball bushing 17 and the rod 5. The loading head is held in the axial direction by a ball bushing 18 in the casing 19, which is fixed to the stand 20. A cylindrical cap 21 is assembled to the loading head through two sets of thrust bearings and by rotating it with the lever 22, the relative position of the loading head to the test cylinder can be adjusted. With this system stated in the foregoing, almost pure torque can be applied to the cylinder without significant effects of bending moments and axil forces. Combined loading of torsion and compression is also possible by attaching a load cell *23* to the lower end of the loading head.

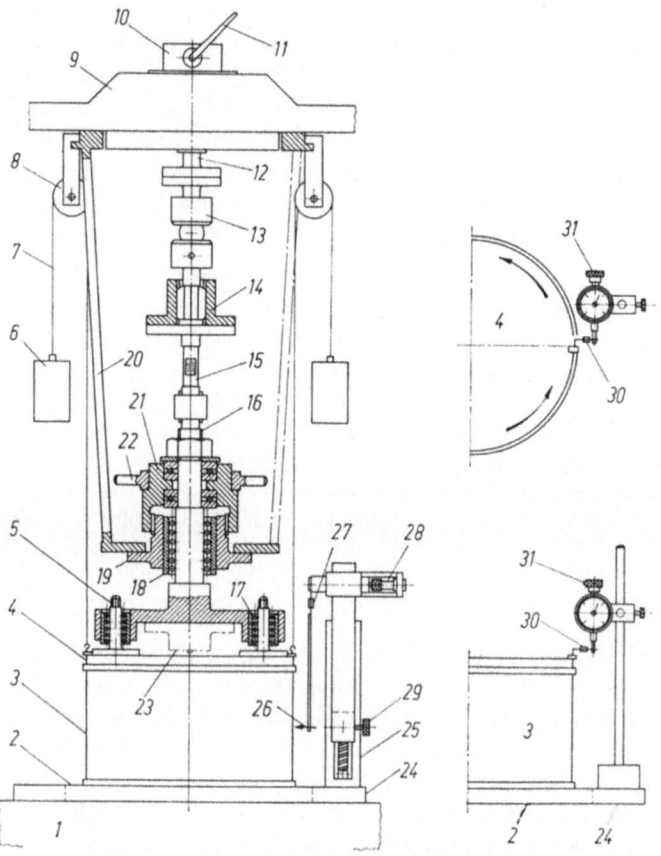

Fig. 1. Schematic diagram of the test setup.

The deflection measuring apparatus, comprising from *24* to *29*, is the same as stated in the previous paper [11], with which both axial and circumferential distributions of the deflection can be measured precisely without affecting the buckled surface. The axial shortening of the cylinder is measured utilizing the strain gauges *30* bonded to a thin lever fixed at the end of a dial gauge *31*, which, in turn, is used for calibration. Two sets of these are used in series to obtain the average value of edge shortenings at diametrically opposite points on the stiffening plate. The same sets are used for measuring the angle of twist by arranging them as shown in the upper right figure. A six-element dynamic strain meter is used for strain measurements while a X-Y recorder is used for recording the results.

3. Typical Test Results

To save the space, only the results for the test specimen No. 500-4 will be stated in detail. The relation between the torque T and the angle of twist ψ is shown in Fig. 2, in which the solid curve $0 - a - b - c - d - e - 0$ represents the actual result. The point a is the critical point, curves ab and de correspond to the unstable dynamic snap-through, while bcd to the stable equilibrium path. The snap-through occurs due to a relatively low torsional rigidity of the loading system, which is estimated to be about $1.9 \cdot 10^3$ Nm/rad, while the hysteresis loop in the range bc indicates a slight friction to the axial movement of the cylinder. (With the test specimen No. 1000-4, experiments are carried out by changing the axial load stepwisely, from which the frictional force is estimated to be less than 1 kg $= 10$ N, which corresponds to about 1% of the critical compressive load.) Assuming that the loading system is stiff enough and the frictional force is constant, we will obtain the curves depicted by dotted lines, from which the result for the ideal case with neither friction nor snap-through can be estimated as shown by the chain line. It will be seen that the equilibrium torque after buckling decreases monotonously with twist in the present range of experiments.

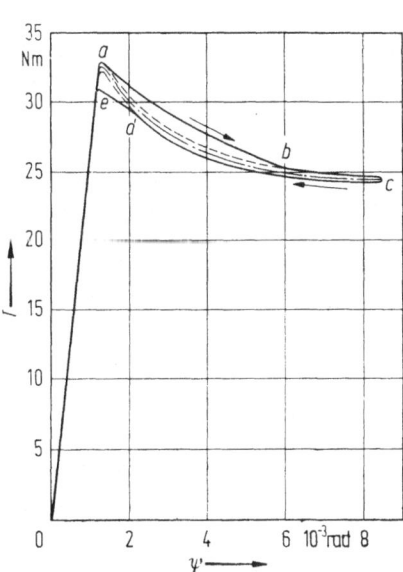

Fig. 2.
Postbuckling relation between torque T and angle of twist ψ (specimen No. 500-4).

The relation between the torque T and the axial shortening δ is given in Fig. 3. As before, the chain line represents the result estimated for the idal case. It will be seen that after buckling, the length of the cylinder decreases monotonously with the development of buckled waves.

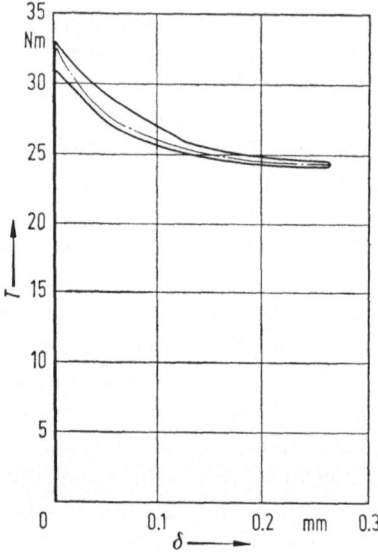

Fig. 3.
Postbuckling relation between torque T and axial shortening δ (specimen No. 500-4).

To see the overall distribution of buckled waves, circumferential distributions of the deflection w (positive inward) are recorded along various circular sections, at a typical postbuckling configuration with $T = 25.1$ Nm and $\psi = 5.43 \cdot 10^{-3}$ rad. The result is shown in Fig. 4, in which x and θ stand for the axial and angular coordinates, respectively. It will be seen that the number of buckled waves n is 15 and that the wave distribution is quite regular irrespective of the longitudinal seam located at the right side blank portion.

In the meanwhile, axial distributions of w are recorded along various angular positions, a part of which extending about one wavelength is shown in Fig. 5. The contour lines for the buckled surface are obtained from these with the results shown in Fig. 6. In the figure, solid and dotted lines correspond to the inward (positive) and outward (negative) deflections, respectively. Axial distributions of w passing through the valley and crest along the central section are also shown. It is to be noted that the waveform is symmetric with respect to the center of each buckle and that the buckled surface is almost covered with inward buckles as the angle of twist increases. It is to be added that the number of buckled waves n remains unchanged in the range of experiments here conducted.

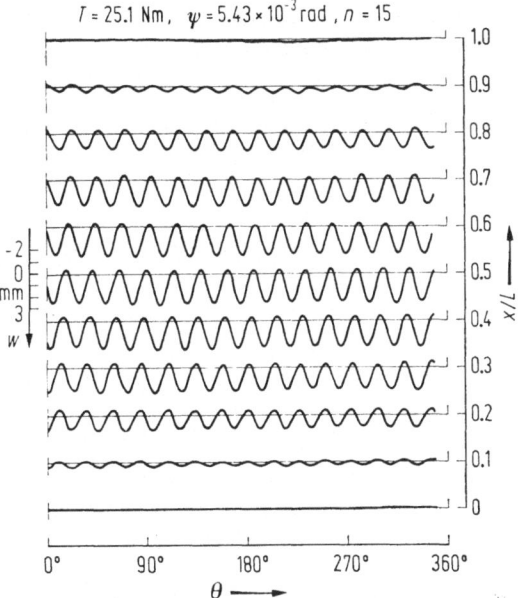

Fig. 4. Circumferential distributions of the deflection at various circular sections (specimen No. 500-4).

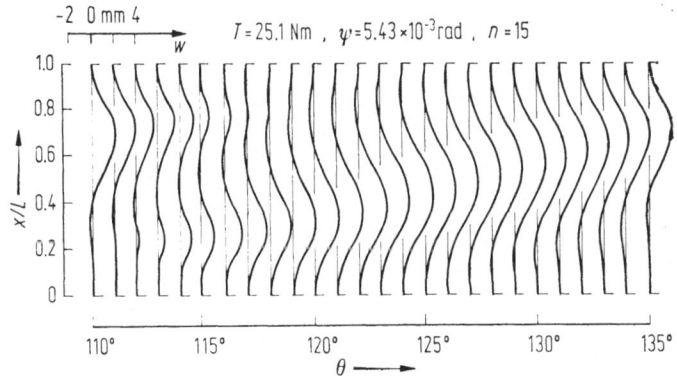

Fig. 5. Axial distributions of the deflection at various angular positions (specimen No. 500-4).

Finally, variations of the maximum inward and outward deflections along the central section of the shell, w_m, with applied torque T, are obtained with the results shown in Fig. 7. The chain line corresponds to the ideal case as before. It will be seen that with the increase in twist after buckling, the inward deflection increases monotonously while the outward deflection soon attains its maximum value and then decreases.

Similar experiments are carried out for each test specimen. It is found that the number of buckled waves remains unchanged during the

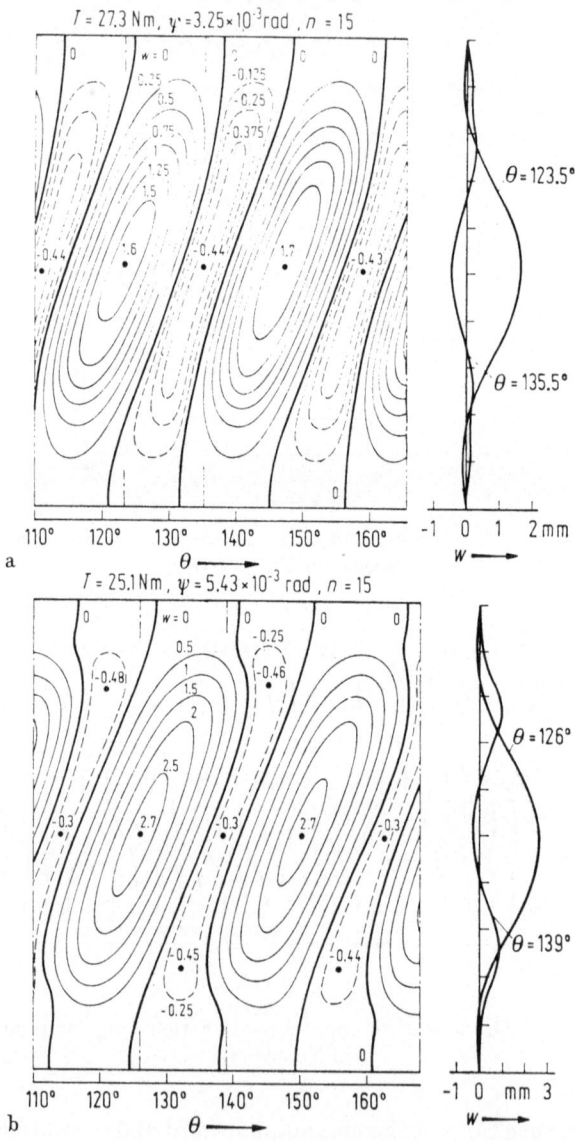

Fig. 6 a and b. Contour lines for typical postbuckling configurations (in mm, specimen No. 500-4).

test, although for relatively long shells with $Z \geqq 200$, the equilibrium states with the smaller wave numbers can be realized artificially in the fully buckled region. In general, the test can be repeated with almost the same result, although slight reductions in the buckling and post-buckling loads are observed in the immediate successive tests.

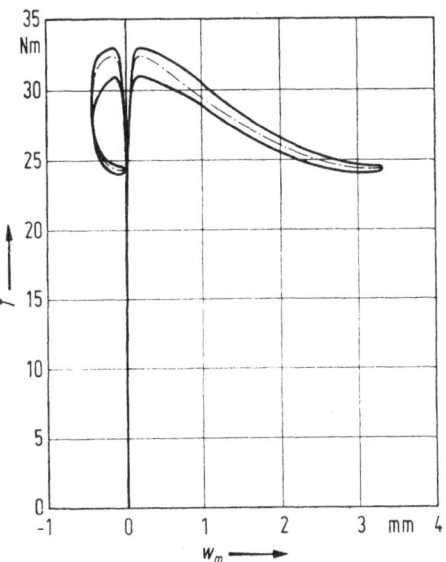

Fig. 7. Postbuckling relation between torque T and maximum central deflections w_m (specimen No. 500-4).

4. Buckling Load and Wave Number

To examine the degree of accuracy of the present experiments, critical loads and the corresponding wave numbers here obtained are compared with those theoretically predicted under the completely clamped boundary conditions, $u = v = w = w_{,x} = 0$ [4]. The results are shown in Fig. 8, in which k_s and β are the nondimensional load and wave number factors, respectively, defined by

$$ k_s = \frac{TL^2}{2\pi^3 R^2 D} = \frac{\tau h L^2}{\pi^2 D}, \quad \beta = \frac{nL}{\pi R}, \quad D = \frac{Eh^3}{12(1-\nu^2)}. \tag{1} $$

Some of the previous experimental results [1, 3] are also shown for reference. It will be seen that critical loads here obtained range from 87 to 94 per cent of theoretical ones, while experimental wave numbers are equal to or slightly less than those theoretically predicted, which seem to indicate a satisfactory accuracy of the present experiments.

320   N. Yamaki

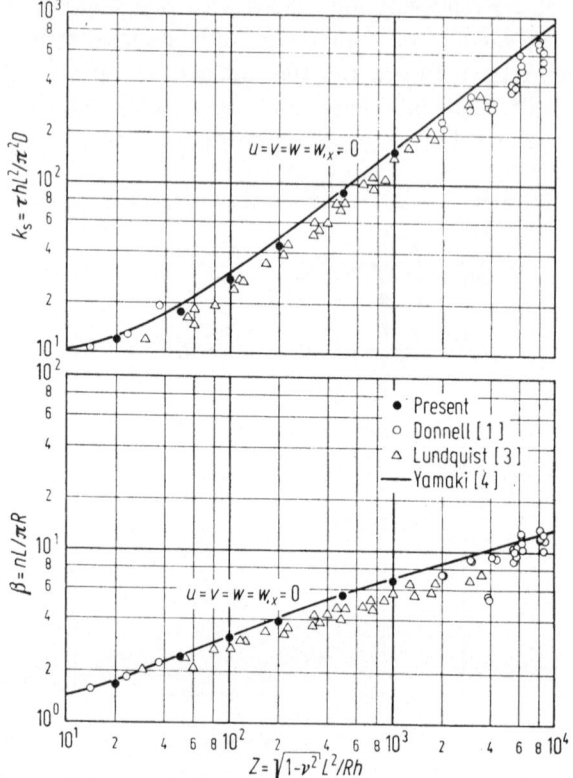

Fig. 8. Comparison between theory and experiment for the buckling of clamped
cylindrical shells under torsion.

5. Summary of the Postbuckling Behavior

To clarify the influence of shell geometries on the postbuckling behavior, the torque-twist relations for shells with various values of Z
are plotted together in Fig. 9. In the figure, T_{cr} is the critical torque
theoretically predicted, while ψ_{cr} is the corresponding angle of twist
expressed by

$$\psi_{cr} = (1 + \nu) LT_{cr}/\pi R^3 hE. \qquad (2)$$

In Fig. 9a, thick solid lines represent the present experimental results,
from which the effect of Z on the postbuckling torque-twist relation
can be clearly observed. The dotted lines correspond to the experimental results obtained by Weingarten [9], which will be seen to lie considerably below the present results. Further, chain lines for $Z = 0$
represent the approximate solution for a clamped infinite strip under
shear previously obtained by the author [15]. In Fig. 9b, approximate
theoretical results for clamped shells by Nash [6] and those for elasti-

cally supported shells by Hayashi [7] are shown by the chain and dotted lines, respectively. It will be seen that both of these are quantitatively in poor agreement with experimental results here obtained.

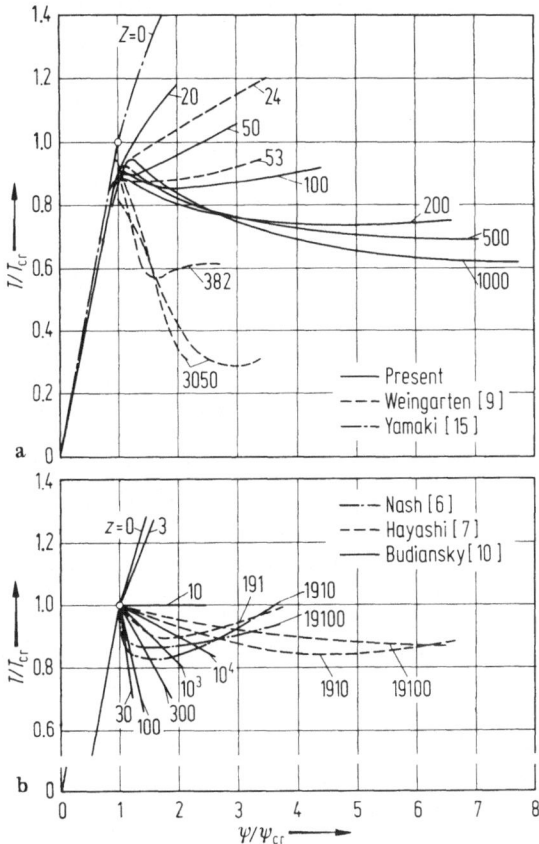

Fig. 9 a and b. Effect of Z on the postbuckling relation between T/T_{cr} and ψ/ψ_{cr}.

On the other hand, the initial postbuckling slopes determined by Budiansky [10] for completely clamped shells are depicted by solid lines, which seem to compare favourably with the present experimental results, at least for large values of Z. However, further theoretical studies will be necessary to achieve a reasonable agreement between theory and experiment.

Postbuckling relations between torque and axial shortening for each test specimen are illustrated in Fig. 10. The numeral in the parentheses denotes the number of buckled waves here observed for the special case when $R/h = 100/0.247 = 405$.

Variations of the maximum inward and outward deflections along the central sections of each cylinder, with applied torque, are shown in Fig. 11. It will be seen that with the increase in twist after buckling, the inward deflection increases monotonously, while the outward deflection soon attains its maximum and then decreases. Further, the inward deflection is much larger than the outward deflection, which exceeds ten times the shell thickness for relatively long shells with Z greater than 200.

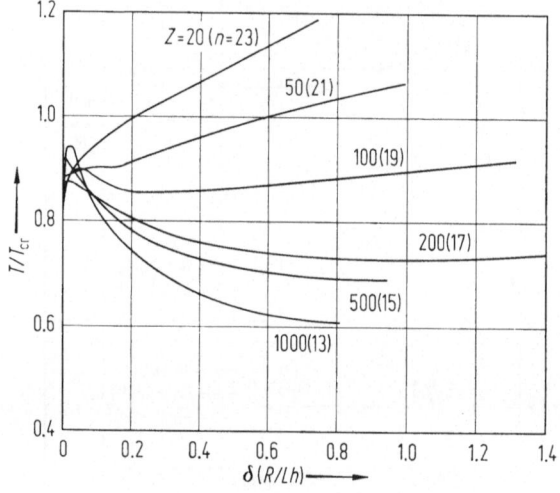

Fig. 10. Effect of Z on the postbuckling relation between T/T_{cr} and $(R/Lh)\ \delta$.

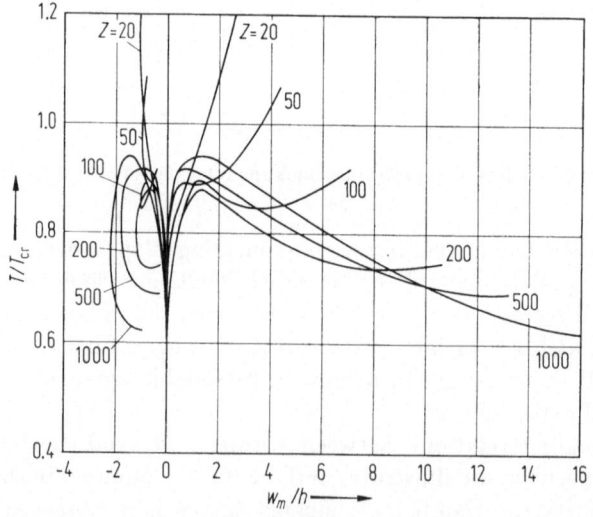

Fig. 11. Effect of Z on the postbuckling relation between T/T_{cr} and w_m/h.

It has been shown that critical loads here obtained are in fairly good agreement with those theoretically predicted. Noting that the effect of initial imperfections on the postbuckling behavior will be small in comparison with that on the critical load, the present experimental results for the postbuckling behavior will be considered to be almost equivalent to those for imperfection-free cylinders, which will be of great use for elaborate theoretical studies in the future.

Acknowledgement

The author wishes to express his thanks to Messrs. K. Otomo and T. Watanabe at the Institute of High Speed Mechanics, Tohoku University, for their assistance during the course of this work.

Appendix

For the sake of completeness, actual experimental results for the specimens with Z other than 500, which were omitted in the text, together with some pictures, are given in the following.

In each of Figs. A1 through A5, the symbols a, b, c, d, and e, attached to the figures, stand for the following specifications:

a Relation between T and ψ,
b Relation between T and δ,
c Relation between T and w_m,
d Distributions of buckled waves,
e Contour lines for the buckled surface (in mm).

324   N. Yamaki

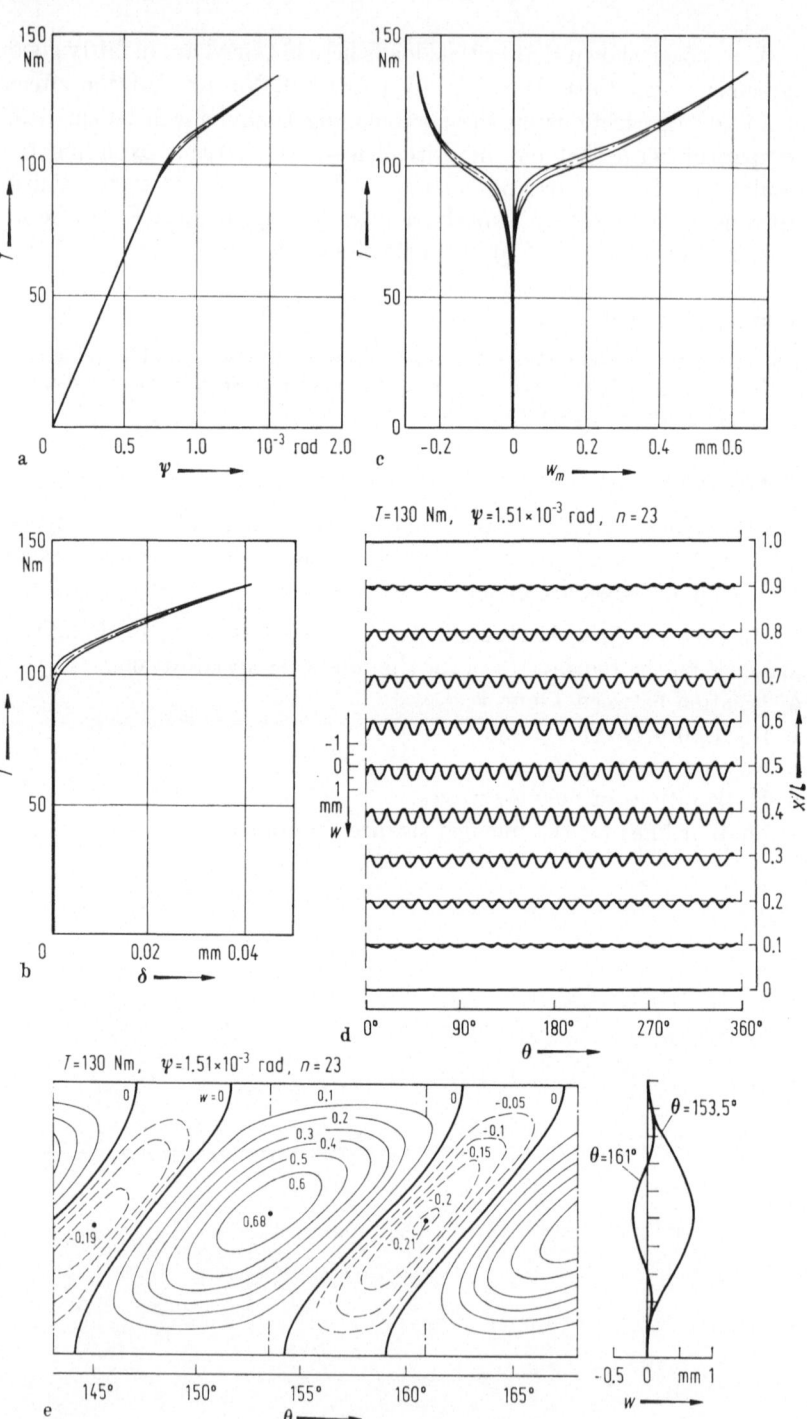

Fig. A1. Specimen No. 20-5.

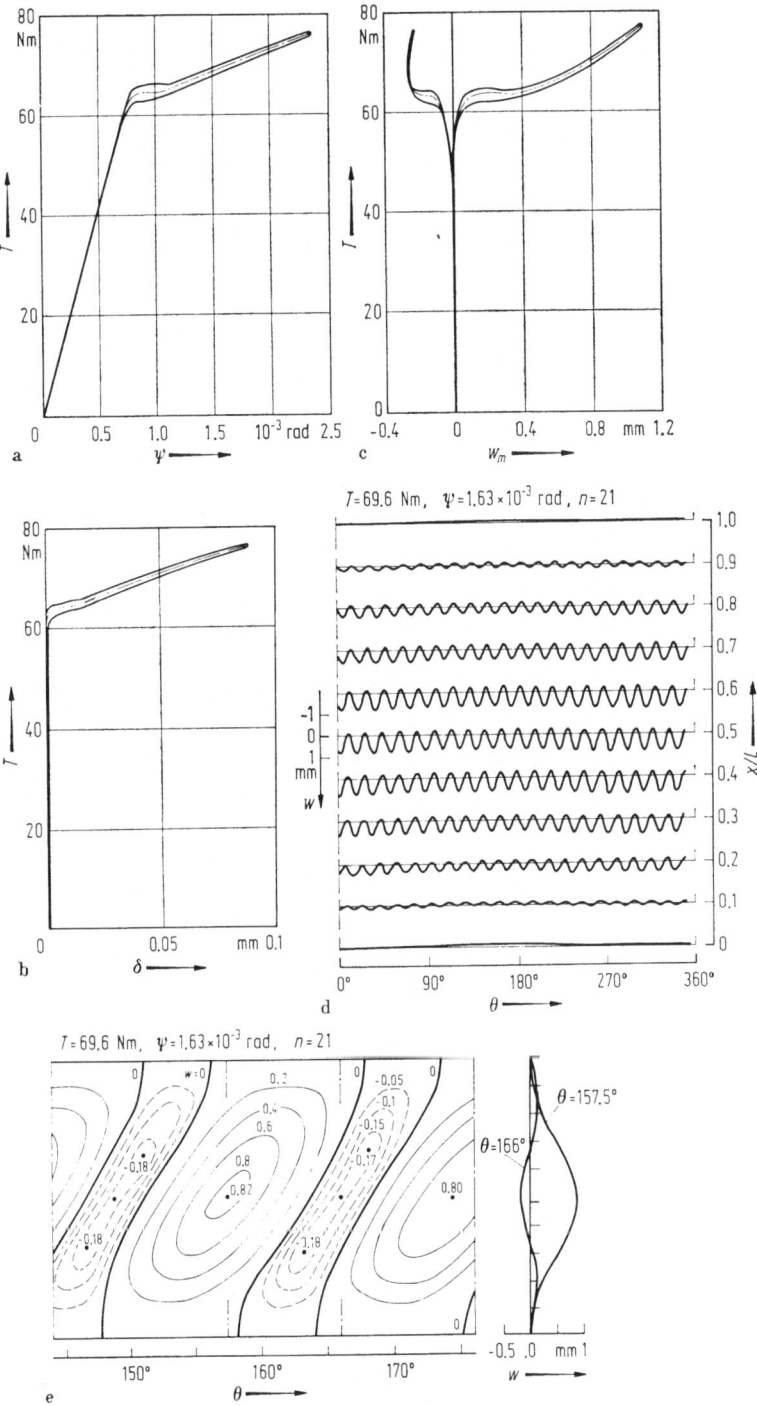

Fig. A2. Specimen No. 50-4.

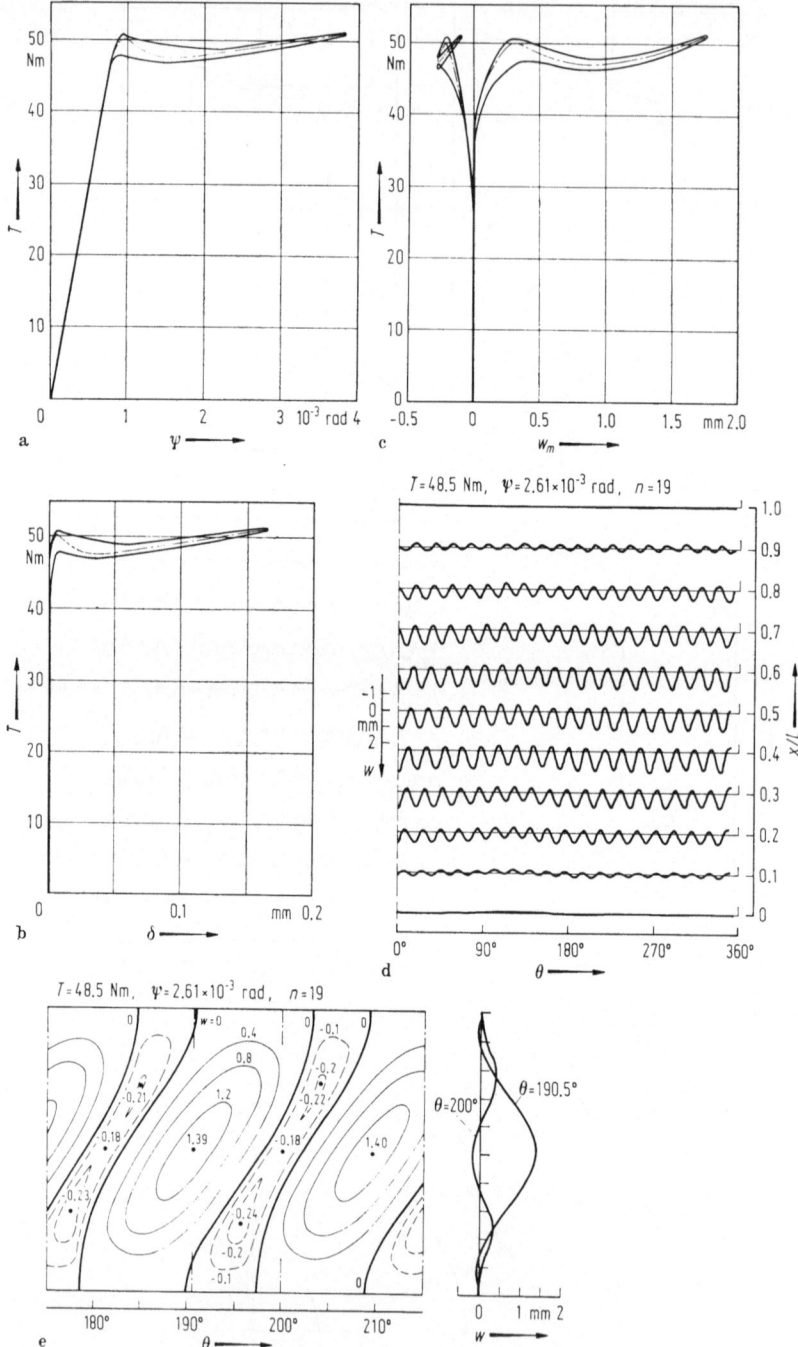

Fig. A3. Specimen No. 100-4.

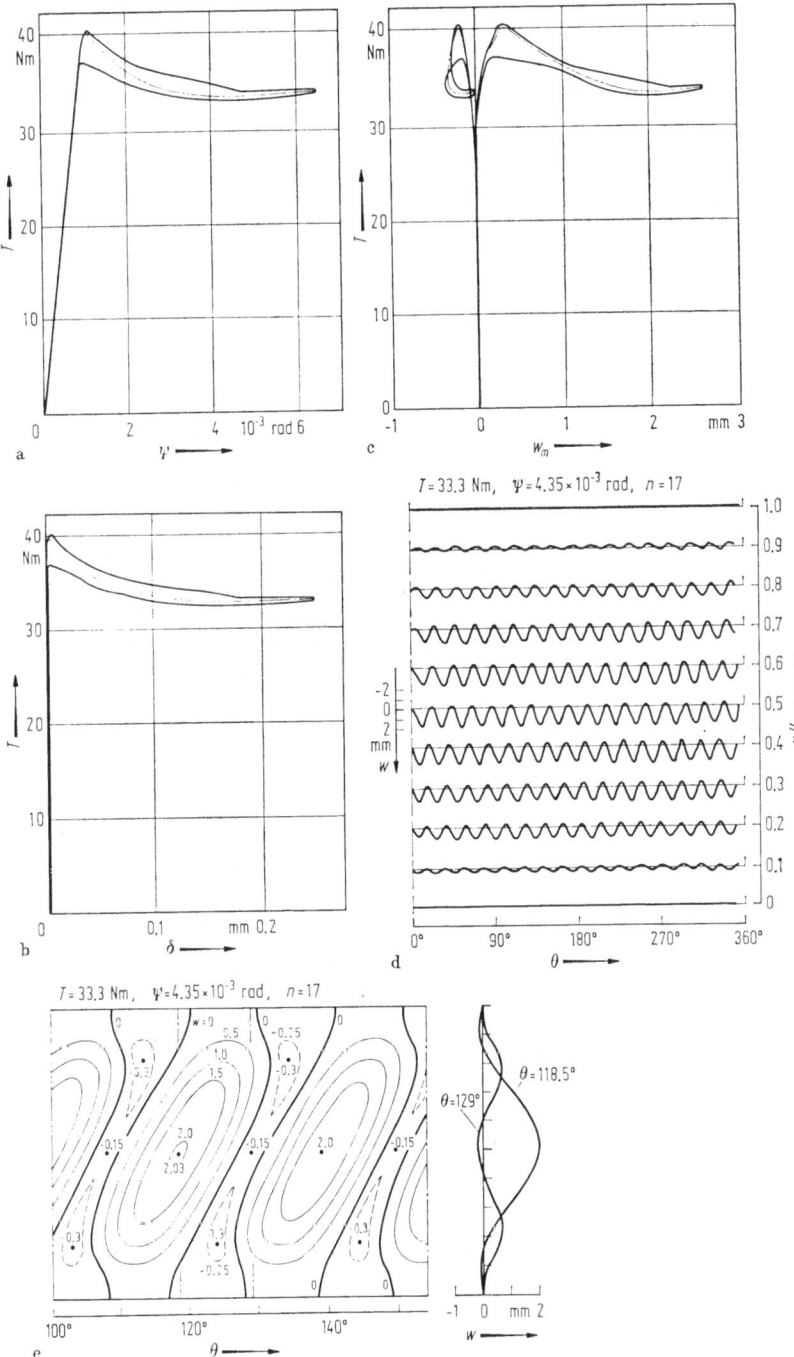

Fig. A4. Specimen No. 200-4.

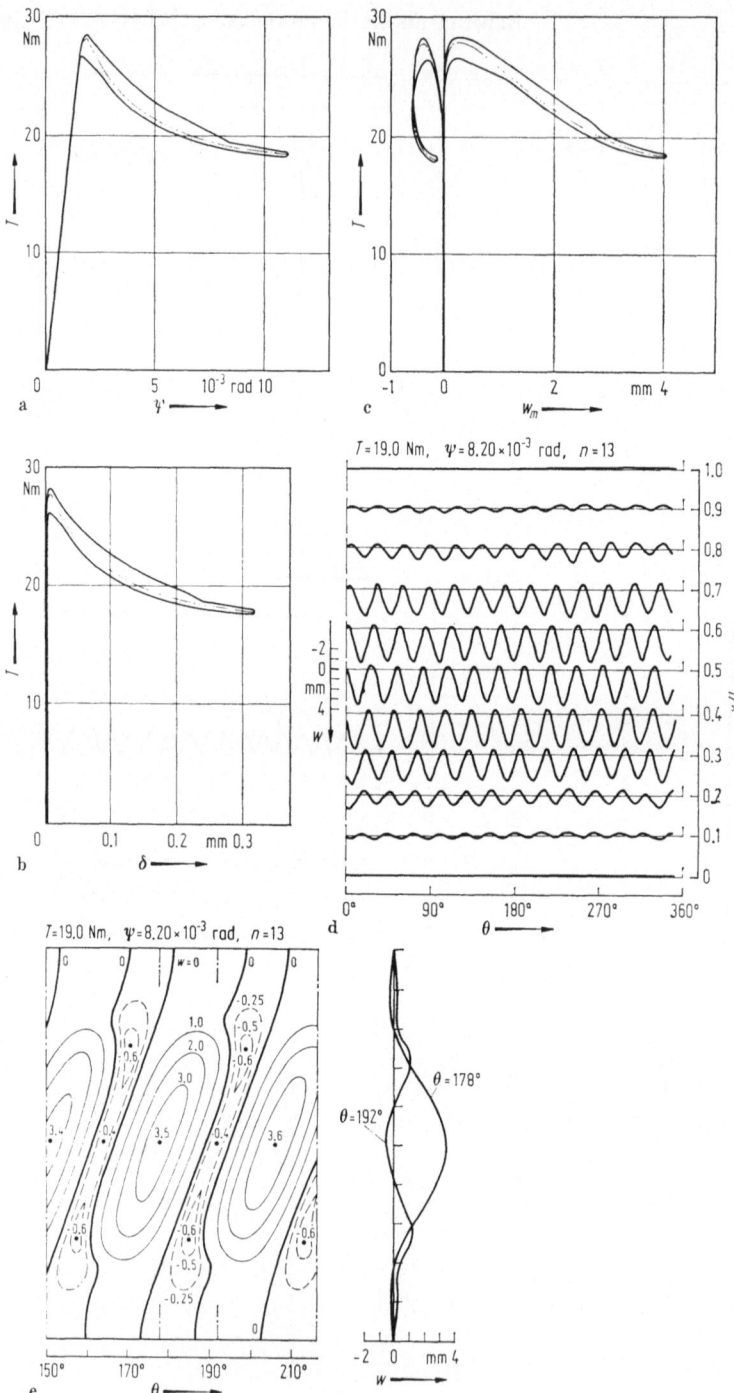

Fig. A5. Specimen No. 1000-4.

Photo 1. Test set-up.

Photo 2.
Buckle pattern for specimen
No. 100-4.

Photo 3.
Buckle pattern for specimen No. 1000-4.

References

1. Donnell, L. H.: Stability of Thin-Walled Tubes Under Torsion. NACA Rep. No. 479 (1933).
2. Batdorf, S. B.: A Simplified Method of Elastic-Stability Analysis for Thin Cylindrical Shells. NACA Rep. No. 874 (1947).
3. Lundquist, E. E.: Strength Tests on Thin-Walled Duralumin Cylinders in Torsion. NACA T.N. No. 427 (1932).
4. Yamaki, N.: Buckling of Circular Cylindrical Shells under Torsion (Rep. 1 and 2). Rep. Inst. High Speed Mech., Tohoku Univ. **17**, 171—184 (1966); **18**, 121—142 (1967).
5. Loo, T. T.: Effects of Large Deflections and Imperfections on the Elastic Buckling of Cylinders Under Torsion and Axial Compression. Proc. 2nd U.S. Nat. Congr. Appl. Mech. 345—357 (1954).
6. Nash, W. A.: Buckling of Initially Imperfect Cylindrical Shells Subject to Torsion. J. Appl. Mech. **24**, 125—130 (1957).
7. Hayashi, T.; Hirano, Y.: Postbuckling Behavior of Orthotropic Cylinders under Torsion. Trans. Japan Soc. Aero. Space Sci. **7**, 47—54 (1964).
8. Nash, W. A.: An Experimental Analysis of the Buckling of Thin Initially Imperfect Cylindrical Shells Subject to Torsion. Proc. Soc. Exp. Stress Anal. **16**, 55—68 (1959).
9. Weingarten, V. I.: The Effect of Internal Pressure and Axial Tension on the Buckling of Cylindrical Shells Under Torsion. Proc. 4th U.S. Nat. Congr. Appl. Mech., 827—842 (1962).
10. Budiansky, B.: Post-Buckling Behavior of Cylinders in Torsion. Theory of Thin Shells. Ed. by F. I. Niordson. Berlin, Heidelberg, New York: Springer 1969, pp. 212—233.
11. Yamaki, N.; Otomo, K.: Experiments on the Postbuckling Behavior of Circular Cylindrical Shells Under Hydrostatic Pressure. Exp. Mech. **13**, 299—304 (1973).
12. Yamaki, N.; Otomo, K.; Matsuda, K.: Experiments on the Postbuckling Behavior of Circular Cylindrical Shells Under Compression. Exp. Mech. **15** 23—28 (1975).
13. Yamaki, N.; Tani, J.: Postbuckling Behavior of Circular Cylindrical Shells under Hydrostatic Pressure. ZAMM **54**, 709—714 (1974).
14. Yamaki, N.; Kodama, S.: Postbuckling Behavior of Circular Cylindrical Shells Under Compression. To be published in the Int. J. Non-Linear Mechanics.
15. Yamaki, N.: Postbuckling Behavior of a Clamped Infinite Strip Under the Action of Shearing Forces. ZAMM **46**, 249—252 (1966).

Design Philosophy in Structural Stability

A. H. Chilver

Cranfield Institute of Technology, Cranfield, England

Summary

The paper aims to show the vital importance of buckling and stability in structural engineering design, the evolution of design attitudes towards the stability of structures, and the fundamental problems the field poses for basic research in stability.

With the growing sophistication of materials and methods of construction, efficient and economic structural forms have become increasingly slender and thin, and thereby more prone to buckling. This has led to a wide range of stability problems in different materials and in different structural forms. Some of these problems have been tractable, while others have remained essentially unresolved. This has led to a continuing stimulation of basic research studies in stability over the years, and, as structural design becomes more sophisticated, so the need for basic research will continue.

The main purpose of stability research is to help develop structural design philosophy generally and, in particular, attitudes towards stability in structural design. Deficiencies in these attitudes have led, in the past, to serious weaknesses in design concepts, and ways are discussed in which attitudes can be influenced and developed to overcome these shortcomings.

Finally, an attempt is made to summarise the main ways in which fundamental research in stability can play a cogent role in the development of design philosophies.

1. The Importance of Buckling and Stability in Structural Engineering and Design

1.1. The main purpose of research into the field of buckling and stability of structures is to extend basic knowledge and apply that knowledge to the solution of real problems of structural engineering. Such a statement of purpose is perhaps self-evident, and is widely understood amongst researchers in the field of stability. However, the *increasing* importance of buckling and stability problems throughout structural engineering is perhaps not so widely understood. Buckling and stability considerations enter increasingly into the design of most modern engineering structures, and it is relevant to consider the reasons for this.

1.2. The earliest problems of structural engineering were related to the *strengths* of massive structures. Such strength problems were concerned mainly with the breakdown of engineering materials by fracture, yielding, and so on. Although material breakdown can lead to an overall "instability" in the broadest sense, such problems did not involve "buckling". Some of the early problems of structural engineering were concerned with the "stability" of massive structures on their foundations; it is interesting that foundation problems have always been regarded as involving a "stability" concept which has not been brought into the main-stream of structural stability thinking.

1.3. But problems of "buckling", as such, became important design problems when, during the 19th century, the introduction of metallic structural materials made it possible to build structures of more extreme geometric *proportions*. During the early 19th century, slender cast-iron columns introduced a new element of clear, wide areas between the columns of a building; this led ultimately to the use, in this century, of mild-steel columns in multi-storey buildings. Again, the use of thin, wrought-iron plates during the 19th century led to a new set of buckling and stability problems of compressed and sheared plates, which became the components of new structural forms, especially in bridges.

1.4. Metals, applied to structural engineering during the 19th century, led essentially to structural forms of more extreme geometric *proportions*: slender columns, thin plates and thin shells. With the emphasis on extreme geometric proportions, it is not surprising that the field of structural stability theory, from the 19th century onwards, was concerned primarily with the study of "buckling". This is the reason for the wide appeal and use, during the latter half of the 19th century and the early part of this century, of Euler's approach to buckling problems: the study of those critical loading conditions at which elastic deformations were indeterminate—at least at the level of "first-order" buckling theory.

1.5. But since the turn of the century, the importance of buckling in structural engineering has gone beyond simple problems of indeterminate elastic deformations. At about the turn of the century, some new stability problems were beginning to appear in structural engineering. Major steps forward had already been made in bridge engineering: long, slender spans were possible in the increasingly strong steels then available, but a number of structural "accidents" had already demonstrated that there would be "stability" limits to the speeds and weights of railway trains using the new, sophisticated bridges. These limits were not of static strength, nor of static buckling, but of overall or dynamic stability.

1.6. Again, in the early part of this century, and particularly during the 1920's, stability problems became critical in a new and rapidly-developing field—aircraft structures. Aeronautics attracted new and sophisticated structural materials, which led to more widespread studies of static buckling, as for example, thin plates, reinforced sheets and cylindrical shells. At the same time, it led to the deeper study of the dynamic stability of the aircraft structure; early monoplanes had a fundamental problem in wing flutter, and the resolution of this problem was a major step forward in structural stability. The related problems of dynamic stability of long-span, flexible, suspension bridges came to a head in the 1940's.

1.7. Since the second world war, a new set of structural problems has emerged in the civil engineering field. In Britain, the Ferrybridge cooling tower problem was a case where overall stability, in the broadest sense, was not adequately considered; the Ronan Point disaster, when a modern "systems" building collapsed after one unit had been seriously damaged, is typical of another class of stability problems where overall stability is critically dependent on structural integrity; more recently, box-girder problems in bridges have become critical and have affected adversely overall structural stability.

1.8. Historically, therefore, we see an increasing importance of *overall* structural stability in structural engineering. Buckling—although an important aspect of structural stability—is not so generally important as overall structural stability itself. The increasing need for economy of materials will lead to geometric proportions which are more extreme. Other desirable economies will encourage increasingly large structures and to experiments with new structural forms, all of which are likely to lead to more, rather than fewer, stability problems. It is clear there is a need in structural engineering for concepts of "overall" structural stability; the purpose of this paper is to direct researchers to this need, and to encourage wider studies of overall stability of structural systems of direct interest to engineers.

1.9. In considering the importance of buckling and stability in structural engineering, a distinction has been implied between stability as an overall concept and buckling as a characteristic of deformation. The early history of stability research shows a strong interest, at that time, in buckling as an undesirable deformation. Much of the early study of buckling was dominated by considerations of "rigid" structural systems of elastic materials. Buckling was seen as a secondary problem once overall rigidity of a structure had already been given. This approach has dominated much of stability research up to the present time, in the study of both static and dynamic problems. Much recent work in dynamic buckling, although of undoubted academic value and interest, is difficult to apply in practice because of limitations in our knowledge of the actual dynamic loads experienced by structures. More readily applicable work in dynamic stability was that in the 1920's on aircraft flutter problems, and this later opened up an understanding of related problems in civil engineering structures. In a history extending over a period of 200 years or so, stability theory has tended to concentrate on the buckling of "rigid" structural systems; for this reason, central interest in stability theory has tended to move away from the problem of overall stability of structures.

2. Evolution of Design Attitudes towards Stability

2.1. Structural engineering designers have tended to approach stability problems in two extreme ways:

(1) The checking of stability mechanisms to ensure structures are "rigid". It is clearly important to ensure any structure is not a near-mechanism, and the avoidance of near-mechanisms has always been a feature of good structural design.

(2) Checking that buckling deformations are within acceptable bounds. This usually involves an assessment of critical loads, and ensuring such critical loads are not encountered in practice.

2.2. The initial concern in structural engineering was with static problems and with buckling as an undesirable change of shape. Design attitudes have tended to isolate areas of buckling of a structure and to analyse these in depth. Thus, even today, it is single-column curves and single-plate buckling curves which are in wide use in design offices, albeit the structures being designed and built are complex, integrated, whole structures. Design attitudes have tended to relate buckling loads to geometric parameters of single components. Thus, we have notions of acceptable slenderness-ratios of columns, thinness-ratios of plates, and so on.

2.3. An important feature of structural design philosophy is the important role in this philosophy of *elastic* buckling. At first sight this seems illogical, since the higher (and presumably, therefore, more efficient) stressed states should involve plastic or creep effects; but it should be remembered that design has been basically concerned with the avoidance of critical buckling states of systems of extreme geometric proportions. The designer has readily accepted the other constraints on structural behaviour—plasticity, creep, etc.—and has tended to combine these, in a largely empirical fashion, to devise design rules.

2.4. The main approaches to the treatment of stability problems in structural engineering, with some examples, can be summarised as follows:

Main approaches to the treatment of stability problems
in structural engineering design, with examples of these approaches

Wholly theoretical	Theoretical base modified by testing	Structure simulated in models	Environment simulated in models	Behaviour simulated in models
Assessment of critical states, under conservative loading	Columns	Soil foundations	Shells under wind-loading	Cylindrical shells
	Beams		Flexible, long-span bridges	Spherical shells
	Framed buildings			Reinforced shells
Estimates of flutter states	Plates		Flutter of plates and shells in high-speed flow	
	Reinforced plates			

Wholly theoretical approaches to design are rare, but are used for the assessment of some critical states of elastic systems, especially when conservative loading is involved. A large number of design approaches is based on a mixture of theoretical concepts and empirical facts; "column-curves" are in this class, involving interaction curves derived by testing, usually of models. In the case of the stability of a soil foundation, design is based essentially on the results of model tests of the soil material. In some of the most difficult stability problems, the environment is not fully understood, as for example the wind-loading of a cooling tower; in such cases the environment is simulated under model conditions. Again, there is an important class of practical problems in which the structure and environment are understood, but the resulting behaviour is complex, as for example, in many shell instability problems; in such cases, design is based on the behaviour observed in structural models. Some attitudes to design involve mixtures of these approaches.

2.5. Design attitudes have evolved in a largely *deterministic* way. A given geometric form of structure has been assumed to have consistent material properties, and a load can be defined as a buckling load for that type of structure.

2.6. The main deficiency in design attitudes has been an over-concentration on buckling deformations and too little attention to real assessments of overall structural stability. This is the basic reason for the lack of understanding of overall stability which permeates so many structural engineering problems of recent times. Dynamic problems, for example, pose difficult problems of overall stability, and design attitudes have found it difficult to grapple with these. Problems such as Ronan Point and Ferrybridge are essentially concerned with the overall stability of degenerate structures, and design attitudes have again been difficult to establish. These trends pose an important problem for stability researchers: how can research best help the engineer and designer in approaching problems of overall stability?

3. A Broad Concept of Stability in Structural Design

3.1. Strong emphasis has been made on the need to take a broader view of stability in structural design; let us examine how this broader view can be built up. The behaviour, B, of a structure, S, is a function of S and of the environment, E, within which the structure operates. Thus, $B = B(S, E)$.

S and E are not themselves continuous variables, and B is not therefore a simple continuous function of S and E. The dependence of B on S and E is illustrated in Fig. 1. At the design stage, the structure,

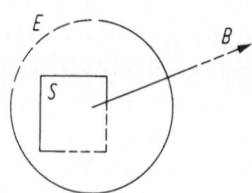

Fig. 1. The actual problem of stability, as it occurs in structural design practice: S, a structure (not built at the design stage) definable in a limited way in statistical terms; E, an operating environment, definable at the design stage in a limited way in statistical terms; B, the estimated behaviour of the structure in its operating environment.

S, is usually only partially defined; a full definition of structure would require a knowledge of the statistical variations of geometric forms and material properties; in practice, only average values of such statistical parameters are used in design. Again, the operating environment, E, is usually only partially defined in practical design; an average loading may be used, and it will frequently be assumed that such loading is likely to be the most critical for stability. At the design stage, the resulting behaviour, B, is estimated in practice either by direct extension of current theoretical and empirical knowledge or an extension based on some new theoretical concepts or model tests.

3.2. In discussing design philosophy, it is important to realise that S, E and B are *conceptual* at the design stage. They do not correspond necessarily to the actual structure built, nor to the actual environment encountered. Thus, the behaviour conceived at design is not necessarily realised in practice. This emphasises the need to define S, E and B in statistical, rather than deterministic, terms.

3.3. The actual environment experienced by a structure cannot be defined simply in a continuous way. For example, in the history of a structure, a static load may be followed by a dynamic load, and the change of environment is not definable in simple terms. By contrast, we have tended in stability research to describe all behaviours in terms of continuously varying parameters. Continuous relations between certain "environment" parameters and certain aspects of "behaviour" are used as a design guide to the likely behaviour of a structure in its operating environment. Some examples of this for structures generally are shown in Fig. 2. In the case of buckling of a compressed column, Fig. 2a, stability theorists assume a continuous variation of static load, (which in this case corresponds to the environment), leading to a continuously-varying buckling deformation, (in this case corresponding to the resulting behaviour). A similar situation holds for the buckling of a highly unstable structure, Fig. 2b. In the case of creep buckling of a column, Fig. 2c, theorists reduce environment to a steady column-load applied for a critical time, leading to a continuous variation of lateral displacement. The behaviour of a multi-storey portal frame, Fig. 2d, is described in terms of the load parameter and lateral deflection at an appropriate point. For a structure whose deformations must be limited, Fig. 2e, we assume again a continuous variation of loads. In the case of a structure subjected to wind-loads, Fig. 2f, we assume that, as the wind-speed is increased slowly, (which is unlikely to be the case), the amplitudes of steady oscillations increase continuously, (again an idealistic behaviour).

3.4. These traditional continuous descriptions of stability (or instability) are very limited. They are indications, at the most, of the likely stability or buckling of a structure in a few specific and academic states. This raises the question: what are we looking for more precisely in defining the stability of a structure? For design, structural stability is a feature we look for in structural systems whereby, under operating conditions, a structural system behaves, throughout its operating life, in a "controlled" way. During its operating life, the structure will be expected to show no deviations leading to uncontrolled behaviour. Instability is essentially the loss of controlled behaviour; as such, it takes many different forms: elastic buckling, plastic buckling, yielding of a component leading to uncontrolled behaviour, fracture leading to uncontrolled behaviour, growing oscillations under wind loading. Thus, this broad definition covers situations which are statical, dynamical, elastic, plastic, creep, and so on. It covers not only elastic buckling, which is a phenomenon which may, or may not, be relevant to structural stability.

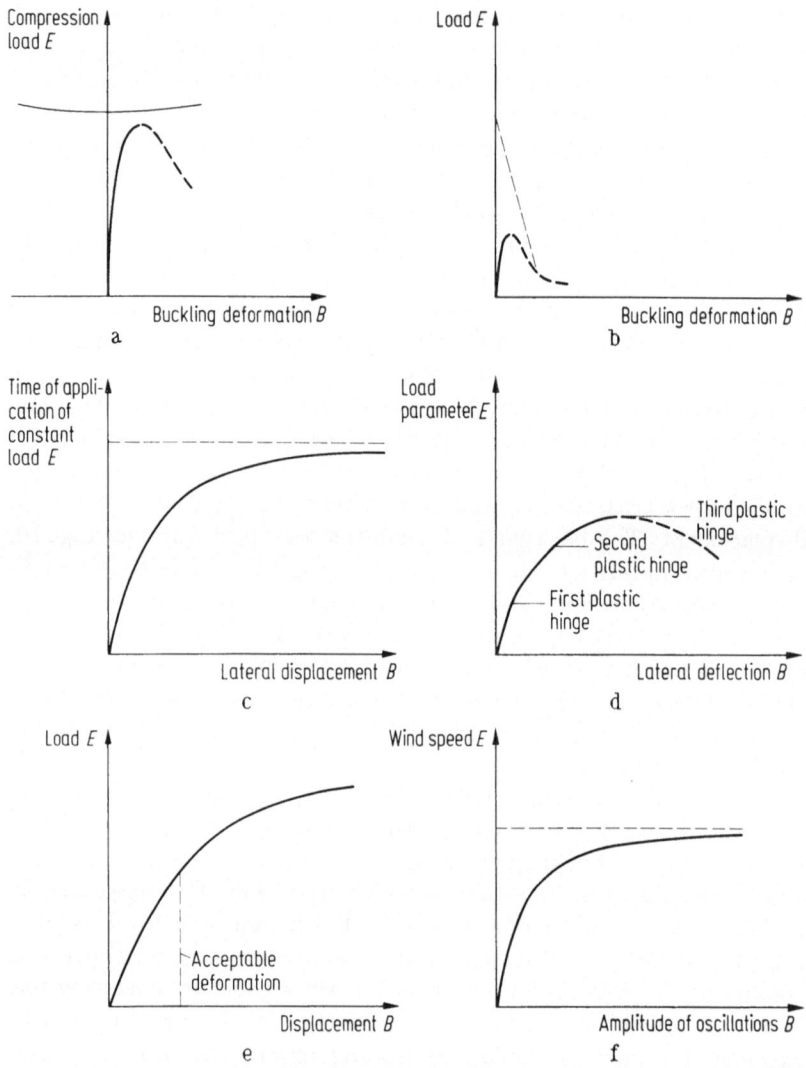

Fig. 2a—f. Simplified and continuous environment-behaviour relations in a range of stability problems.

a) Buckling of a compressed column; b) Buckling of a highly-unstable structure; c) Creep buckling of a column; d) Formation of plastic hinges in a framed-structure; e) Structure constrained by limiting deformations; f) Critical wind-speed for a tall structure.

4. Structure, Environment and Behaviour

4.1. This definition involves a complex system: (1) a structure, (2) an operating environment, (3) behaviour either "stable" or "unstable". Within this broad definition the designer may be interested in, for example, compressed columns, sheared plates, pressurised shells, cooling towers under wind loads, or oscillations of tall chimneys, in any of which structural instabilities may set in. Some of these instabilities will have "micro" sources: for example, a crack in a structure, propagation of which becomes unstable and catastrophic; or a local buckle which reduces overall load-carrying capacity. But these micro sources are structurally important at the macro level, at which overall structural instability may occur.

4.2. A structure, either at the conceptual design stage or when built, cannot be defined in every respect. Moreover, whereas a "built" structure is determined, at the design stage a structure can only be conjectured. At the design stage, there is only a certain probability of achieving a given structural form. Thus, statistical measures of shape and material properties must play a fundamental role in any comprehensive theory of stability.

4.3. The environment within which a structure will be expected to behave in a stable fashion is not a simple load axis, since not all the loads on such a continuous load axis are experienced by the structure. The style of definition of operating environment depends on the nature of the problem; for example, gust loads on an aircraft are usually defined in terms of up and down gust velocities, rather than by a meteorological description of the weather. The definition of environment is bound to involve statistical measures based on previous engineering experience; the operating environment is essentially a time history of the environment, and this will imply random loading. The operating environment consists strictly of an infinite number of "loading" vectors; the designer reduces these to a relatively simplified set of loads.

4.4. It has been suggested that in structural engineering practice, "controlled" structural behaviour is aimed at. We can enlarge on this by considering a number of different examples: in a tall building the designer looks for a long, continuous history of an essentially rigid structure, that is, he looks for a stable geometric form; a long-span, flexible bridge is designed essentially to have a dynamic behaviour within acceptable bounds, that is, controlled dynamic movements; an aircraft structure is designed to achieve a controlled flight path, usually in a repeated fashion; by contrast, a space vehicle may be expected to perform perhaps only one mission. A common feature of

340   A. H. Chilver

these examples is the expectation of a controlled behaviour within presented operating conditions; this will usually require repeated histories, and therefore repetition to an acceptable degree is part of expected behaviour. For engineering purposes the predicted behaviour must imply controlled (or stable) behaviour; "instability" is the occurrence of a phenomenon which leads to loss of control. The phenomena of buckling are all relevant to this, but form only part of the subject.

4.5. The essence of good stability design is the use in practice of structural forms relatively "insensitive" to mild deviations from "expected" environment. This is almost a generalisation of the "infinitesimally-small disturbance" concept of static stability. There is little use in structural engineering for highly-sensitive structures, that is, structures whose stability is critically sensitive to any single parameter, such as geometric imperfections or small dynamic loads. This may sound surprising, with so much work undertaken in recent years on highly unstable buckling problems, and it reflects on the usefulness of such work. Structurally-useful forms of "stability" may be summarised as in Fig. 3. Structures which are mildly-sensitive to buckling over a wide-range of imperfections are useful in structural design; structures in this class include columns and flat plates. Structures which show continously *increasing* sensitivity with imperfections over possible values of such imperfections are *not* structurally useful; very flat arches are in this class. Another class of structures shows an initial sensitivity to small imperfections, but this sensitivity diminishes with increasing imperfection; this is an important class of structures, and includes practical spherical and cylindrical shells; although these are

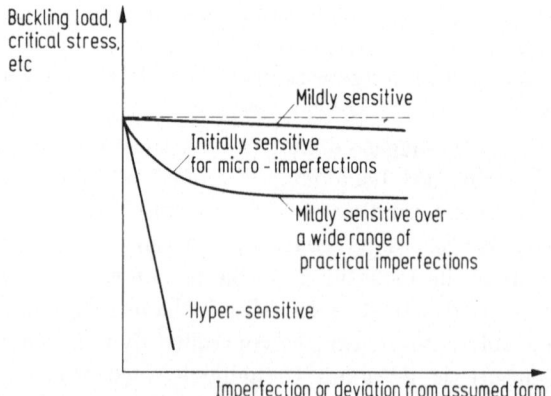

Fig. 3. Forms of sensitivity to deviations from assumed geometric forms, etc.

initially very sensitive to imperfections, for a wide range of practical imperfections a "plateau" of buckling loads exists.

4.6. The "plateau" concept for useful structural design is important. If a structure is only mildly sensitive at all imperfections, then the structure can be useful in engineering. If a structure, highly unstable for small imperfections, has a "plateau" over a wide range, then it is of equal structural use. This explains why many apparently highly-unstable forms are useful in practical structures; over the range of actual imperfections, a plateau of buckling loads exists, and can be relied on in design.

4.7. For similar reasons, there is a need to re-appraise the criticality of optimal structural forms. Optimal forms—for example, those exhibiting simultaneous buckling—can be unacceptably critical and less optimal forms may be useful in engineering practice.

5. Fundamental Problems the Field Poses for Basic Research in Stability

5.1. The pre-requisites of design for structural stability are: (1) an adequate knowledge of the structure, (2) an adequate knowledge of the operating envirnoment, and (3) an ability to analyse adequately the resulting behaviour. We consider each of these in turn.

5.2. An adequate knowledge of structure: this requires an adequate knowledge of material characteristics, (and their variation in time and throughout the structure), geometry (and especially micro-geometry) and integrity (by which is meant the ability of the structure to remain whole and not degenerate in use). We are always tempted to define these qualities precisely but, in fact, all these properties (even for a given structure) can be defined only approximately. This is particularly so of micro-imperfections of geometry, which are generally not accurately measurable. For a structure yet to be built, these properties can be given only with probabilities (rather than certainties) of occurrence. Any knowledge of structure, for design purposes, must therefore be probabilistic in form.

5.3. An adequate knowledge of environment: for an understanding of the stability of a structure we need to know not only the "median" environment but the micro or macro perturbations which disturb this median situation. Both the median environment and the perturbations need to be defined statistically. Very little information about stability is conveyed if a static buckling load, suitably factored, is used as a "safe load" in design. The test of stability is whether a structure can withstand its environment and still behave in a "controlled" way, and we must always test stability by this criterion.

5.4. An ability to treat adequately the interaction between structure and environment: within the framework of the definition outlined above of what constitutes structural stability, the analytical study of the interaction of a structure and its environment is crucial, and is essential if any quantitative assessment of stability is to be made. In this context the phenomena of buckling are important and the work which researchers are doing in this field can only do good in building up basic knowledge.

5.5. Two important fields in structural stability, which this discussion brings out, are the study of structure and environment. Although they are important, they are, in many ways, neglected fields of study. There are relatively few comprehensive researches into variations of material properties (except perhaps the effects of residual stresses). There are no systematic studies of the statistical laws governing geometric imperfections. There are few studies of the probability of occurrence of static loads, and there is little study of the statistical description of actual dynamic loads.

5.6. Most research in structural stability has been concerned with the analysis of the interaction of a load and a structure. First, the structure is approximated to in deterministic terms. Second, the environment is represented only very crudely, sometimes in stochastic terms, but generally in deterministic terms. In many cases the interaction of these two artificial systems is tractable. This tractability leads to a whole field of theoretical study, and in this the main interest in the fundamental problem of structural stability is frequently lost.

5.7. Clearly, the study of structure in isolation or, for that matter, the study of environment in isolation, does not lead to any advance necessarily in structural stability. Again, it is clearly vital to study the interaction of the two systems. But we tend to idealise the two systems so extensively as to make very few of our theoretical studies of much practical value in design. Although we have made considerable progress in the last 25 years in the basic understanding of the mechanics of a number of stability problems, the impact on design methods and philosophy has been small. In practice, stability is still very much in the "column-curve" stage of the turn of the century.

5.8. The most important problems of structural stability at the present time are those involving more accurate descriptions of structure and environment and at the same time the development of more powerful analytical tools to study their interaction. We are limited at present in making real assessments of stability because we have essentially two statistically-defined systems interacting and our analytical techniques for dealing with this situation are themselves limited.

6. The Role of Model-Testing of Structures

6.1. The sheer complexity of many structural stability problems means that wholly-theoretical solutions will not always be found. A systematic and soundly-based approach to the model-testing of structures would be very helpful to the whole field of structural engineering.

6.2. As an example, let us consider the general problem of the "buckling" load of an elastic structure. For an homogeneous, isotropic, elastic material, the elastic properties can be defined by two constants, Young's modulus, E, and Poisson's ratio, ν. If we envisage that geometric scaling is preserved, the actual scale is defined by a single geometric parameter, say L. Then the critical buckling load, P, may be written

$$P = F(E, \nu, L),$$

where F is a function of E, ν and L. Dimensional analysis, applied to this functional relationship, gives

$$P/EL^2 = f(\nu).$$

If a critical stress σ, is introduced, then $\sigma \alpha P/L^2$, and

$$\sigma/E = g(\nu),$$

where g is a function of ν. Thus, we have the interesting result that the critical stress parameter (σ/E) of geometrically-similar homogeneous elastic systems is a function of only Poisson's ratio, ν. In practice, the dependence on ν is usually weak, and so model-tests using most elastic materials will give useful results over a wide range of materials.

6.3. Buckling is usually present in structural systems when one geometric dimension is of a smaller order than another dimension; for example, slenderness ratios of columns and thinness ratios of plates. In such cases, two geometric parameters L and l, where $L \ll l$, are needed to define the geometry. Then

$$P = F(E, \nu, L, l).$$

Dimensional analysis then gives

$$P/El^2 = f(\nu, l/L),$$

or

$$\sigma/E = g(\nu, l/L),$$

since σ is usually associated with the smaller linear dimension, l. If, as is usually the case, the dependence on ν is weak, then

$$\sigma/E = g(l/L).$$

Some common examples of this relationship are: columns, for which $\sigma/E \alpha (l/L)^2$; flat plates, for which $\sigma/E \alpha (l/L)^2$; and compressed cylinders, for which $\sigma/E \alpha l/L$. Indeed, many structures fall into the class

$$\sigma/E \; \alpha (l/L)^n,$$

where n takes simple values, such as 1 and 2. It is interesting to speculate that in many cases these relationships could have been determined by simple testing of models.

6.4. Model-testing methods have been most widely used in situations where environmental conditions can be simulated but where no analytical tools exist at present. This is particularly so in cases where loading is derived from fluid flow, which may generate loads of turbulent flow. In such cases we can make progress in structural design if we use more effectively the methods of dimensional analysis.

6.5. Another important practical consideration in theoretical analysis is that most practical structures have a number of nearly-simultaneous critical buckling loads. Recent research studies have shown that, in such cases, the buckling path, (that is, the buckling-load/deformation relation) is not well-defined and is highly sensitive to imperfections. Coincidence of buckling loads appears to generate buckling paths which are coupled forms of the basic paths. The analytical study of such problems is attractive and a challenge to researchers, but it is unlikely to be fruitful in design. This complication suggests, again, that our best resort is to the study of the behaviour of realistic models. Such models should show, first, that the structural form is not a hypersensitive one, and, second illuminate the variation of the buckling loads (stresses) with the relevant geometric and other parameters.

7. Conclusions

7.1. The structural designer is becoming increasingly concerned with problems of *overall* structural stability. Although problems of buckling are an important part of structural stability, the primary need in structural engineering is for a philosophy of the overall stability of structures.

7.2. In developing this philosophy, there is need for deeper understanding of the physical make-up of structures themselves and of the environments within which they operate. In any comprehensive philosophy of stability, structure and environment will be expressed in statistical terms.

7.3. The designer is concerned most, of course, with the behaviour of the structure within its environment. In some of the most difficult problems, the behaviour cannot be deduced analytically. For this reason, model testing will always play a vital role in the development of any stability design philosophy.

7.4. The structural designer is most concerned with achieving a "controlled" behaviour of a structure. This is what is meant broadly by the stability of structures. In this sense, the field of structural stability is fundamental and all-pervading in structural design.

Design of a Mars Entry "Aeroshell"

R. W. Leonard, M. S. Anderson, W. L. Heard, Jr.

NASA Langley Research Center, Hampton, Virginia, U.S.A.

Summary

The external shell of the Project Viking capsule, which will provide atmospheric deceleration for the 1976 landing of an unmanned spacecraft on the surface of Mars, is a wide-angle, stiffened cone subject to buckling under entry aerodynamic pressure. Complex, highly optimized structural prototype and flight designs were evolved through the application of relatively advanced buckling analysis methodology. Both designs were evaluated through tests and analysis with improved shell-of-revolution computer programs. Deviations between the analyses and experiments were resolved only by modeling the thin-walled rings as shell branches. The results illustrate the great complexity of shell behavior and the designer's need for reliable analysis tools capable of representing detailed structural behavior with greater accuracy than is current practice.

1. Introduction

The use of advanced buckling analysis technology in the design of real structures is still surprisingly limited. Even in the aerospace industry, which has generally led other product industries in employment of advanced technology, plate- and shell-like structures are still frequently designed on the basis of classical buckling formulas for idealized geometries (see, for example, [1]), with heavy reliance on development tests to account for deviations of the actual structure from the assumed ideal model, for probable imperfections of the fabricated structure, and for defects in classical buckling analysis methodology. Not only is this process expensive but the resulting designs may be far from optimum in their proportions and mass.

In this paper, we will describe an unusually demanding, real-life shell design problem where relatively advanced buckling analysis and test technology was applied and which required unusually precise analysis techniques for resolution. We will try to assess the lessons implied by this application as to the design engineer's need for buckling analysis tools and for understanding of shell buckling behavior to produce successfully and efficiently such designs.

2. Symbols

b	Width of ring web	t	Skin thickness
C	Correlation factor for general instability	$\bar{\eta}$	Ratio of weighted average of stiffened-shell-wall bending stiffness to skin bending stiffness
E	Young's modulus of structural material		
h	Ring web thickness	δ	Initial deviations of shell surface from conical
L	Meridional length of truncated cone between support or payload ring and base ring	ϱ	Density of structural material
		σ_s	Meridional stress
n	Circumferential wave number	σ_θ	Circumferential stress
p	Buckling pressure	Br, Rg	Subscripts designating branched shell and ring theory calculations, respectively
$\bar{R}$	Average circumferential radius of curvature of truncated cone		

3. Aeroshell Design Requirements

For man-made planetary entry bodies, the name "aeroshell" has been given to the forward-facing external shell which must withstand the aerodynamic pressures and temperatures of atmospheric deceleration. The general configuration and structural design requirements of aeroshells are dependent on the characteristics of the entry mission and the properties of the planetary atmosphere.

3.1. The Mars Entry Mission

In 1975, the United States will launch two identical "Project Viking" spacecraft toward the planet Mars with the objective of landing two

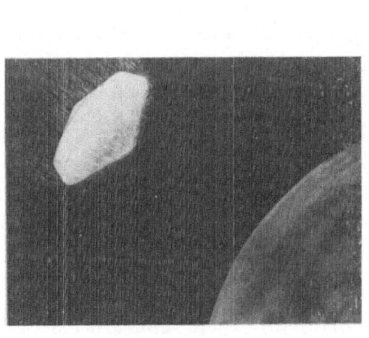

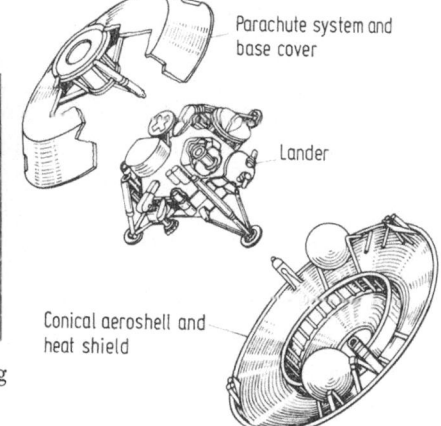

Parachute system and base cover

Lander

Conical aeroshell and heat shield

Fig. 1. Viking lander capsule entering Martian atmosphere.

automated scientific laboratories on the Mars surface [2]. Each space-craft will consist of two parts, an Orbiter and a Lander capsule. On reaching the vicinity of Mars in mid 1976, the two parts will separate, the Orbiter entering an elliptical orbit about the planet to maintain a communication link with the Earth, and the Lander capsule entering the Martian atmosphere as depicted in the artist's drawing of Fig. 1. As also shown in Fig. 1, each Lander capsule consists of three main assemblies: the aeroshell protected by a thin ablative heat shield to provide the first stage of atmospheric deceleration, a base cover containing a parachute for a second stage of deceleration in the lower atmosphere, and the lander with retrorockets and shock-absorbing legs for final deceleration and landing.

The configuration of the Viking aeroshell is partly dictated by the functional requirements of this system design. The deceleration force of the large lander and base cover/parachute system mass is transmitted to the aeroshell through a deep "payload ring" (Fig. 1) about half way between the aeroshell apex and outer edge. No restraint is applied by the base cover to the outer edge of the aeroshell which is therefore a free edge with a stiffening "base ring".

The remaining major configuration characteristics are due to the low density and pressure of the Martian atmosphere which at the surface are about 1 percent of Earth sea-level conditions or about the equivalent of conditions at 30 km altitude (see, for example [3]). The aeroshell diameter, 3.5 m, is therefore the largest consistent with the launch vehicle dimensions and the shape, a wide-angle (140°) cone with a shallow spherical nose, was chosen for its high drag coefficient in a low density stream (see [4]). Configuration details will be given later.

3.2. Structural Design Criteria

The center of gravity of the Viking Lander capsule has been deliberately positioned off the center line so the capsule will enter the Mars atmosphere at a slight angle of attack and generate lift to achieve the desired entry performance. The pressure loading on the aeroshell is, thus, slightly asymmetric; the expected distribution of pressure is shown schematically in Fig. 2 on a cross section of the aeroshell. The maximum expected (design limit) pressure at the stagnation point is 12.5 kPa. This pressure is multiplied by the factor of safety 1.25 to obtain the design ultimate pressure, 15.6 kPa, which the aeroshell design is required to withstand without buckling or other failures.

Also illustrated in Fig. 2 is the concentration of the reaction loading from the decelerating lander and base cover masses through the payload ring and the absence of reaction loading at the base ring. This unique disposition of loads affects the requirement of design for buckling strength in two ways. First of all, the resulting prebuckling membrane stresses in the shell are biaxial compression in the region of the aeroshell forward of the payload ring but a combination of meridional tension with circumferential compression aft of the payload ring (Fig. 2). This meridional variation of prebuckling stresses needs to be taken into

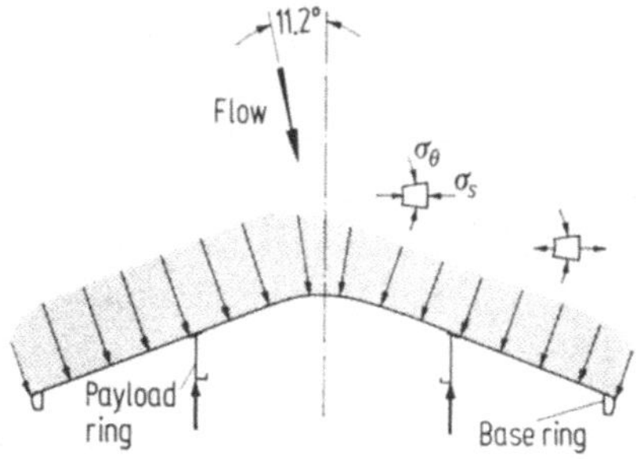

Fig. 2.
Distribution of pressure
and reaction loads on the
Viking aeroshell.

account in both aeroshell designing and subsequent evaluation to achieve a reliable, minimum mass structure. Second, for low values of base ring stiffness, the lack of support at the aeroshell outer edge leads to failure in an inextensional bending ("ring") mode, with two waves around the circumference [5]. Minimum total shell weight [6] corresponds to the minimum base ring weight required to enforce a node at the base ring-shell interface and force buckling in a general instability "shell" mode.

Finally, structural designs of Viking aeroshells have been required to take account of such practical constraints as ease and reliability of fabrication, use of available sheet gages, and so forth. In particular, because the aeroshell is relatively lightly loaded by the low-density Martian atmosphere, practical ring-stiffened designs can be achieved with no more mass than sandwich aeroshell designs which are dominated by minimum available sheet gages [7]; for this reason, the designs described in this paper are ring-stiffened shells.

4. Aeroshell Designs

Two separate Viking aeroshell designs have been produced. The first was a structural prototype which was designed, analyzed, fabricated, and tested at the Langley Research Center to establish the reliability of a somewhat advanced design process. The second was the flight aeroshell produced by the Project Viking prime contractor, which was designed by the same general process but with different fabrication constraints.

4.1. The Design Method

The design rationale was based on the premise that all possible buckling modes be approached except where prevented by practical constraints such as minimum gage. Buckling modes considered were general
buckling, local buckling of skin and rings, and inextensional modes of
buckling at low harmonic numbers (usually $n = 2$) where the largest
amplitude is at the base ring.

Synthesis computer programs usually require many analysis iterations to achieve a minimum weight design. To keep computation times
reasonable, these procedures must be quite simple. In [7] and [8], a
procedure was developed for near optimum design of ring-stiffened
cones that utilized suitably simple formulas with a few accurate analyses used to correct the results. Details of the procedure are given in
the references and only the main steps are outlined herein. For general
buckling, a generalization of the formula developed by Baruch and
Singer [9] was used:

$$p = \frac{0.92CEt^{2.5}}{L\overline{R}^{1.5}} \left[(1 + \bar{\eta})^{0.75} - \frac{(\overline{R}t)^{0.5}}{L} \bar{\eta} \right],$$

where $\bar{\eta}$ ([7] or [8]) is a weighted average of the stiffened shell-wall-
bending stiffness divided by skin bending stiffness and C is a correlation factor that is adjusted to bring the equation into agreement with
a more accurate shell analysis. (The suitability of such an approach is
indicated by the fact that C varied from unity no more than ± 10 percent for a very wide parameter range and the variation is usually much
less for one given design condition.)

For local buckling, the stress resultants in the shell and rings were
determined from membrane theory and hoop strain compatibility
between rings and skin. The buckling condition in the skin was determined using the classical formula for an infinitely long rectangular plate
subject to biaxial stress; for the ring, appropriate buckling formulas
that have been published for the ring cross sections were used. In [8],
practical manufacturing constraints were imposed. These included
restricting the number of ring sizes to a specified number and selecting
ring and skin thicknesses from a table of standard available gages. In
addition, the portion of the shell forward of the payload ring was
included in the design process. Though general buckling was not a
problem in this area, adequate rings to prevent local buckling were
required. The procedure of [8] was used for design of both the prototype
and flight aeroshells.

The procedure followed the outline in Fig. 3. The design program,
with a trial value of C, generated all required dimensions which were

then input into a comprehensive shell of revolution stress and buckling analysis program. The SALORS system ([10] and [11]) was used for this purpose. In the SALORS analysis, each ring was modeled discretely with a ring theory embodying a complete set of bending, extensional, and torsional stiffnesses, but not permitting ring cross-section distortions; nonlinear axisymmetric prestress was calculated and bifurcation buckling from this stress state determined. The SALORS cal-

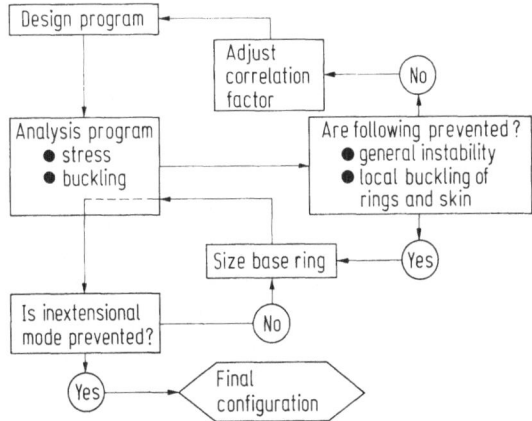

Fig. 3. Flow chart for design process.

caluated stresses were compared with the allowable local buckling stresses determined in the design program and the overall buckling strength was compared with the required value. By adjusting C and the pressure level in the design program, a new design was generated which satisfied all buckling requirements based on the previous SALORS analysis. Two or three iterations were usually sufficient to achieve a converged solution. The design at this point was based on obtaining the equivalent of simple support at the payload ring and base ring; these rings were then designed [8] to prevent the inextensional modes from occurring.

4.2. Prototype and Flight Designs

Details of the designs resulting from this process are shown in Fig. 4a for the prototype and Fig. 4b for the flight article. A photo of the flight article in Fig. 5 indicates the overall arrangement which was typical of both aeroshells. Slight differences, such as those in the payload and base ring configurations, are due to the fluid state of the design in its early stages (with the prototype approximately 1 year ahead of the flight configuration) and to differing manufacturing constraints by the two different fabrication organizations. The small rings for the prototype were fabricated by rolling which led to the selection of a channel section having a cylindrical web and outward-facing flanges maintained in tension, during rolling to avoid buckling. A slightly more efficient Z-section, fabricated by stretch forming, was used for the flight article.

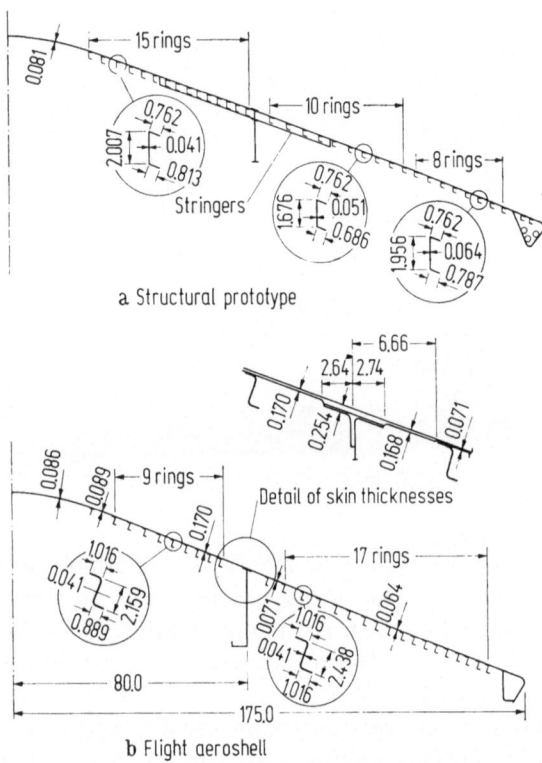

Fig. 4a and b. Aeroshell final design configurations (dimensions in cm).
a) Structural prototype; b) Flight aeroshell.

The most notable difference in the two aeroshells was the method used to accommodate the very high stresses in the vicinity of the payload reaction. Longitudinal stringers were used for the prototype while manufacturing considerations led to the choice of a series of stepped lands for the flight article. No attempt has been made to define the complex details of the stringers in Fig. 4a; however, a photograph of this region is shown in Fig. 6.

Fig. 5. Construction details of flight aeroshell.

Fig. 6. Construction details of prototype aeroshell.

5. Aeroshell Evaluation Methods

The evaluation of the aeroshell was based on rather sophisticated analysis and on testing full-scale specimens of both the prototype and flight designs. Testing was considered to be especially important because of the unprecedented low mass design that was optimized to a high degree using skin gages approaching the minimum that could be fabricated uccessfully. In addition, characteristics of the base ring necessary to prevent inextensional buckling had very little prior experimental confirmation.

5.1. Analysis Computer Programs

The design was based on shell-of-revolution analysis (SALORS) that was at the forefront of the state of the art at the time. General shapes with discontinuities, variable material properties and cross sections, discrete rings, and arbitrary loads, could be considered. Thus, all the features exhibited by the designs in Fig. 4 can be accurately modeled as long as there is no geometric or material property variation in the circumferential direction. Subsequently, further advances were made in shell-of-revolution programs and some programs (for example, [12—14]) now have the capability of branched shell analysis (three or more shell-of-revolution segments intersecting at one meridional location). The SRA system [13] was used to study the effect of ring deformations on stress and buckling of the Viking aeroshells by modeling the payload and base rings and the webs of the remaining rings as branched shells.

The SRA system of programs is actually six separate programs as follows with some programs providing the input for others:

SRA 100—linear asymmetric stress analysis,
SRA 101—bifurcation buckling from a linear asymmetric prestress state (input from SRA 100),
SRA 200—nonlinear axisymmtric stress analysis,
SRA 201—bifurcation buckling from a nonlinear axisymmtric prestress state (input from SRA 200),
SRA 202—imperfection sensitivity (input from SRA 200 and SRA 201),
SRA 300—vibration including effects of axisymmetric prestress state (input from SRA 100 or SRA 200).

The shell theory on which the stress, buckling, and vibration programs are based is that due to Novozhilov [15]. The imperfection sensitivity program is based on the theory of Koiter [16] as extended by Budiansky [17] and represents a unique capability for general shells of revolution.

All of the programs beyond the linear stress analysis involve an iteration procedure, each step of which is a linear analysis. Thus the basic solution technique for all programs is similar. A forward integration by the Runge-Kutta method is used to generate a set of solutions which, though incompletely specified, satisfy boundary conditions at one edge of the shell. Boundary conditions at the other edge completely determine the solution. The shell is divided into sufficiently small intervals to prevent numerical problems that occur with the growth of exponential terms in the solution of the differential equation.

5.2. Test Technique

Both aeroshells were tested under uniform pressure using fixtures depicted schematically in Fig. 7. The aeroshell is supported on the payload ring; the base ring is free except for a membrane pressure seal arranged to apply only a negligible force to the ring. The volume in the sealed cavity is partially evacuated to produce the external pressure loading. Several hundred strain gages were used on each test, and deflections were measured along meridional and circumferential lines. More complete details of the test technique are given in [8].

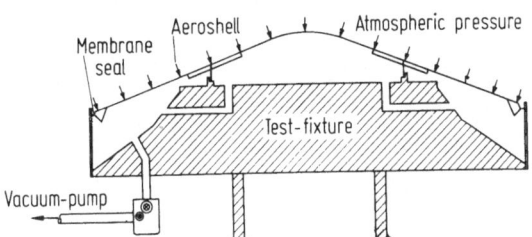

Fig. 7. Schematic of test setup.

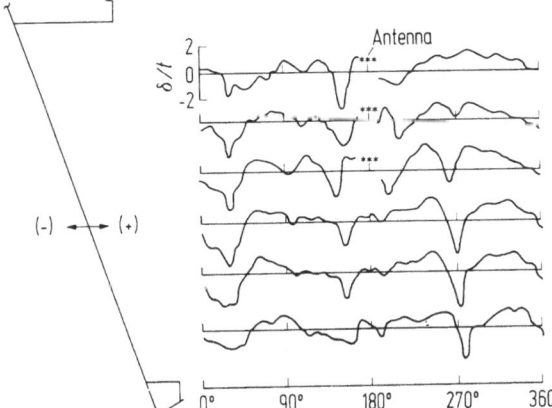

Fig. 8. Initial imperfections in test model of flight aeroshell.

Prior to test, contours of the aeroshell were measured to assess the amount of initial imperfections present. Results for the flight model tested are shown in Fig. 8. The circumferential variations at several meridional stations, corresponding to the sketch on the left, are shown. This design was fabricated from three circumferential segments and tends to exhibit a strong $n = 6$ component in the initial imperfection shape.

6. Results and Discussion

Extensive analyses with the computer programs of [13] have been carried out for both the prototype and flight designs and compared with the test measurements. Selected results are presented and discussed in the following paragraphs.

6.1. Prebuckling Stresses

Calculated and measured outer surface strains in the prototype aero-shell are presented in Fig. 9. The upper and lower plots are the variations along the shell meridian of circumferential and meridional strain, respectively, for a uniform pressure load of 13.8 kPa. While this non-linear calculation was made with the SALORS program with the channel rings modeled as discretely located rings, the results are essentially identical to calculated outer surface strains obtained with the program SRA 200 with the rings modeled as branched shells. Except in the vicinity of the stringers, large discontinuities in predicted

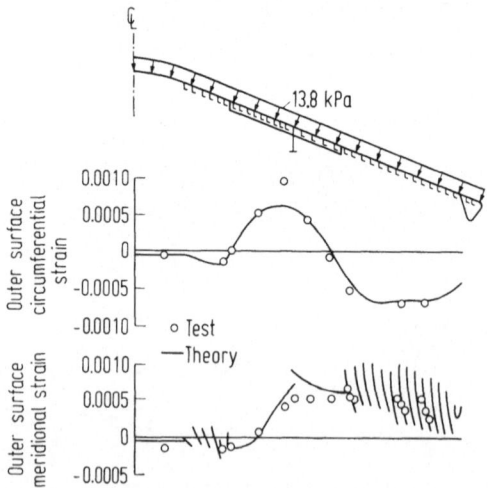

Fig. 9. Circumferential and meridional skin strains at outer surface of structural prototype aeroshell.

skin meridional strain are evident at the ring locations. In the three bays that were instrumented with three meridional strain gages each, the agreement of the measured with the calculated strains is seen to be remarkably good. In fact, agreement is everywhere good except in the region where local circumferential variations due to the stringers cannot be modeled by a shell-of-revolution analysis (which requires that the stringer properties be averaged and uniformly distributed around the circumference).

Obviously, precision in the prediction of such complex stress fields requires great care and facility in the analytical modeling of structural details. Yet, great precision is clearly mandatory in highly optimized designs where local skin and stiffener web or flange failure modes are approached. It will be illustrated in the following section that the accurate prediction of general instability may sometimes also require great precision in structural modeling.

6.2. Buckling Strengths

Effect of Load Asymmetry. To verify and determine the extent of the conservatism implied by the assumption of uniform pressure loading in the design and subsequent evaluation of the Viking aeroshells, the influence of entry pressure asymmetry on buckling strength was evaluated analytically for the flight aeroshell design. The analyses were made with computer programs SRA 100 and 101 and with all the ring webs modeled as shell branches. In Fig. 10, the assumed uniform pressure load is compared with the expected asymmetric distribution of pressure. The resulting buckling mode shapes and relative values of the buckling loads are also compared. For asymmetric pressure, the maximum pressure at buckling exceeds the uniform buckling pressure by only 8 percent; the circumferential buckling shape is dominated by the $n = 6$ component which is the critical harmonic for uniform pressure. On the high-pressure side of the shell, the two modes are very similar.

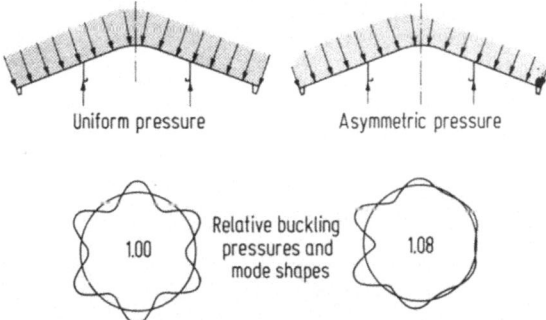

Fig. 10. Buckling results for uniform and asymmetric pressure distributions.

The uniform-pressure approximation is, therefore, regarded as both realistic and safe.

Strength of Viking Aeroshell Designs. Calculations of the uniform-pressure buckling strength of the prototype aeroshell are compared with the test results in Fig. 11 where buckling pressure is plotted as a function of circumferential wave number of the buckling mode. The dashed and solid curves for perfect shells were obtained with the programs SRA 200 and 201 from analytical models and processes that are identical except for modeling of the small rings. The dashed curve results when each ring is modeled entirely with ring theory while the solid curve is obtained when the ring webs are modeled as shell branches for increased accuracy. The branched shell analysis agrees with the test results in both load and critical wave number ($n = 7$); the buckled test shell is shown in Fig. 2. For comparison, additional branched shell

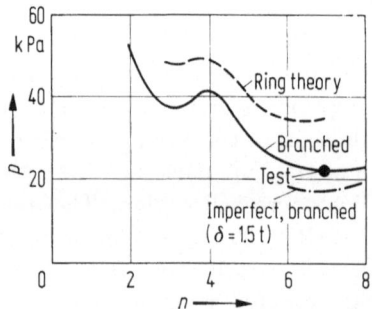

Fig. 11.
Buckling pressure as function of circumferential wave number for structural prototype aeroshell.

Fig. 12. Buckled prototype aeroshell test model.

calculations were made at wave numbers 6, 7, and 8 with the use of program SRA 202 and assumed initial imperfections having the shape of the buckle mode and a maximum amplitude of one and one-half times the skin thickness; this amplitude approximates the largest measured on the test shell. Note in Fig. 11 that the influence of the assumed imperfection is enough to reduce the calculated buckling strength below the test value but is considerably less than the influence of accurately modeling the ring webs as shell branches which changed not only the load but the critical wave number from 6 to 7.

Figure 13 is a corresponding plot for the flight aeroshell design. As might be anticipated from the greater width-to-thickness ratios of the ring webs for this design (Fig. 4), the importance of modeling the webs as shell branches is overriding for the flight design and the assumed imperfection, which again had the shape of the buckle mode with amplitude approximately the maximum measured (Fig. 8), has little effect. The test result agrees well with both branched shell calculations.

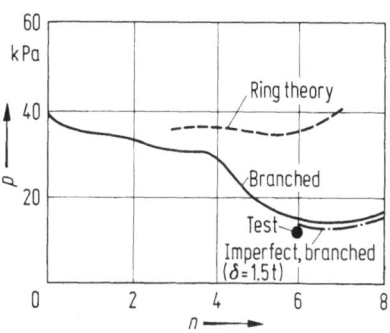

Fig. 13.
Buckling pressure as function of circumferential
wave number for flight
aeroshell.

Figure 14 shows the meridional shapes of the critical buckling modes, including the substantial deformations of the rings as determined by the branched shell analyses with program SRA 201. These plots show defor-mations amplified to achieve an arbitrary unit maximum deflection of the skin and hence are only approximately comparable; however, the greater deformations of the flight aeroshell rings are quite evident[1].

These results show the importance of ring deformations during buckling on the strength of conical aeroshells. Similar ring and stringer deformation effects were noted for other lightly stiffened structures in [18].

Effect of Payload Ring Restraint. The importance of ring deformations can be further illustrated with another comparison. The outstanding

---

[1] To achieve the target strength of the flight aeroshell design, these ring deformations were restrained with small metal clips in subsequent test and flight articles.

flange of the large payload ring is elastically restrained by the lander payload itself. The analytical results of Fig. 13 were obtained assuming the outstanding flange fixed, and the test was conducted with the actual constraint provided by the lander structure. To assess the importance

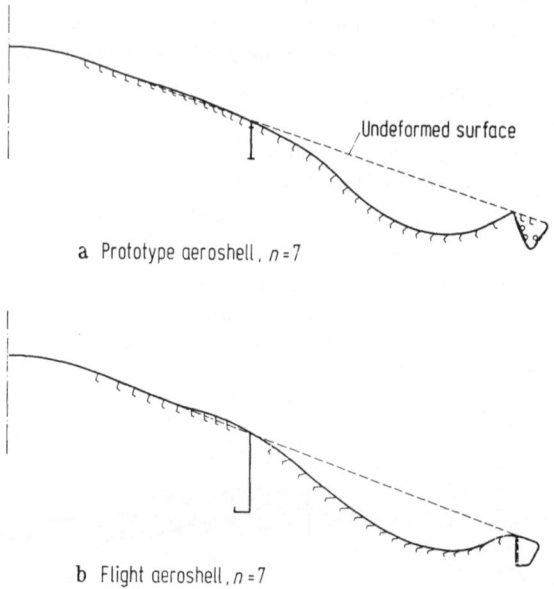

a Prototype aeroshell, $n = 7$

b Flight aeroshell, $n = 7$

Fig. 14a and b. Meridional buckling mode shapes (branched shell calculations). a) Prototype aeroshell ($n = 7$); b) Flight aeroshell ($n = 7$).

of this restraint, a branched shell calculation was made with the outstanding flange of the payload ring free. The result is compared with the fixed-flange calculation and test in Fig. 15. As shown by the dashed curve, the buckling strength is not only reduced but the inextensional ($n = 2$) mode becomes critical. The meridional shapes of the flange-fixed and flange-free $n = 2$ modes are also shown in Fig. 15 for comparison. Clearly, design of shells with such rings modeled by ring theory (as even this shell was) could lead to serious errors since ring theory does not admit deformations that adequately distinguish between flange-free and flange-fixed designs.

6.3. Effect of Ring Flexibility on Optimum Design

The preceding results show substantial differences in calculated buckling strengths, depending on the theory used, and indicate that for reliable design one must account for local ring deformations. The question arises as to what effect ring deformations would have on the mass of an optimally designed conical shell. Figure 16 gives a mass ver-

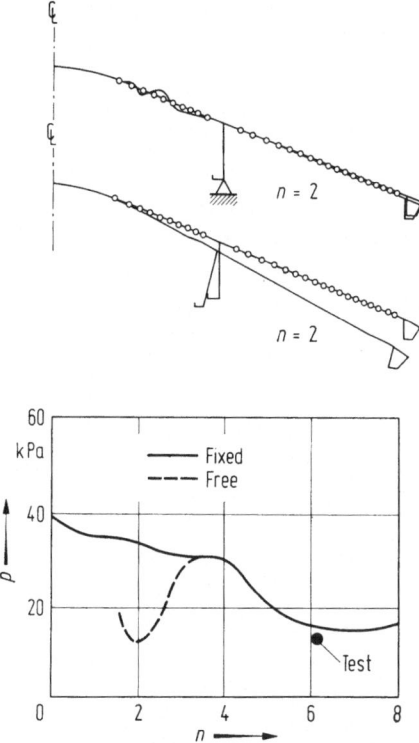

Fig. 15. Influence of payload ring restraint on buckling strength of flight aeroshell (calculated with branched shell theory).

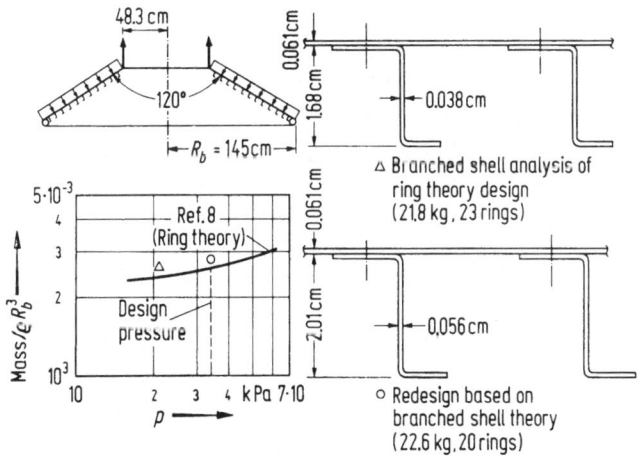

Fig. 16. Comparison of ring theory and branched shell optimum designs.

sus strength curve for 120° ring-stiffened truncated cones subject to
uniform external pressure and supported at their inner edge. The mass
parameter includes the mass of the skin and small rings, but no payload
ring or base ring mass. This curve was first presented in [8] and is based
on modeling with ring theory. A typical design at a pressure of 34 kPa
was analyzed with branched shell theory and the result is shown by the
triangular symbol, a reduction in strength of 40 percent from the ring
theory calculation. A new design was determined by the method
described earlier except branched shell theory was used to establish
the correlation factor. This design, shown by the circle symbol is only
4 percent heavier than the original design, indicating that ring deformations may impose only a small penalty if properly taken into account
during design. Sketches of the original design and the revised design
are shown on the figure and at first glance they appear very similar.
Note the revised design has thicker gage rings; the effect on weight,
however, is partially offset by the fact there are only 20 rings as compared to the original design which has 23 rings.

At this time, only a limited number of calculations have been made
for buckling of conical ring-stiffened shells using branched shell theory.
Though not enough parameters have been systematically varied to
develop a complete picture, it is natural to assume that the ratio of
buckling pressure based on branched shell theory to that calculated
from ring theory is a function of the width/thickness ratio of the web
of the ring. A plot of these quantities is shown on Fig. 17 showing all
the points calculated to date. Although the data represent two radii,
two cone angles, and two different ring cross sections, they fall in a
fairly narrow band. These preliminary results suggest that rings need

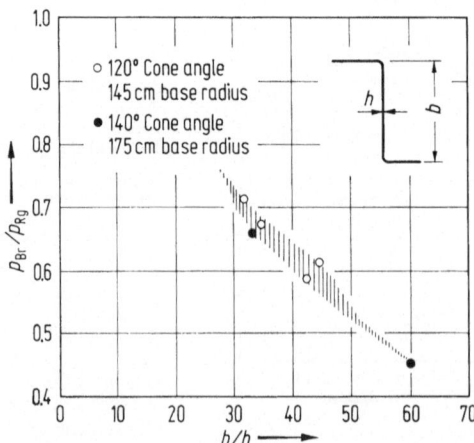

Fig. 17. Effect of ring proportions on buckling strength of stiffened cones.

to have rather low depth-to-thickness ratios if ring theory is to be applicable for wide angle cones loaded by external pressure. Development of general curves such as this would be of great benefit to designers since it is still quite expensive and difficult to make an analysis where all the rings are modeled as branched shells.

7. Concluding Remarks

Complex, highly optimized structural prototype and flight designs of the Project Viking Mars entry aeroshell were evolved through the application of relatively advanced buckling analysis methodology. These designs have been subsequently evaluated through a variety of more comprehensive analyses with improved shell-of-revolution computer programs capable of modeling thin-walled rings as shell branches and of accounting for the influence of initial imperfections through a variation of the method of Koiter. The analytical results have been compared with the results of careful tests of a full-scale specimen of each design.

The results indicate that necessary precision in stress calculations for near optimum designs requires not only very comprehensive analysis tools but careful modeling of structural details as well. Moreover, general buckling strength may be adversely affected by ring cross-section deformations that can only be represented adequately in analysis by modeling both large and small rings as shell branches. In the same way, this effect can be accounted for in design, probably at little cost in added mass.

Thus, while imperfections remain an important factor to be taken into account in design and analysis of shell structures (especially unstiffened shells), correct analytical modeling of structural details may be more important for lightly loaded, stiffened shell structures. Designers therefore need analysis tools capable of representing detailed structural behavior with much greater accuracy than in the best current practice. Since the few current computer programs adequate to this task are new and very complex, they require much sophisticated knowledge of computer analysis and shell behavior. The challenge, therefore, is to deliver this needed powerful analysis capability in a form which is simple and reliable for designer's confident use.

References

1. Anonymous: NASA Space Vehicle Design Criteria (Structures). Buckling of Thin-Walled Circular Cylinders. NASA SP-8007, revised August 1968. Buckling of Thin-Walled Truncated Cones. NASA SP-8019, September 1968. Buckling of Thin-Walled Doubly Curved Shell. NASA SP-8032, August 1969. Buckling Strength of Structural Plates. NASA SP-8068, June 1971.

2. Martin, J. S., Jr.; Soffen, G. A.: The Viking Mission. Science and Technology **31** (1973). The American Astronautical Society, Tarzana, California.
3. Anonymous: NASA Space Vehicle Design Criteria (Environment). Models of Mars Atmosphere (1967). NASA SP-8010, May 1968.
4. Cohen, G. A.: Evaluation of Configuration Changes on Optimum Structural Designs for a Mars Entry Capsule. NASA CR-1414, August 1969.
5. Cohen, G. A.: The Effect of Edge Constraint on the Buckling of Sandwich and Ring-Stiffened 120 Degree Conical Shells Subjected to External Pressure. NASA CR-795, 1967.
6. Dixon, S. C.; Carine, J. B.: Preliminary Design Procedure for End Rings of Isotropic Conical Shells Loaded by External Pressure. NASA TN D-5980, 1970.
7. Cohen, G. A.: Structural Optimization of Sandwich and Ring-Stiffened 120 Degree Conical Shells Subject to External Pressure. NASA CR-1424, 1969.
8. Heard, W. L., Jr.; Anderson, M. S.; Anderson, J. K.; Card, M. F.: Design Analysis and Tests of a Structural Prototype Viking Aeroshell. AIAA J. of Spacecraft and Rockets **10**, No. 1, 56—65 (1973).
9. Baruch, M.; Singer, J.: General Instability of Stiffened Circular Conical Shells Under Hydrostatic Pressure. Aeron. Quart. **16**, 187—204 (1965).
10. Heard, W. L., Jr.; Anderson, M. S.; Chen, M. M.: Computer Program for Structural Analysis of Layered Orthotropic Ring-Stiffened Shells of Revolution (SALORS)—Linear Stress Analysis Option. NASA TN D-7179, October 1973.
11. Anderson, M. S.; Fulton, R. E.; Heard, W. L., Jr.; Walz, J. E.: Stress, Buckling, and Vibration Analysis of Shells of Revolution. Computers and Structures **1**, No. 1/2, 157—192 (1971).
12. Bushnell, D.: Stress, Stability and Vibration of Complex Branched Shells of Revolution: Analysis and User's Manual for BOSOR 4. NASA CR-2116, 1972.
13. Cohen, G. A.: Computer Analysis of Ring Stiffened Shells of Revolution. NASA CR-2085, February 1973. User Document for Computer Programs for Ring-Stiffened Shells of Revolution. NASA CR-2086, March 1973.
14. Svalbonas, V.: Numerical Analysis of Stiffened Shells of Revolution. NASA CR-2273, September 1973.
15. Novozhilov, V. V.: The Theory of Thin Shells. P. Noordhoff, Groningen, 1959, Chap. 1.
16. Koiter, W. T.: On the Stability of Elastic Equilibrium (In Dutch With English Summary). Thesis, Polytechnic Institute at Delft, Amsterdam: H. J. Paris 1945; English translation. NASA TTF-10833, 1967.
17. Budiansky, B.; Hutchinson, J. W.: Dynamic Buckling of Imperfection-Sensitive Structures. Proceedings of the XI International Congress of Applied Mechanics. H. Görtler, ed., Berlin, Göttingen, Heidelberg: Springer 1964, pp. 636—651.
18. Bushnell, D.: Evaluation of Various Analytical Models for Buckling and Vibration of Stiffened Shells. AIAA J. **11**, No. 9, 1283—1291 (1973).

Interaction between Local Plate Buckling and Overall Buckling in Thin-Walled Compression Members — Theories and Experiments

R. Maquoi, Ch. Massonnet

University of Liège, Belgium

1. Practical Importance of the Problem

One of the imperatives of modern steel construction is the use of thin plates and the development of more efficient shapes in order to better compete with reinforced and prestressed concrete. For hot rolled shapes, the thinness b/t of their constituting walls is such that local buckling is excluded; on the contrary, an economical use of cold rolled light-gage steel requires that local buckling of the walls be tolerated in service and thus, that design be based on effective width of these buckled elements. These work in service in the postcritical range.

The expansion of the metal, which constitutes an efficient means of valorizing the material, has mainly been encouraged by G. Winter, whose first tests have been made nearly twenty years ago. In 1968, the IABSE Congress in New-York adopted general conclusions aiming at extending to "heavy" steel construction the use of wall postcritical strength [1]. Such a tendency has been adopted in the 1969 AISC Specifications [2] for steel building design, which contain special rules inspired by the light-gage steel AISI Specifications [3].

On the other hand, any up to date design method of thin-walled members must necessarily comply with the principles of the modern semi-probabilistic theory of safety, for which elements of ideally perfect shape do not exist. Indeed, industrial structural elements always contain random imperfection distributions and the above design method must lead to a sufficiently small risk of collapse—of the order of 10^{-5}—even if all of these imperfections reach simultaneously their upper characteristic values in the statistical sense.

Until recently, the design of thin-walled compression members was based on the naive theory of optimization, mainly developed by F. Bleich [4], according to which:
(a) the column axis is perfectly straight and the walls are perfectly plane,
(b) local plate buckling cannot precede general buckling of the member,

366   R. Maquoi and C. Massonnet

(c) the optimum design is therefore based on the inequality
$$\sigma_{cr}^{\text{plate buckling}} \gtreqqless \sigma_{cr}^{\text{(overall) buckling}}.$$

2. Criticism of the Naive Theory

The actual behaviour of thin-walled members does not agree with the above theory. Due to geometrical and structural imperfections, the column axis bends under increasing axial load and there is a stress redistribution in the cross section so that the theoretical initial uniform stress distribution becomes more and more disturbed (Fig. 1). It is equivalent to say that the effective area of the cross-section is progressively

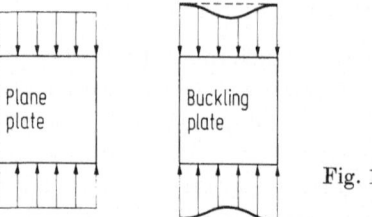

Fig. 1

reduced, so that stresses and strains are increased. So, for industrial elements, which always contain unavoidable imperfections, local plate bending and overall bending occur from the beginning of loading and lead to a permanent but increasing interaction. Such a problem may be evidently solved with a computer, by using either analytical methods, or numerical methods, the more refined of which is a large displacement elastoplastic finite element computer program.

However, practice requires less sophisticated approaches and this is the reason why several authors have proposed simplified design methods whose results are hopefully accurate enough for a practical design. Most known design methods have been proposed by the 1969 AISC Specifications, by the SERCOM [5] in Belgium, by Skaloud [6, 7] in Czechoslovakia, by Klöppel and his team [8, 9] in Germany and by Bergfelt [10] in Sweden.

On the other hand, these last years, several theoretical papers, due to Koiter and Kuiken [11], van der Neut and al [12, 13, 14], Thompson [15] and others have emphasized the danger presented by structural elements which can perish by two different instability modes, for which the critical stresses are nearly the same. All these theoretical works are based on idealized mathematical models which lead to a tractable analytical discussion and tend to show that:

(a) due to imperfections, the actual collapse strength is highly inferior to the optimum strength given by Bleich's calculations;

(b) the equilibrium at collapse is unstable;

(c) the design given by the naive theory ($\sigma_{cr}^{pb} = \sigma_{cr}^{b}$) leads to too thin wall thicknesses.

These academic papers, however, are based on the consideration of an idealized model which disregards the presence of webs and their inter-action with the other walls constituting the cross section. Besides, no plastic strength reserve is considered; the column fails by pure general instability triggered by the deterioration of the effective area and moment of inertia.

The objections and conclusions made by Koiter, van der Neut and Thompson are not supported by tests, the main of these having been made by Skaloud, Bergfelt, Graves-Smith and by Klöppel, Schmied an Schubert.

3. Aim of Present Paper

Because one of the authors of the present paper has some responsibility in the drafting of structural codes at the Belgian and at the European level, he feels that a clarification of the above problem is urgently needed. Both authors believe that the best way would be to perform a large number of carefully planned tests on thin-walled columns with box-type cross section—the latter being the most used in practice. However, such tests are so lengthy and costly that above way does not seem tractable. It appears that Klöppel and Schubert [9] have developed a rather elaborate analytical design method whose accuracy has been proved by a sufficiently large number of tests. Indeed, as will be seen further on, this method gives for the ratio $P_{experiment}/P_{theory}$ a mean value of 0.968 and a standard deviation of 0.13. It is true that Klöppel's design method remains too complicated for daily use in the design office; therefore, a small ALGOL program has been written to make calculations much easier and speedier.

The authors of the present paper feel that such a program constitutes a good tool and recommends itself for numerical simulation of experiments on a computer with varying values of initial imperfections and, in addition, shows the effect of the yield stress on the ultimate strength.

The main aim of the present paper is to examine the validity of the academic criticisms presented in the literature against the classical naive thin-walled column design method in the light of the Klöppel-Schubert model.

4. Skaloud's Researches

The numerous tests executed by Skaloud [7] have clearly shown that it is not possible to define a range where plate buckling would predominate and another one where buckling alone would be observed. These experiments have illustrated the interaction between both phenomena and emphasized the fact that the collapse strength depends always on two parameters: the column slenderness ratio λ and the wall thinness b/t.

For columns of small slenderness ratio, the effect of general buckling is negligible and the ultimate load is given by the ultimate load of the walls. The ultimate strength exceeds very often the critical load of local buckling because of the reserve of postcritical strength.

If, however, the slenderness ratio is such that general buckling influences substantially the behavior, the ultimate load, according to Skaloud, is definitely below the value given by the buckling curve. The values of the buckling critical stresses, calculated for a bar whose walls are replaced by their "effective" parts to take account of local buckling, are often near the experimental results. They may however differ from the values of the experimental ultimate stresses, because of the imperfections of the geometry and the loading, of the unavoidable residual stresses, and of the fact that the redistribution of the stresses due to the bending of the bar is neglected.

Finally, in the case of very short columns with thick walls, for which general and local buckling may be neglected, the bearing capacity may exceed the "squash load" $A\sigma_y$ because of the strain hardening of the material. A is the gross area of he cross section.

The basic assumptions of the design method put forward by Skaloud [6] are:

(1) The axis has a sinusoidal initial deflection of magnitude:

$$m = 0.3 \,(\lambda /100)^2 i \tag{4.1}$$

where i is the core radius.

(2) The maximum stress may be determined by the second order theory

$$\sigma_{\max} = \sigma_m \left[1 + \frac{m}{1 - (\sigma_m/\sigma_{cr})} \right] \tag{4.2}$$

where $\sigma_m = P/A$ is the mean normal stress and σ_{cr}, the critical buckling stress.

(3) The geometrical properties of the mid-span section of the member are derived from the consideration of the effective widths of the various walls, depending on the wall thickness b/t, the ratio of the edge stresses and the yield stress. The ratio

$$C_L = A/A_{eff} \tag{4.3}$$

is given in tables.

(4) The collapse criterion is purely elastic:

$$\sigma_{\max} = \sigma_y. \tag{4.4}$$

According to this theory, the mean collapse stress $\sigma_{m,u}$ is given by the equation

$$\sigma_{m,u} = \sigma_y/C_L C', \tag{4.5}$$

where C_L is the coefficient that takes care of the wall buckling (see above), and C' is a coefficient that takes care of general bar buckling.

The basic idea of an interaction between general and local buckling is good. According to Klöppel ([9], p. 45), the method gives, however, a too large collapse strength, because:

(a) assumptions (1) and (2) furnish too small values for the coefficient C' and therefore a too large value for $\sigma_{m,u}$;
(b) the method used for calculating the effective width is slightly unsafe.

As we shall see in Section 5, the Klöppel-Schubert method is simply an improvement of the Skaloud method.

5. Klöppel's Experiments and Theories

Klöppel and Schubert [9] have developed an iterative method enabling the calculation of the maximum tensile and compressive stresses at the four corners of a thin-walled bar of hollow rectangular section.

Based on measurements on industrial members, the following initial imperfections are recommended by the above authors:

(a) for the walls, a relative initial deflection:

$$f_0'/t = 0.0004(b/t)^2 \qquad \text{if } b/t < 150,$$
$$= -0.90 + 0.012\,(b/t) \quad \text{if } b/t > 150. \tag{5.1}$$

(b) For the whole member, an initial imperfection axis:

$$f_0 = i/20 + l/500 \tag{5.2}$$

with i radius of gyration of the gross cross section, l buckling length.

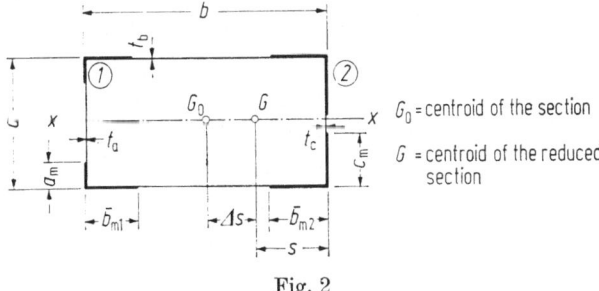

Fig. 2

Suppose we consider a cross section of a compressed thin-walled member deforming in plane Gx and presenting its concavity to the left. Figure 2 represents the reduced cross section. It gives the following relations:

$$A_a = a_m t_a, \quad A_{b1} = b_{m1} t_b, \quad A_{b2} = b_{m2} t_b, \quad A_c = c_m t_c,$$

$$A_{red} = (A_a + A_{b1} + A_{b2} + A_c)\,2;$$

$$s = 2[A_{b1} b_{m1}/2 + A_{b2}(b - 0.5 b_{m2}) + A_c b]/A_{red}, \qquad (5.3)$$

$$s_1 = s, \quad s_2 = s - 0.5 b_{m1}, \quad s_3 = b - s - 0.5 b_{m2}, \quad s_4 = b - s,$$

$$I_{red} = (A_a s_1^2 + A_{b1} s_2^2 + A_{b1} b_{m1}^2/12 + A_{b2} b_{m2}^2/12 + A_{b2} s_3^2 + A_c s_4^2)\,2;$$

$$\Delta s = s - b/2, \quad W_{red1} = I_{red}/s, \quad W_{red2} = I_{red}/(b - s).$$

The design of the section is requested for stress resultants M and N, which are attached to the initial centroid of the full box section. For a bending moment in plane Gx, we have, however, different effective widths on the concave (5) and convex (2) sides, so that the centroid of the reduced section is displaced towards the convex side (Fig. 2). The extreme stresses are given by

$$\sigma_1 = \frac{P}{A_{red}} + \frac{M + \Delta s\,P}{W_{red1}}, \quad \sigma_2 = \frac{P}{A_{red}} - \frac{M + \Delta s\,P}{W_{red2}} \qquad (5.4)$$

which, solved for P and M, gives

$$P = \frac{(\sigma_1 W_{red1} + \sigma_2 W_{red2})}{W_{red1} + W_{red2}} A_{red}, \qquad (5.5)$$

$$M = \left(\sigma_1 - \frac{P}{A_{red}}\right) W_{red1} - P\Delta s. \qquad (5.6)$$

Klöppel and Schubert take as point of departure the well known formula giving the maximum edge stress of an eccentrically compressed bar

$$\sigma_{max} = \sigma_m \left(1 + \frac{m_0}{1 - (\sigma_m/\sigma_{cr})}\right), \qquad (5.7)$$

where $\sigma_m = P/A$ and m_0 is the global initial eccentricity of the axial load at midheight expressed by taking the core radius i as unity.

Let us remark that, because of wall buckling, A and P_{cr} must be calculated for the reduced section, thus

$$\sigma_{max} = \frac{P}{A_{red}} \left(1 + \frac{m_{0 red}}{1 - (P/P_{cr\,red})}\right). \qquad (5.8)$$

They use the same collapse criterion as Skaloud, namely

$$\sigma_{max} = \sigma_y. \qquad (5.9)$$

As the geometrical properties of the mid-span cross section depend on the effective widths a_m, b_{m1}, b_{m2}, c_m (Fig. 2), which depend themselves on the compression force P and bending moment M, it is clear that the determination of the edge stresses σ_1, σ_2, given by formulae (5.4) is a complicated non linear task which can only be solved by iteration and for which the use of a computer is highly recommended.

Klöppel and Schubert give, as appendix of their paper [9], an ALGOL program which performs automatically the computation of σ_1, σ_2, when the geometrical dimensions a, b, t_a, t_b of the cross section and the length of the column are given, as well as the load P and bending moment M. This program accepts the wall imperfections defined by (5.1) and any axis imperfection defined by formula (5.2). To avoid extensive use of it, the German researchers have proposed approximate design methods which, compared with experimental results of a lot of collapse tests, give safe values of the ultimate load.

From our point of view, the main interest of the German work lies in that the refined analytical design method, controlled by test results, is summarized by a computer program constituting a good tool for numerical simulation. This ALGOL program has been translated by us in FORTRAN IV and somewhat modified to enable a direct calculation of the ultimate load, as characterized by above collapse criterion.

6. Comparison between Various Theories by Means of Thompson's Efficiency Chart and Conclusions

6.1. General

Whereas weight has in general a rather small value as optimizing criterion of a whole civil engineering structure, it represents a good merit function in the search for the optimal shape of an isolated compression member. Thus, from a practical engineering view point, it makes sense to *design a thin-walled compression member of definite length for minimum weight*. This leads *to maximize the load carrying capacity of a column subject to the constraint that the weight shall be held constant*.

Concentrating on an axially loaded square tube of width b and of thickness t, it is seen that, if the area of the cross section (Fig. 3)

$$A = 4bt, \tag{6.1}$$

and the length L of the column are kept constant—to have a constant weight—, the design will only depend on the dimension b. It is then

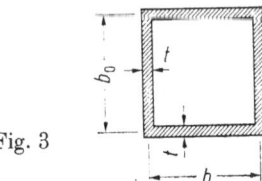

Fig. 3

possible to discuss the variation of the collapse strength P with b in the diagram of Fig. 4.

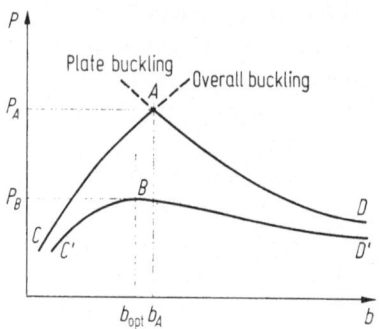

Fig. 4

If the walls are perfectly plane and free from residual stresses, and if, in addition, the column axis is perfectly straight, the optimum design is given by point A at the intersection of Euler's buckling curve and local plate buckling one. This is the design method according to the so-called naive theory. However, it is well known that because of unavoidable initial imperfections, there is an interaction between both buckling modes and that the two curves CA and AD will be replaced by a unique one $C'BD'$ showing a maximum P_B in B, for a value of the thickness which differs from that given by the naive theory.

In their paper [15], Thompson and Lewis have introduced a non dimensional diagram, which will be called in what follows *efficiency chart* and which is particularly suitable for discussing our problems.

6.2. Efficiency Chart of the van der Neut's Model

Van der Neut [12] has studied the behaviour of an idealized column, whose cross section is constituted by two load-carrying flange-plates and two unspecified fictitious webs which simply serve to maintain the structural integrity of the column without contributing to the transmission of axial stresses (Fig. 5). The flanges are therefore assumed to be simply supported along their longitudinal edges. For such a model, the critical loads P_E for Euler column buckling and P_{LB} for local flange

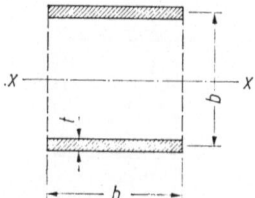

Fig. 5

buckling are respectively:

$$P_{\mathrm{E}} = \pi^2 EI/L^2 = \pi^2 Eb^3 t/2L^2 \qquad (6.2)$$

and

$$P_{\mathrm{LB}} = 2bt\sigma_{\mathrm{cr}}^{\mathrm{pb}} = 2\pi^2 Et^3/3(1-\nu^2)\,b. \qquad (6.3)$$

Thompson and Lewis introduce the characteristic ratio:

$$x = P_{\mathrm{E}}/P_{\mathrm{LB}} = 3(1-\nu^2)\,b^4/4t^2L^2, \qquad (6.4)$$

or, with (6.1):

$$x = 3(1-\nu^2)\,b^6/A^2L^2, \qquad (6.5)$$

whence:

$$b = \left[x\,\frac{A^2L^2}{3(1-\nu^2)} \right]^{1/6}. \qquad (6.6)$$

As the weight remains constant, the latter formula shows that the non-dimensional factor x can be used as a measure of b. In addition, let us introduce another non-dimensional ratio

$$y = P^x/P_{\mathrm{LB}}, \qquad (6.7)$$

where P^x represents the load at which the column axis collapses, whether or not this is associated with buckling of the flanges.

It is very easy to see from (6.7), (6.1), (6.3) that the load P can be written as:

$$P^x = \frac{\pi^2 EA^{5/3}L^{-4/3}}{4[3(1-\nu^2)]^{1/3}}\,yx^{-2/3} = Kyx^{-2/3}. \qquad (6.8)$$

As K is a constant for given values of A and L, we can see that, instead of discussing the variation of P^x as a function of x, it is equivalent to study the variation with x of a function

$$\varrho(x) = yx^{-2/3}. \qquad (6.9)$$

This may be called the efficiency of the shape because its maximum value 1, is obtained from the naive theory of buckling ($P_{\mathrm{E}} = P_{\mathrm{LB}} = P^x$).

The last paper of van der Neut [14] contains five diagrams giving—with our notations—the function $y - f(x)$ for five values (1.25%, 2.5%, 5%, 10%, 20%) of the flange imperfection parameter:

$$\alpha = f_0'/t, \qquad (6.10)$$

where f_0' is the magnitude of flange wave imperfection, and for six values (0%, 1%, 2%, 4%, 7%, 15%) of the column axis imperfection parameter:

$$\beta = 2f_0/b \qquad (6.11)$$

where f_0 is the wave amplitude of the column axis imperfection.

Starting from these diagrams, it suffices to calculate $\varrho(x)$ for various values of x, α and β to be able to construct the corresponding efficiency charts. These are reproduced on Figs. 6 to 11 and show that:

(a) even for very small values of the imperfection parameters α and β, the loss of collapse strength is appreciable;

(b) for all values of the imperfections, the optimum corresponds to values of x less than unity, and leads therefore to walls thicker than the value given by the naive buckling theory.

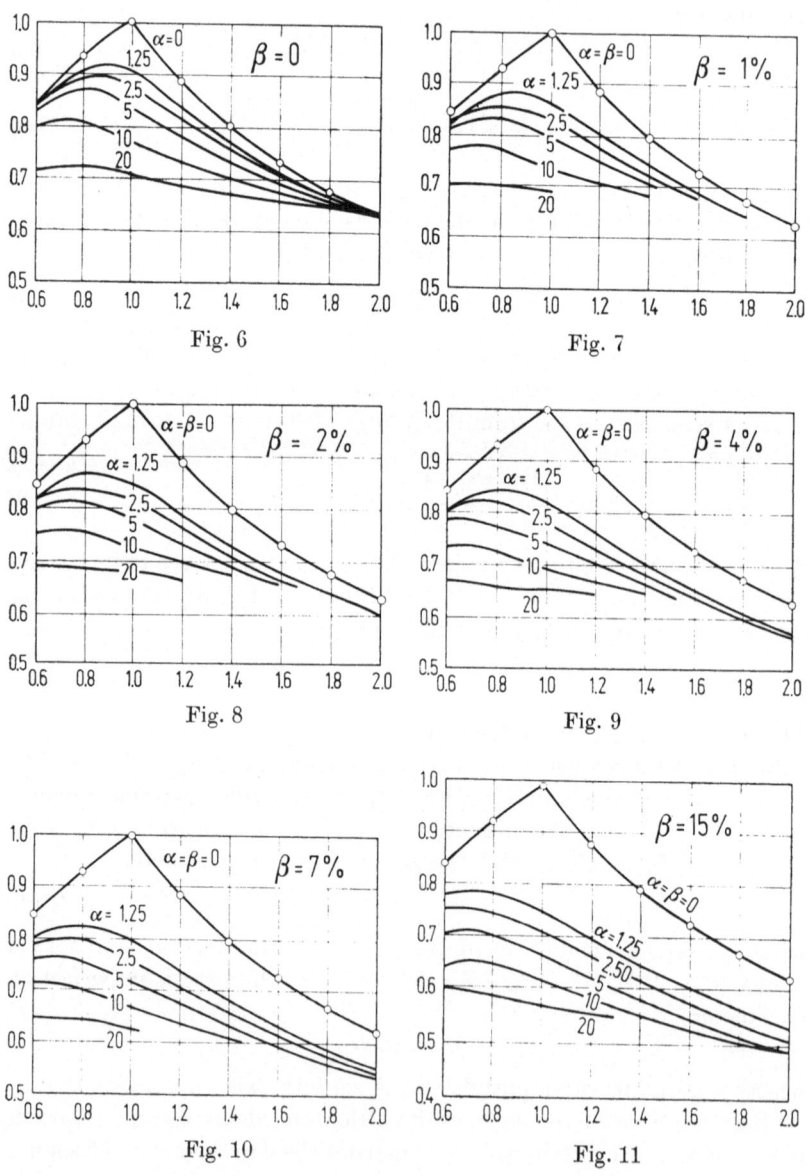

Fig. 6

Fig. 7

Fig. 8

Fig. 9

Fig. 10

Fig. 11

It is important to remark that this theory does not consider the influence of the yield strength of the material, which, when it is reached in some points, can strictly limit the ultimate strength of a member.

6.3. Study by Koiter and Kuiken

In their paper [11], Koiter and Kuiken establish approximate formulae showing how the collapse strength varies with a certain plate imperfection parameter μ. These formulae are based on the general non-linear theory of elastic stability, taking into account the imperfection sensitivity of the column, and are only valid for the assumed idealized cross section which corresponds with van der Neut's model. Main notations and results are summarized hereafter for the case of a square cross section.

As in Section 6.2, critical loads are:

$$P_{\mathrm{E}} = \pi^2 EI/L^2 = \pi^2 Eb^3 t/2L^2 \tag{6.12}$$

and

$$P_{\mathrm{LB}} = 2\pi^2 Et^3/3(1 - \nu^2)\, b. \tag{6.13}$$

A parameter λ_{E} is defined as the value of λ^x for which the collapse load $P^x = \lambda^x P_{\mathrm{LB}}$ is equal to the Euler's critical load, so that:

$$\lambda_{\mathrm{E}} = 3(1 - \nu^2)\, b^4/4L^2 t^2. \tag{6.14}$$

The notations used by Thompson and Lewis are thus here:

$$x = P_{\mathrm{E}}/P_{\mathrm{LB}} = \lambda_{\mathrm{E}} \tag{6.15}$$

and

$$y = P^x/P_{\mathrm{LB}} = \lambda^x \tag{6.16}$$

and corresponding efficiency charts, giving

$$\varrho(x) = y x^{-2/3} \tag{6.17}$$

are also curves for

$$\varrho(\lambda_{\mathrm{E}}) = \lambda^x \lambda_{\mathrm{E}}^{-2/3}. \tag{6.18}$$

In [11], Koiter and Kuiken propose approximate formulae which, with above notations and an imperfection parameter μ related to the van der Neut's one α by

$$\alpha = \sqrt{\frac{32}{3(1 - \nu^2)}}\, \mu, \tag{6.19}$$

take the following form:

for $x < 1$:

$$\varrho(x) = x^{1/3} - \frac{8 x^{7/3} \mu^2}{(1 - x)^3},$$

for $1 < x < \dfrac{5}{3}$:

$$\varrho(x) = x^{-2/3} - \frac{(5 - 3x)\, x^{-2/3} \mu^{2/3}}{(x - 1)^{1/3}\, (2 - x)^{2/3}}.$$

$$\text{(6.20 a, b)}$$

This study only considers the flange imperfection and disregards the effects of column axis imperfection. It does not mention what can occur when the yield stress is reached in some points of the cross section; it may thus be criticized in the same way as van der Neut's approach.

In (6.20), the last terms of the second member represent the discrepancy of the efficiency calculated in the case of a perfect column member with a perfectly straight axis and plane flanges.

Above formulae are only valid for the case $\gamma = 1$; that corresponds to the boundary conditions of Marguerre and Trefftz, for which the longitudinal edges remain straight.

For the case $\gamma = 0.845$, i.e. if the longitudinal edges are completely free from stresses (boundary conditions of Hemp), Koiter and Kuiken propose other asymptotic formulations, which after the same transformation as above, may be written:

for $x < 1$: $\varrho(x) = x^{1/3} - \dfrac{8x^{7/3}\mu^2}{(1-x)^3}$,

for $x = 1$: $\varrho(x) = x^{-2/3} - 2^{3/4}x^{-2/3}\mu^{1/2}$,   (6.21a, b, c)

for $1 < x < 1.96$: $\varrho(x) = x^{-2/3} -$

$$- (4.07 - 2.07x)(x - 1)^{-1/3}(1.69 - 0.69x)\,x^{-2/3}\mu^{2/3}.$$

The case $\gamma = 0.845$ corresponds to the case discussed by van der Neut.

From calculations we have made, it appears that the approximate formulae of Koiter and Kuiken which are formally valid for very small values of μ agree very well with curves (for $\beta = 0$) derived from van der Neut's model. When the imperfection parameter becomes rather important, asymptotic formulae are no more valid and more complex formulations, which are not reproduced here, must be used.

6.4. Study by Graves-Smith

In his Ph. D. Thesis [16] and in further papers [17, 18] Graves-Smith studied in detail the behaviour of a square tube of constant thickness. As this shape is commonly used in practice and, furthermore as the theory takes into account the plastic strength reserve, the results should be highly significant. Unfortunately, the author only gives a single diagram from which an efficiency chart could be built.

We have for a square cross section

$$A = 4bt = \text{const}, \quad I = (2/3)\,b^3t, \qquad (6.22\text{a, b})$$

$$\sigma_{\mathrm{cr}}^{\mathrm{pb}} = 4\pi^2 D/b^2 t = \pi^2 E t^2/3(1 - \nu^2)\,b^2, \quad \sigma_{\mathrm{E}} = \pi^2 E/\lambda^2 = \pi^2 E b^2/6L^2.$$

$$(6.23\text{a, b})$$

As in Section 6.2, we set

$$x = \frac{P_E}{P_{LB}} = \frac{\sigma_E}{\sigma_{cr}^{pb}} = \frac{1 - \nu^2}{2} \left(\frac{b^2}{Lt}\right)^2, \tag{6.24}$$

or, with (6.22 a)

$$x = 8(1 - \nu^2)\, b^6/L^2 A^2, \tag{6.25}$$

whence

$$b = \left(\frac{x}{8} \frac{A^2 L^2}{1 - \nu^2}\right)^{1/6}. \tag{6.26}$$

As x varies monotonously with b, it may be taken as a measure of b. With Thompson's notations, the collapse load is:

$$P^x = y P_{LB} = y A \sigma_{cr}^{pb} = y A \frac{\pi^2 E}{3(1 - \nu^2)} \left(\frac{t}{b}\right)^2 =$$

$$= y \frac{\pi E A^3}{48(1 - \nu^2)\, b^4}, \tag{6.27}$$

or, replacing b by its value (6.26),

$$P^x = K' y x^{-2/3}, \tag{6.28}$$

where K' is the constant

$$K' = \frac{\pi^2 E A^{5/3} L^{-4/3}}{12(1 - \nu^2)^{1/3}}. \tag{6.29}$$

This shows that, as in the case of the van der Neut shape, we have to discuss the variation, with x and with the magnitude of the imperfections, of the so-called efficiency factor

$$\varrho(x) = y(x)\, x^{-2\,3}. \tag{6.30}$$

In his paper [17], Graves-Smith gives curves of y as function of a parameter $(L/r)/(L/r)_{cr}$, equal to the ratio of the actual slenderness to the critical one. With

$$L/r = \sqrt{\pi^2 E/\sigma_E}, \quad (L/r)_{cr} = \sqrt{\pi^2 E/\sigma_{cr}^{pb}}, \tag{6.31a, b}$$

above parameter may be expressed as

$$(L/r)/(L/r)_{cr} = \sqrt{\sigma_{cr}^{pb}/\sigma_E} = x^{-1/2}. \tag{6.32}$$

To obtain the efficiency chart of the square tubular section, we simply have to derive from the four curves of equation $y = f(x^{-1/2})$ given by Graves-Smith in Fig. 15 of [17], the curves

$$\varrho(x) = y x^{-2/3}. \tag{6.33}$$

The chart so obtained is given in Fig. 12. The only trouble is that the curves of Graves-Smith are given for values of the parameter

$$\gamma = f_0'/b \qquad (6.34)$$

where f_0' is the magnitude of flange imperfection instead of $\alpha = f_0'/t$ as on preceding charts. It is easy to see that

$$\gamma = \alpha t/b \qquad (6.35)$$

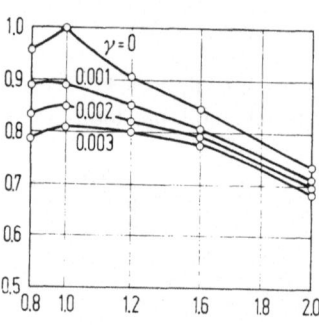

Fig. 12

and that it is impossible to deduce t/b from the diagram of Graves-Smith. From the study of Graves-Smith, it can be concluded that, for columns of square section:

(a) experimental results corroborate very well the theoretical interaction curve;

(b) the ultimate strength remains constant for slenderness ratios up to $0.5(L/r)_{\mathrm{cr}}$ and this implies that the overall buckling is not significant for the range $x > 4$;

(c) for flange imperfections γ, ranging from 0.01 to 0.03, the ultimate stress is not seriously affected by quite large initial overall imperfections.

The model of Graves-Smith is much more realistic than that of van der Neut because of the presence of webs and the consideration of plastic deformations. Figure 12 shows that the ultimate strength given by the naive theory is reduced by the effect of initial geometrical imperfections.

However, contrary to Thompson's conclusions, the actual ultimate strength for a given imperfection reaches its maximum near $x = 1$, and this maximum is rather very flat, so that a design following the naive theory is qualitatively valid. We believe that the contradiction between the conclusions of Thompson and Graves-Smith must be explained by the fact that the model of the last author is much more realistic. Indeed, we shall see that the point of view of Graves-Smith is corroborated by our numerical simulation computations.

7. Numerical Simulation of a Square Column up to Collapse

As mentioned before, the program established by Klöppel and al. is used to study the efficiency at collapse of a prismatic column with square box cross-section (Fig. 3). The ends are supposed perfectly hinged and the column is axially loaded.

7.1. Procedure Adopted

We chose a length L and an area $A = 4bt$ of the cross section so that, for a given length, the weight remains constant. If we let vary the wall thickness t, the width b is determined by the value of A. It is thus possible to examine some range of flange thinness b/t and slenderness ratio λ.

To be found is the value of t which produces the maximum collapse strength for the bar considered, taking into account the initial flange and axis imperfections, the non dimensional parameter x and the steel grade.

The values of $\alpha = f_0'/t$ given by (5.1) have been obtained by Klöppel and al. on the basis of a statistical study, and we shall adopt them in the present paper. About the axis imperfection, it depends not only on the fabrication procedures but also indirectly on the unavoidable small end eccentricities of applied loads. For this reason, we consider in present study a range of values for this parameter.

The non-dimensional ratio x simultaneously takes into account the thinness b/t of the four identical flanges and the slenderness ratio λ of the column member. A range from 40 to 120 is examined for the slenderness ratio, to cover "short" members as well as "long" ones.

The third parameter to be examined is the steel grade; indeed, it is clear that the ultimate strength of a given column depends on this parameter when the collapse criterion $\sigma_{\max} = \sigma_y$—which is very realistic—is adopted. Two steel grades will be studied: AE 24 and AE 36 with European symbols. The first is the regular mild steel with a yield stress of about 240 N/mm² (34.8 ksi). The second is the regular high strength steel, with a yield plateau at $\sigma_y = 360$ N/mm² (52.2 ksi).

For the numerical simulation, following values are chosen:

$L = 5000$ mm ($\triangleq$ 16.5 ft),
$A = \quad 40$ cm² ($\triangleq$ 6.2 sq. in),
$\sigma_y = \quad 240$ N/mm² for AE 24 ($\triangleq$ 34,810 psi),
$\sigma_y = \quad 360$ N/mm² for AE 36 ($\triangleq$ 52,210 psi).

7.2. Computer Results and Discussion

The main computer results are summarized in Fig. 13 and 14. They are presented just as the efficiency chart proposed by Thompson and

Lewis. The upper curve with a peak for $x = 1$ represents the behaviour of a box column with perfectly plane walls and straight axis; it corresponds thus to the naive theory.

It is seen that the efficiency of an imperfect member is always lower than unity for $x = 1$.

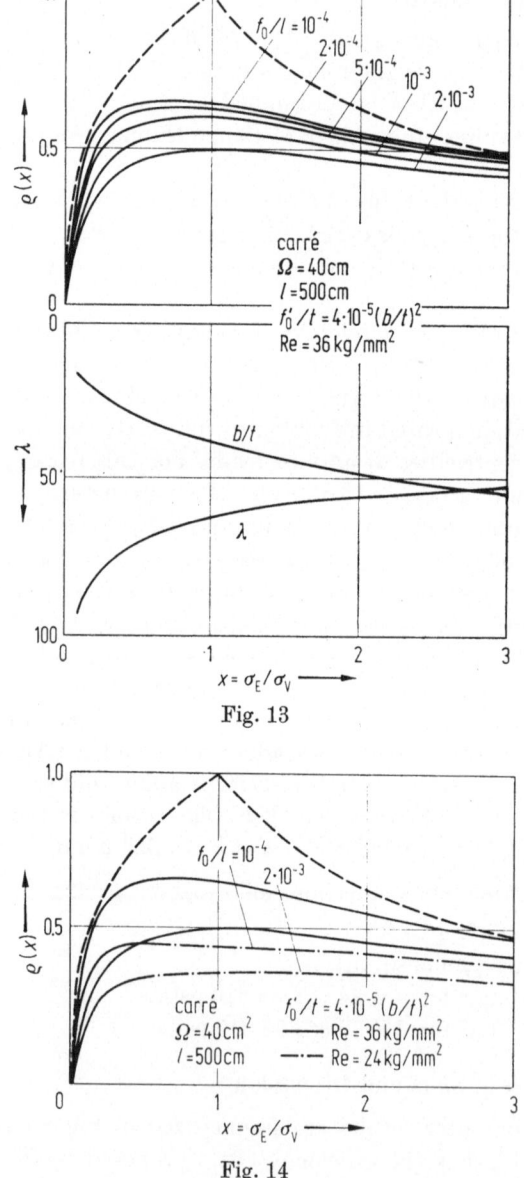

Fig. 13

Fig. 14

This loss of strength depends:

(a) on the value of the yield strength,
(b) on the imperfection of the walls, f_0'/t,
(c) on the imperfection of the axis f_0/l.

Indeed, if the critical stress according to Bleich's theory exceeds the value of the yield strength, the maximum corner stress will reach σ_y before Bleich's ultimate strength is exhausted. This reduction of efficiency is substantial, as shown by the comparison of steels AE 24 and AE 36 on Fig. 14.

On the same figure, it can be seen that the maxima of curves are very flat, especially for steel AE 24. It seems thus that, especially for mild steel, for the range of examined slenderness and thinness ratios, the value of maximum efficiency does not depend very much on the value of x.

It must also be observed that the maximum efficiency occurs either for $x < 1$ or for $x > 1$ according to the magnitude of the axis imperfection. Such a conclusion is in contradiction with the predictions of Thompson and Lewis who found that the maximum was always slipping to $x < 1$. It is in agreement with Fig. 12, which represents theoretical results of Graves-Smith. Such apparent paradox can be explained by the difference between the basic assumptions of the theoretical solutions, as has been explained above.

The efficiency evidently diminishes also with increasing axis imperfection and it seems that the decrease is nearly proportional to the amplitude f_0/l. The main conclusion of the present study is probably the fact that it is practically impossible for a column of normal slenderness to benefit from a large postcritical strength, which should be represented by points beyond curves of the naive theory. This is due to the fact that the loss of rigidity resulting from the presence of imperfections has a most severe effect, which is not compensated by the postcritical reserve of strength.

Such a conclusion does not apply for sufficiently large values of x, which correspond to small slenderness ratios and very thin walls. In this case, overall buckling does not occur and the ultimate strength of the column depends on the ultimate strength of the constituting plate walls, which can individually develop their postcritical strength. However, this type of columns presents only a very small interest in construction pratice.

382 Interaction between Local Plate Buckling and Overall Buckling

8. General Conclusions

The main conclusions are as follows:

1. The strength erosion due to interaction between buckling modes is confirmed.

2. Because of the choice of the collapse criterion ($\sigma_{max} = \sigma_y$), the results are quite different for mild steel and high strength steel.

3. Curves in efficiency charts are very flat in the vicinity of $x = 1$, so that the naive theory still constitutes an acceptable tool for the choice of plate thickness.

References

1. IABSE, 8th Congress, New York, 1968, Final Report, pp. 497–498.
2. Specifications for the Design of Steel Buildings. American Institute of Steel Construction, 1969.
3. Light-Gage Cold formed Steel design manual. American Iron and Steel Institute, 1962.
4. Bleich, F.: Buckling Strength of Metal Structures. McGraw-Hill 1952.
5. Baar, S.; Hick, F.: Construction Métallique. No. 3, 1972 (in French).
6. Skaloud, M.: Acta Technica, 1962, pp. 52–86.
7. Skaloud, M.; Zörnerova, M.: Acta Technica, C.S.A.V., No. 4, 1970.
8. Klöppel, K.; Schmied, R.; Schubert, J.: Der Stahlbau **35**, 321 (1966); **38**, 9 (1969) (in German).
9. Klöppel, K.; Schubert, J.: The Calculation of the Carrying Capacity in the Postbuckling Range of Thin-Walled Box Columns Loaded by Concentric and Excentric Compressive Force (in German). Publications of the Institute for Statics and Steel Construction, Darmstadt, 1971.
10. Bergfelt, A.: Bulletin Technique de la Suisse Romande. No. 17, 363–368 (1973) (in French).
11. Koiter, W.; Kuiken, G.: Report WTHD 23, Laboratorium voor Technische Mechanica, Delft, Mei 1971.
12. van der Neut, A.: Report VTH of the Delft University of Technology, 1968.
13. van der Neut, A.; Meijer, J.: Report VTH-160, Delft Universty of Technology, April 1970.
14. van der Neut, A.: Report VTH 172, Delft University of Technology, August 1972.
15. Thompson, J. M. T.; Lewis, G. M.: J. Mech. Phys. Solids **20**, 101–109 (1972).
16. Graves Smith, T. R.: Ph. D. Thesis, Cambridge University, 1966.
17. Graves Smith, T. R.: Thin Walled Steel Constructions. Symposium 1967 at Univ. Coll. Swansea. Crosby Lockwood and Son Ltd., London.
18. Graves Smith, T. R.: 8e Congrès de l'A.I.P.C. New York, 1968.

Buckling Design in Ship Structures

E. M. Q. Røren, H. R. Hansen

Det norske Veritas, Oslo, Norway

Abstract

Ship hull structures yearly consume in the order of 15 million tons of steel. An appreciable amount of this goes into parts of the structure which are either designed on the basis of buckling criteria or a combination of stress and buckling criteria, or it is included with no other function than that of preventing buckling of the structure.

Most of the ship structures consist of a fairly complex system of stiffened plates which are subjected to hydrostatic and dynamic lateral loads as well as in plane stresses which may be both tensile and compressive in a more or less bi-axial fashion. The theoretical buckling strength of stiffened plates in various bi-axial stress fields is of course reasonably well defined. But there are still some problems related to the influence of initial deviations, the corrections for elasto-plastic behaviour, the influence of lateral loading, and the evaluation of "realistic" boundary conditions.

Slender unstiffened or lightly stiffened shells have not been typical for ship structures. However, the present developments of LNG-carriers with very thin shell of revolution type tanks have necessitated development of buckling design criteria both for unstiffened spherical shells and for lightly stiffened cylinders. The importance of the relationship between shape imperfection and reduction (knock down) factor is for both types of shells rather obvious. For circular cylinders even light stiffening gives large increases in the theoretical buckling strength. The extent to which this constitutes a real increase is an important factor for the designer. Typical for the cylinders is also that they operate in the elasto-plastic region.

For these and for several other important aspects of buckling, design decisions have had and have to be made. Regrettably the basic theoretical and experimental knowledge is uncomfortably sketchy pointing to a need for further basic and design-oriented research.

Symbols

A_s	Cross sectional area of stiffener	E	Young's modulus of elasticity
$D = Et^3/12(1-v^2)$	Plate flexural stiffness	E_x, E_y	Young's modulus in x- and y-direction
D_x, D_y	Plate flexural stiffnesses in x- and y-direction	e_s	Eccentricity of stiffener

F	Airy's stress function	λ	Elastic buckling eigenvalue
H	Torsional plate stiffness	ν	Poisson's ratio
h_x, h_y	Effective thickness in x- and y-direction	ν_x, ν_y	Poisson's ratio in x- and y-direction
h	Shear carrying thickness	ϱ	Imperfection sensitivity factor
I	Stiffener moment of inertia	σ_{cp}	Critical plastic buckling stress
$l = 1.73\sqrt{Rt}$	Meridional natural buckling half wave length of unstiffened cylinder	σ_{cE}	Critical elastic buckling stress
		σ_y	Yield stress of material
q	Lateral load	**Vectors**	
R	Radius of shell	$\boldsymbol{F}$	Stress function vector
s	Stiffener spacing	$\boldsymbol{P}_b$	Load vector in bending
t	Shell thickness	$\boldsymbol{P}_m$	Membrane load vector
w	Lateral deflection		
w_0	Initial lateral deflection (imperfection)	$\boldsymbol{w}$	Vector of lateral deflections
$\gamma = EI/Ds$	Ratio between bending stiffness of stiffeners and plate	**Matrices**	
$\delta = A_s/st$	Degree of stiffening. Area ratio	$\boldsymbol{K}_B$	Bending coefficient matrix
		$\boldsymbol{K}_G$	Geometric coefficient (stiffness) matrix
δ	Initial deviation of shell	$\boldsymbol{K}_m$	Membrane coefficient matrix

1. Introduction

Ship hull structures are some of the largest, most complex and costly man-made structures. Ships are now on order in the range exceeding half a million tons of carrying capacity, with a length of more than 400 metres. The failure of such structures will have considerable consequences both for the people on board and for the environment. Consequently, by any system of values it is obvious that the reliability of these structures is very important.

The steel weight of a large ship is on the order of one eighth to one seventh of the load carrying capacity. Thus for example a 250 000 t.dwt. tanker has a fully loaded displacement of about 285 000 tons. The structures are quite slender; typical thicknesses of longitudinal bulkheads are about 16 mm, hull plating 20—35 mm, spacing of stiffeners (typical) 1 000 mm, length, breadth and depth of hull approximately 320, 60 and 30 m, respectively. Design against buckling is indeed very important, keeping in mind, however, that stiffness is also important for other reasons and that fatigue, brittle fracture etc. also play a role in the design.

The normal ship structures are heavily stiffened plate panels of traditional configurations. Even with these traditional structures several aspects of the buckling behaviour are not sufficiently well known to satisfy the designers completely. And as novel types of marine structures have recently been developed it has become apparent to the designers that for other types of structures than those pertaining to ships buckling design is even less well defined. The main purpose of the following paper has been to point out, rather than to attempt to solve, some of the still unanswered, or only partially answered questions related to the design against buckling of marine structures.

2. Stiffened Plate Panels

Ship structures consist mainly of flat or slightly curved plate panels stiffened on one side. For most parts an orthogonal stiffening system with primary stiffening in the direction of the main in-plane loads and deeper girders in the other (or both) directions is employed, see Fig. 1.

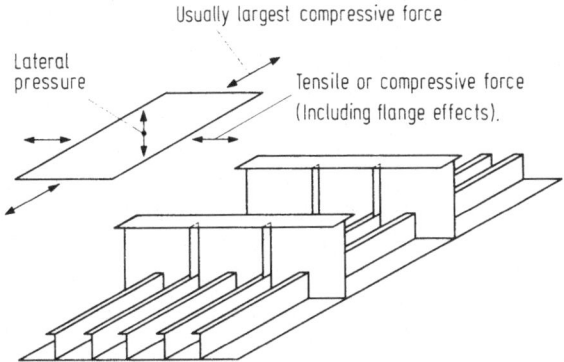

Fig. 1. Typical stiffened plate sub-assembly as employed in most ship structures.

The loads are due to static loading (still water), dynamic loading (wave induced) and in some instances temperature changes. In most cases both tensile and compressive in-plane loads will be present with the same order of magnitude in a given direction. The in-plane loads will often be higher in one direction than the other, but for certain structures both may be of the same order of magnitude. In many cases large in-plane loads are present, together with considerable lateral loads. The lateral loads will usually be acting in both directions depending upon the load conditions. It should be clear that all loads depend upon the handling of the ship at sea. Classification rules, however, presume "normal" handling and care. The relative importance of the different load components varies with time, handling and trade of the ship.

However, even when assuming the loads to be defined the calculation of the structural capability is not at all simple. The design as it is presently defined in our rules [1] employs a margin of "safety" against the critical stress varying with slenderness. This is based upon the familiar assumption and common experience that the ultimate load carrying capacity of a plate panel is appreciably above the critical buckling strength. This is true for slender panels having low critical stresses, but it is not equally relevant for the present ship structures having a characteristic slenderness range from 25 to 80.

In the higher part of the slenderness range elastic buckling will occur for the normal hull steels with yield stress from 25 to 40 kp/cm². In the lower part, and particularly with the low yield steels, the plastic effects will, however, be pronounced, and for this purpose the Johnson-Ostenfeld correction is generally employed:

$$\sigma_{cp} = \sigma_y[1 - (\sigma_y/4\sigma_{cE})].\tag{1}$$

For plane and unwelded plates the Johnson-Ostenfeld relation is considered to be somewhat conservative, but for the practical structures with weld stresses and distortions it may be quite reasonable. However, good experimental proofs are rather scarce.

Another aspect of stiffened panel buckling is the sensitivity to imperfections. It is normally assumed that plates are insensitive to imperfections, but this is regrettably a questionable truth. For all welded ship structures the deviations of the plates are pronounced for all but the heaviest plates where buckling in itself is unimportant. Typical deviations are from 0.1 to 0.5 or more of the thickness, and usually with a fairly regular shape due to the welding contraction on the side of the stiffeners, see Fig. 2a. This regular shape is normally

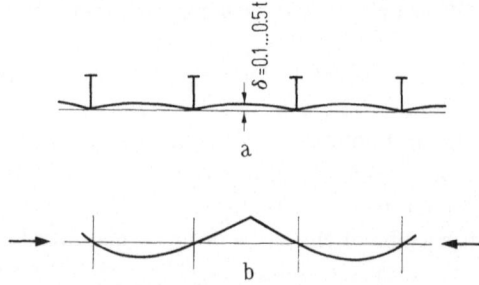

Fig. 2. a) Typical shape of deviation due to welding contraction at stiffeners; b) Irregular post buckling shape due to deviations as shown above.

expected to give a slight increase of the theoretical buckling strength similar to that arising from the influence of lateral pressure. Such effects are, however, never accounted for in the design, and indications are that the deviations may actually reduce the load carrying capacity to some extent. The one-sided deviations have been shown not to change the basic buckling pattern, but the influence on the post-buckling shape may be quite pronounced. Fig. 2 b shows an example of the clearly irregular post-buckling pattern recently experienced in the deck of a transversely framed ship.

Some tests on stiffende panels [2] appear to contradict the normal assumption previously referred to that stiffened plates have an ulti-mate strength well above their critical panel stress level. For example the plates shown in Fig. 3 had a b/t of 55 and δ/t in the 0.3 range. For

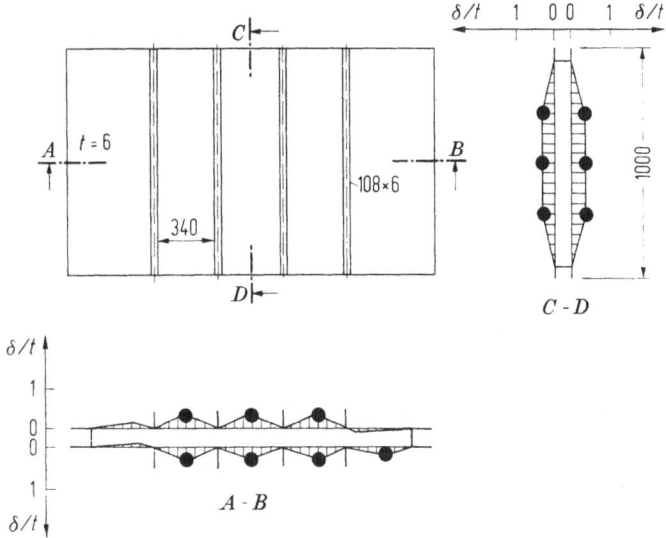

Fig. 3. Geometry and initial deflection of testpanel (from [2]).

the slab stiffened specimens non-linearities became appearent from about 0.6 of the critical panel buckling strength, see Fig. 4, with the ultimate strength approximately 0.85 of the critical (or 0.9 when correcting for plasticity according to Johnson-Ostenfeld). The slabs had a torsional buckling strength close to the panel buckling and the principal reason for the low strength appears to be the interaction of the panel buckling and stiffener tripping. This is in good agreement with the theoretical considerations of Tvergaard [3] who refers to the imper-fection sensitivity of stiffened plates when the plate buckling strength and Euler collapse load of the stiffener coincide. Such coincidence is not usual, but in cases of buckling design interaction effects are obviously important. In such cases the now fairly usual optimization methods will be particularly dangerous if interaction is not accounted for.

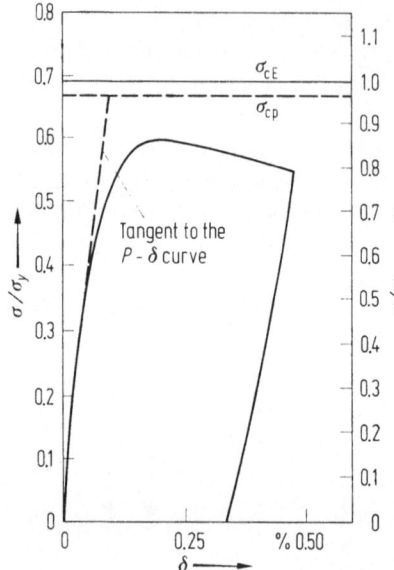

Fig. 4. Load-deflection relationship of slab stiffened panels (from [2]).

As previously discussed several important aspects of buckling behaviour of different types of stiffened ship type plate structures are not considered to be sufficiently well established. On that background it has been decided to develop a general computer program taking into account combinations of lateral and inplane loads, initial deviations, plasticity effects and interaction of the different buckling modes. An outline of this is found in Appendix 1.

3. Spherical Shells

For the special purpose of carriage of Liquified Natural Gas (LNG) a spherical tank design has been developed consisting of an unstiffened sphere supported at the equator by a stiffened cylinder. The sphere is subjected to compressive forces in the shell, both overall and locally viz.:

(a) Overall compressive forces due to a small external design pressure when empty.

(b) Compressive circumferential forces in areas close to and above the water surface due to partial filling.

(c) Local compressive forces due to interaction between tank system and hull structure ("forced deflections").

The spheres have R/t ratios of from 350 (in Aluminium) to 800 (in 9% Nickel steel). At present about 75% of the sphere is designed to meet buckling strength requirements and the remaining 25% to meet tensile strength and fatigue requirements.

As far as the uniform external pressure is concerned the design is fairly simple. The design is based on the simple classical buckling strength formula:

$$\sigma_{CL} = 0.606 \, Et/R. \qquad (2)$$

The design employs a safety generally corresponding to the API requirements [4] and slightly below the ASME requirements [5].

The other load conditions give local biaxial stress conditions consisting of membrane compressive stress in one direction and smaller or larger tensile stresses in the opposite direction.

An example of the circumferential compressive stresses existing in the lower hemisphere when heeling at a partial filling is found in Fig. 5. The same condition gives considerable shear, and thus a compressive principal force, in the upper hemisphere which is not included in the figure.

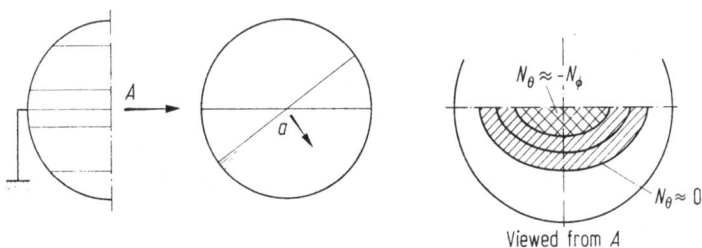

Fig. 5. Compressive circumferential stress picture due to nonsymmetric load (heeling) during partial filling.

A design basis is considerably more difficult to find for these load conditions, ASME for instance, gives no guidance. It is to be expected that considerably larger compressive stresses than for the overall compression case may be allowed in such conditions due to several reasons:

1. The imperfection sensitivity of a unidirectional and local compressive stress is expected to be less than for a bidirectional and global compressive stress.

2. The shape deviations in certain local areas will probably have smaller amplitudes than the maximum allowable, and not conform as well to the preferred buckling pattern of the shell.

3. The "forced deflection" nature of some components of the total stress, and the possibility for re-distribution of stresses around any local shape deviations means that prebuckling non-linearities reduce the possibility of buckling.

The tests performed by Yao [6] seem qualitatively to confirm the importance of points 1 and 2 above, but the limited number of tests and the lacking knowledge of the deviations of the tested shells makes it difficult to draw any quantitative conclusions. API has definite guidance for the various biaxial stress conditions, as shown in Fig. 6. However, the background for the API requirements as described in [7] is not wholly convincing.

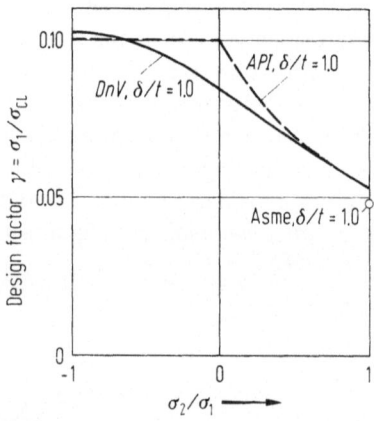

Fig. 6. Biaxial stress requirements for shells with $R/t \geq 300$.
σ_1 Principal compressive stress; σ_2 Principal tensile or compressive stress.

DnV has adopted a similar design curve also shown in Fig. 6, partly based on the Yao results. In addition to the basic curve for $\delta/t = 1$, Koiter's formula for cylindrical shells in axial compression has been adopted as a basis for extension to other design deviations:

$$\varrho = [4/27(1 - \nu_a^2)]^{1/2} (1 - \varrho)^2 \, t/\delta. \tag{3}$$

All of the spheres built to these requirements so far have been hydrostatically tested giving a compressive stress level roughly corresponding to the design compressive stress. Five steel spheres with an R/t of 800 have been subjected to a test stress level of 0.13 of the classical buckling stress, without any apparent deflections.

One sphere with an initial deviation exceeding the allowable ($\delta/t \approx 1.0$) was specially instrumented to follow the deflections during the test. No non-linearity implying an incipient buckling behaviour was experienced.

4. Circular Cylindrical Shells

Cylindrical shell type structures have been used in ships mainly for unstiffened pressure vessel type tanks in which buckling is usually not particularly important, and for submarine hulls which will not be discussed here. The cylindrical supports for the previously described spherical tanks, and for example large-diameter tubular members of floating offshore drilling rigs are recent examples of the use of thin shells as structural elements subject to considerable compressive forces.

With the variety of stiffening systems utilized and loading types present, design against buckling is fairly complex for the structures mentioned. The new rules for offshore fixed structures [8] describe reasonably completely the present thinking. In the following the reasoning behind these rules will not be generally commented. A few aspects of the cylinder buckling design have, however, turned out to be somewhat controversial, and will be further discussed.

Tennyson [9] has made the general statement that ring-stiffened cylinders in axial compression appear less weight effective than unstiffened ones. This conclusion is based upon tests with very small imperfection amplitudes ($\delta/t < 0.1$) compared to the practical fabrication tolerances which are ten times as large. Nevertheless ring-stiffened cylinders are quite extensively employed to carry axial loads. The above mentioned statement is thus not necessarily quite true.

For such cylinders the buckling design is fairly simple and well defined in the two extremes of the stiffening spacing range. With the well known theoretical preferred buckling half wave length:

$$l \approx 1.73 \sqrt{Rt}, \tag{4}$$

short cylinders ($L \ll 1.73 \sqrt{Rt}$) behave as a wide column and long cylinders ($L \gg 1.73 \sqrt{Rt}$) behave as an unstiffened cylinder. However the intermediate range ($1 < L\sqrt{Rt} < 3$) has been utilized extensively for design purposes.

A formula for the intermediate range is given by Harris [10] and is adopted also in [11] and [8]. However, one of the major US steel designers and builders has apparently for a some time utilized a design formula which for example for $L/\sqrt{Rt'} = 2$ gives twice as high a buckling stress. Some recent tests on large scale steel cylinders appear to confirm the shape of the curve as shown in Fig. 7, but with the limited number of tests the conclusion is uncertain and further cylinders are being tested without the results available yet.

The cylindrical support skirts for the spherical LNG-tanks are short (10—12 m), large diameter (30—37 m) thin lightly stiffened shells. With R/t ratios in the 300 to 800 range they are built in steel or

392   E. M. Q. Røren and H. R. Hansen

aluminium and steel, and are subjected to both an axial, a bending and
a shear load. With the large R/t ratio the axial and bending loads are
considered together as an axial compression and the shear is treated
as a torsion of the cylinder, both slightly conservative assumptions.

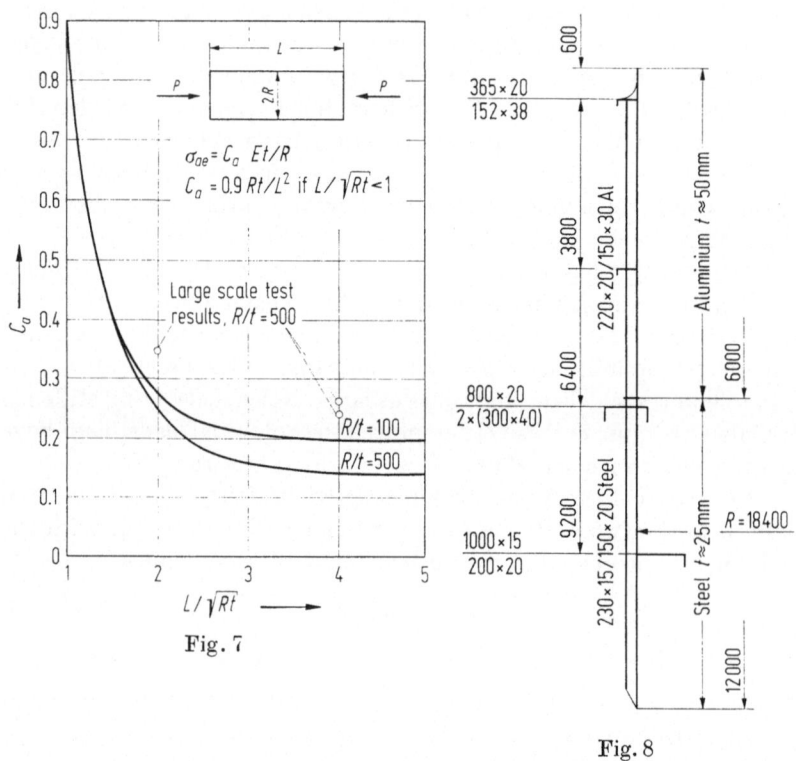

Fig. 7

Fig. 8

Fig. 7. Design elastic buckling stress for unstiffened cylinder.

Fig. 8. Typical shell geometry. Approximate dimensions only, spacing =1,070 m.

The stiffening system consists basically of primary longitudinal stiffeners (stringers) having a small spacing, supported by a limited number
of heavier rings, see Fig. 8. With the present designs the stiffening area
ratio is 15 to 25 per cent.

The calculation of a theoretical classical buckling stress even for a complex stiffened shell is fairly simple using a program like for example BOSOR [12]. Typical results from this type of calculation for the shell in Fig. 8 are shown in Fig. 9.

The classical buckling stress calculated easily becomes very high even with quite low stiffening ratios, see Fig. 10. To what extent this large increase in theoretical buckling strength may be considered in the design is a controversial subject. Hutchinson [13] has shown that

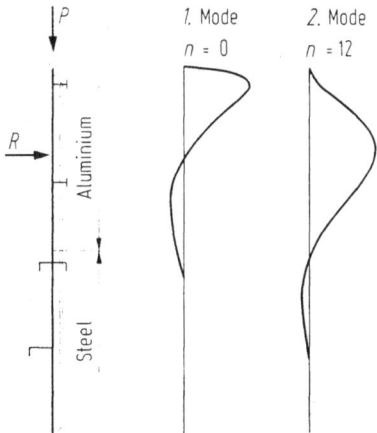

Fig. 9. Example of geometry and two lowest buckling modes for axially loaded cylinder.

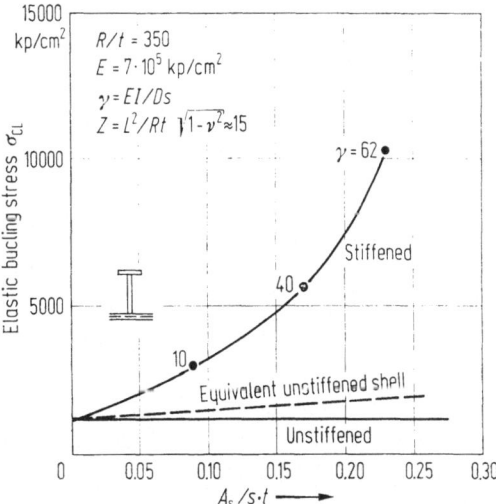

Fig. 10. Theoretical buckling stress for axially loaded stringer stiffened cylinder

lightly stiffened cylinders particularly with external stiffening have a considerable sensitivity to imperfections within the length range considered here. Singer [14] has reported tests which even for relatively low stiffening ratios show high reduction factors ϱ, but the test cylinders appear to be too accurate to reflect the fabrication tolerances for the structures discussed here. Alternative design methods have been either to consider the stiffener material as additional shell thickness but increase the reduction factor somewhat, or to consider the theoretical stiffened shell buckling strength employing a reduction factor similar to or lower than that employed for an unstiffened shell. The latter method has been chosen since it appears to give a better basis for evaluating different stiffenings. Within the very light stiffening ratio range a decreasing reduction factor has been employed apparently confirming reasonably well to a few recent large scale steel model tests, see Fig. 11. Regrettably these tests are noticeably influenced by plasticity so that the corresponding elastic reduction factor is only an estimate.

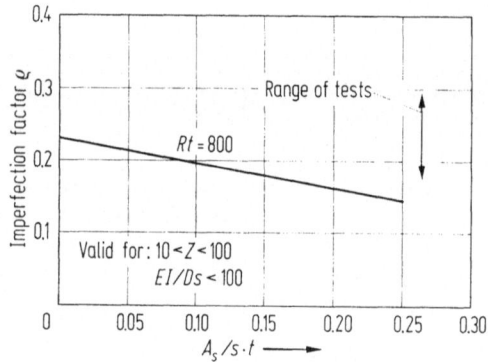

Fig. 11. Assumed design imperfection factor for longitudinally stiffened shell.

5. Conclusion

The buckling design of ships and other marine structures is not nearly as well established as it could be hoped. This has prompted Det norske Veritas to start several investigations with the specific intention of developing more reliable design information. Presently work is going on along several lines:

1. Development of a non-linear method for stiffened plates using Finite Differences, including initial deviations and plasticity (Appendix 1).
2. Testing of medium to relatively large scale metal spheres and lightly stiffened cylinders to investigate the imperfection sensitivity (Appendix 2).

3. Development of Finite Element program for large deflection analysis of general shell structures.

It is hoped that in the future more of the proprietary information existing within this area will be released, and that more of the high-quality research going on could be directed towards developing more reliable design information.

Appendix 1. Large Deflection FDM Program
for Elasto-Plastic Behaviour of Stiffened Plates

1. Scope of Program Development

Rectangular, orthogonally stiffened orthotropic plates with initial deformations are to be considered. The stiffeners are assumed to be parallel to the global coordinate axes (x, y) and may be handled as smeared out or discrete ones.

In-plane bi-axial normal and shear loads together with lateral load may be taken into account. All loads may be arbitrarily varying over the structure (spacewise).

The boundary conditions may be taken as simply supported, fixed or elastically built in with a rotational spring.

It will be possible to determine initial buckling loads of imperfect structures together with the load-deflection behaviour in the elastic as well as plastic range. For the latter purpose, it is possible to use an initial stress iteration technique with constant stiffness similar to that outlined by Nayak and Zienkiewicz [15].

2. Governing Equations

The differential equations for the membrane and bending behaviour of the plate are similar to those developed by Timoshenko [16, 17].

2.1. Membrane

$$L^4(F) = C(w_{,xy}^2 - w_{,xx}w_{,yy}),\qquad(1)$$

where

$$L^4 = E_x \frac{\partial^4}{\partial x^4} + 2B \frac{\partial^4}{\partial x^2\,\partial y^2} + E_y \frac{\partial^4}{\partial y^4} \quad \text{(differential operator)},$$

$$C = E_x E_y, \quad B = [C/G - (\nu_x E_y + \nu_y E_x)]/2,$$

w lateral deflection, F stress function.

2.2. Bending

$$D^4(w) - (h_x F_{,yy}w_{,xx} + h_y F_{,xx}w_{,yy} - 2hF_{,xy}w_{,xy}) =$$
$$= q + h_x F_{,yy}w_{0,xx} + h_y F_{,xx}w_{0,yy} - 2hF_{,xy}w_{0,xy},\qquad(2)$$

396   E. M. Q. Røren and H. R. Hansen

where

$$D^4 = D_x \frac{\partial^4}{\partial x^4} + 2H \frac{\partial^4}{\partial x^2 \partial y^2} + D_y \frac{\partial^4}{\partial y^4} \quad \text{(differential operator)},$$

q lateral load, w_0 initial lateral deflection (imperfection),

h_x, h_y, h thicknesses, D_x, D_y, H rigidities.

Discrete stiffener relations similar to those developed by Tvergaard [3] are used. These equations include bending and torsional stiffness. The tangential bending stiffness, however, is not taken into account.

3. Numerical Formulation

The differential equations (1) and (2) will be solved by a finite difference method (FDM) linearization. A rectangular mesh with constant mesh spacing in each direction will be used.

Equation (1) may be written in the form

$$K_m \cdot F = P_m, \tag{1a}$$

where

K_m coefficient matrix from boundary conditions and the operation of L^4 on each mesh point within the boundary,
F stress function vector,
P_m membrane "load" vector due to lateral deflections.

Similary Eq. (2) get the form:

$$(K_B - K_G) w = P_B, \tag{2a}$$

in which

K_B coefficient matrix from boundary conditions and operation of D^4,
$K_G = f(F)$ geometric stiffness matrix taking the membrane stress state into account,
P_B bending load vector taking account of the effect of lateral loads q together with in-plane stress state and imperfections,
w vector of lateral deflections.

For initial buckling Eq. (2a) transforms to

$$(K_B - \lambda K_G) w = 0. \tag{2b}$$

4. Solution Methods

The linearized equations (1a) and (2a) may be solved during an iteration loop. As a starting point Eq. (2a) will be solved first with $F \equiv 0$, giving a solution for lateral deflections w_1. This is substittuted into Eq. (1a) and a F_1 is found which again is used in Eq. (2a) and so on. The iteration is stopped when $F_{n+1} - F_n \leqq \varepsilon$ and $w_{n+1} - w_n \leqq \varepsilon$ where ε is about 1 per cent.

4.1. Solution Routines

Solution routines available at Det norske Veritas will be used [18, 19]. For Eq. (1a) and (2a) either a Gauss solution or a Crout or Cholesky algorithm may be used.

For the solution of Eq. (2b) we are going to use either a standard eigenvalue routine or a search for zeros in the determinant. This depends on whether the matrices K_B and K_G are symmetric or not.

Appendix 2. Testing of Stringer and Ring Stiffened Cylindrical Shells under Axial Compression

Most of the tests known from the literature cover cylindrical shells where the stringers (vertical stiffeners) are machined out from the shell or riveted to the shell and the stiffener size relative to the shell is much larger than used for structures of the present type.

In order to obtain a more satisfactory design basis for the design of large and thin cylindrical shells with very light stiffening ($A_s/st \leqq 0.2$) experiments with stringer and ring stiffened cylindrical shells under axial compression have been initiated.

In order to make it easier to evaluate the influence of the initial deviations, the test models will be made of a material with a relatively high yield strength where plasticity will not appreciably influence the results.

The cylindrical shell will be clamped to the test equipment. The stringers are tapered in both ends. Both the stringers and the rings will be welded to the shell.

Due to practical limits on the maximum compressive load an aluminium alloy with a yield stress of about 2500 kp/cm² and a proportional limit of about 2000 kp/cm² has been chosen instead of steel with a yield stress of about 7000 kp/cm².

The main characteristics for some typical models are:

Radius $R = 1200$ mm,

Thickness $t — 2.5$ mm,

Length $L =$ abt. 1000 mm,

Number of stiffeners, outside or inside abt. 80,

Stiffener spacing $s =$ abt. 99 mm,

$A_s/st =$ abt. 0.19,

$EI/Ds =$ abt. 15,

$e_s/t =$ abt. $\pm$ 6,

Number of rings $= 2$,

Theoretical classical buckling stress $\sigma_{cl} =$ abt. 3500 kp/cm²,

Theoretical classical buckling load $P =$ abt. 530 Mp,

Maximum compressive load available abt. 450 to 500 Mp.

References

1. Det norske Veritas: Rules for the Construction and Classification of Steel Ships. Oslo, 1974.
2. Kmiecik, M.: The Load-Carrying Capacity of Axially Loaded Longitudinally Stiffened Plate Panels Having Initial Deformations. SFI Report R80. Trondheim, May 1970.
3. Tvergaard, V.: Imperfection-Sensitivity of a Wide Integrally Stiffened Panel under Compression. Int. J. Solids Structures 9 (1973) No. 1, pp. 177—192.
4. API Standard 720. Recommended Rules for Design and Construction of Large, Welded Low-Pressure Storage Tanks. 4th Edition. Wash., D.C., Febr. 1970.
5. ASME Boiler and Pressure Vessel Code. Sect. VIII, Div. 2. New York 1971.
6. Yao, J. C.: Buckling of a Truncated Hemisphere under Axial Tension. AIAA J. 1 (1963) No. 10, pp. 2316—2319.
7. Dvorak, J. J.; McGrath, R. V.: Biaxial Stress Criteria for Large Low-Pressure Tanks. Welding Research Council Bull. Ser. no. 69, June 1961, pp. 14—24.
8. Det norske Veritas: Rules for the Design, Construction and Inspection of Fixed Offshore Structures. Oslo, July 1974.
9. Tennyson, R. C.; Muggeridge, D. B.; Caswell, R. D.: New Design Criteria for Prediction of Cylindrical Shells under Axial Compression. AIAA-Paper No. 71—145, 1971.
10. Harris, L. A., et al.: The Stability of Thin-Walled Unstiffened Circular Cylinders under Axial Compression Including the Effects of Internal Pressure. J. Aeronaut. Sci. 24 (1957) pp. 587—596.
11. Baker, E. H., et al.: Shell Analysis Manual. NASA Contractor Report 912. Wash., D.C., April 1968.
12. Bushnell, D.: Stress, Stability and Vibration of Complex Branched Shell of Revolution: Analysis and User's Manual for BOSOR 4. Lockheed Missiles and Space Co., Palo Alto, Calif., Report LMSC-D243605, March 1972.
13. Hutchinson, J. W.; Amazigo, J. C.: Imperfection-Sensitivity of Eccentrically Stiffened Cylindrical Shells. AIAA J. 5 (1967) No. 3, pp. 392—401.
14. Singer, J.; Arbocz, J.; Babcock, C.: Buckling of Imperfect Stiffened Cylindrical Shells under Axial Compression. AIAA J. 9 (1971) No. 1, pp. 68—75.
15. Nayak, G. S.; Zienkiewicz, O. C.: Elasto-Plastic Stress Analysis. A Generalization for Various Constitute Relations Including Strain Softening. Int. J. Num. Meth. Engng. 5 (1972) No. 1, pp. 113—135.
16. Timoshenko, S. P.; Woinowski-Krieger, S.: Theory of Plates and Shells. 2nd Edition. New York, McGraw-Hill 1959.
17. Timoshenko, S. P.; Gere, J. M.: Theory of Elastic Stability. 2nd Edition. New York: McGraw-Hill 1961.
18. Det norske Veritas: SESAM-69: User's Manual. Oslo, 1974.
19. Bell, K., et al.: NORSAM—A Programming System for the Finite Element Method. User's Manual, Part I: General Description. Trondheim, Febr. 1973.